Advances in Energy Infrastructure Construction Technology

Advances in Energy Infrastructure Construction Technology

Jianyong Han
Li Tian

MDPI

Basel • Beijing • Wuhan • Barcelona • Belgrade • Novi Sad • Cluj • Manchester

Editors
Jianyong Han
School of Civil Engineering
Shandong Jianzhu University
Jinan
China

Li Tian
School of Civil Engineering
Shandong University
Jinan
China

Editorial Office
MDPI AG
Grosspeteranlage 5
4052 Basel, Switzerland

This is a reprint of articles from the Special Issue published online in the open access journal *Energies* (ISSN 1996-1073) (available at: www.mdpi.com/journal/energies/special_issues/EIC_T).

For citation purposes, cite each article independently as indicated on the article page online and as indicated below:

Lastname, A.A.; Lastname, B.B. Article Title. *Journal Name* **Year**, *Volume Number*, Page Range.

ISBN 978-3-7258-2408-3 (Hbk)
ISBN 978-3-7258-2407-6 (PDF)
doi.org/10.3390/books978-3-7258-2407-6

Contents

About the Editors vii

Preface ix

Jianyong Han and Li Tian
Review of Construction Technology of Advanced Energy Infrastructure
Reprinted from: *Energies* **2024**, *17*, 4157, doi:10.3390/en17164157 1

Chaozhe Zhang, Jianyong Han, Songyu Liu, Zhenglong Cao, Chen Jiang and Xuhan Diao et al.
A Review of Efficient and Low-Carbon Pile Technologies for Extra-Thick Soft Strata
Reprinted from: *Energies* **2023**, *16*, 2836, doi:10.3390/en16062836 11

Hong Yu, Hao Li, Zhi-Qiang Zhang, Gui-Feng Zhang, Da-Hai Wang and Hua-Dong Zheng
Failure Patterns of Transmission Tower-Line System Caused by Landslide Events
Reprinted from: *Energies* **2022**, *15*, 7155, doi:10.3390/en15197155 34

Junkuo Li, Fan Gao, Lihuan Wang, Yaning Ren, Chuncheng Liu and Aiquan Yang et al.
Collapse Mechanism of Transmission Tower Subjected to Strong Wind Load and Dynamic Response of Tower-Line System
Reprinted from: *Energies* **2022**, *15*, 3925, doi:10.3390/en15113925 51

Hao Liu, Kaizhi Liu, Yiqi Xiao, Peng Zhang, Meixia Zhang and Youzeng Zhu et al.
Properties of Lightweight Controlled Low-Strength Materials Using Construction Waste and EPS for Oil and Gas Pipelines
Reprinted from: *Energies* **2022**, *15*, 4301, doi:10.3390/en15124301 69

Tianshuo Xu, Le Wang, Peng Zhang, Yuheng Zhou, Kaixin Liu and Xin Feng et al.
Key Techniques for Rapid Jacking and Laying of Pipelines: A Case Study on 'Jingshihan' Gas Pipelines in China
Reprinted from: *Energies* **2022**, *15*, 2918, doi:10.3390/en15082918 80

Sifeng Zhang, Guozhang Ren, Guojian Zhang, Ziyin Ren, Chong Xia and Yuan Gao
Seepage Performance of Fibre Bundle Drainage Pipes: Particle Flow Simulation and Laboratory Testing
Reprinted from: *Energies* **2022**, *15*, 7270, doi:10.3390/en15197270 96

Shidan Chi, Yan Liang, Weixi Chen, Zhen Hou and Tao Luan
Numerical Simulation of Tail Over-Fire Air Supply of a Grate Biomass Boiler
Reprinted from: *Energies* **2022**, *15*, 7664, doi:10.3390/en15207664 114

Duc-Vu Ngo, Jong-Kwon Lim and Dong-Hyawn Kim
Reliability Analysis and Life Prediction of Aging LNG Unloading Arms Based on Non-Destructive Test Data
Reprinted from: *Energies* **2022**, *15*, 9408, doi:10.3390/en15249408 125

Peinan Li, Zeyu Dai, Xi Wang, Jun Liu, Yi Rui and Xiaojun Li et al.
A Study of the Segment Assembly Error and Quality Control Standard of Special-Shaped Shield Tunnels
Reprinted from: *Energies* **2022**, *15*, 2512, doi:10.3390/en15072512 144

Yunjuan Chen, Tao Gao, Fuqiang Yin, Xiaozhi Liu and Jun Wang
Study on Influence of Joint Locations and Hydraulic Coupling Actions on Rock Masses' Failure Process
Reprinted from: *Energies* **2022**, *15*, 4024, doi:10.3390/en15114024 . . . . **174**

About the Editors

Jianyong Han

Dr. Jianyong Han, Master's Supervisor, is an Associate Researcher at Shandong Jianzhu University. He is a member of the Shandong Provincial Science and Technology Experts, a council member of the Shandong Urban Renewal Society, and the Deputy Secretary-General of the Geotechnical Engineering Professional Committee of the Shandong Society of Civil Engineering and Architecture. Dr. Han received his BS and PhD degrees from Northeastern University. His research focuses primarily on deep excavation theory, trenchless technology, and geotechnical anchorage technology in urban underground space engineering. He has led numerous research projects funded by the government and various companies. Dr. Han has published 71 papers in total, including 30 as the first author or corresponding author, with 2 highly cited papers of ESI, 3 hot papers of ESI, and 3 high-impact papers. He has co-authored one textbook, been granted over ten Chinese invention patents, and registered two software copyrights. He has also received three awards for outstanding construction methods in Shandong Province. Dr. Han has won a second prize for science and technology from the Jiangsu Society of Underground Space, a first prize for outstanding achievements from Jinan City, and two Top Ten Innovation Awards from Shandong Province. He serves as a reviewer for over 30 journals indexed by WOS. He is an Editorial Board Member for *Geofluids*, *American Journal of Civil Engineering*, and *Advances in Civil Engineering*, as well as an Excellent Reviewer for the *Chinese Journal of Underground Space and Engineering*. Dr. Han has successfully hosted seven Special Issues and has been recognized as an Excellent Reviewer by the Web of Science system.

Li Tian

Dr. Li Tian is currently a professor and the director of the Department of Civil Engineering at Shandong University. He also serves as the Director of the Key Laboratory of Civil Engineering Intelligent Construction in Shandong Province. Before joining Shandong University, Prof. Li Tian received his BS degree from Northeastern University in 2005 and his PhD from Dalian University of Technology in 2011. His research focuses on the multi-hazard risk and resilience analysis of infrastructures, testing and simulation of steel structures, and structural dynamics and vibration control. To date, he has generated more than 170 peer reviewed papers in well-known international journals, several of which were designated by ESI as "Highly Cited Papers". Additionally, Dr. Li Tian has also made contributions to structural design, performance evaluation, and intelligent maintenance for numerous significant practical engineering projects and has hold over 80 invention patents licensed from China and the United States. He has been the principal investigator of numerous competitive research projects funded by NSFC and the National Grid. He has also served as a reviewer for several journals, including ASCE-*Journal of Structural Engineering*, *Engineering Structures*, and *Reliability Engineering and System Safety*, among others.

Preface

Energy infrastructure is a cornerstone of sustainable economic growth and societal development worldwide. It plays a critical role in ensuring the rational use of resources, facilitating the smooth functioning of cities, and supporting environmental protection by enabling the efficient utilization of energy. Reliable energy infrastructure is essential not only for providing energy resources but also for shaping resilient systems that contribute to the global economy's stability and progress. Without robust and effective energy systems, the pursuit of environmental sustainability and the optimization of resource use would be significantly hampered.

The advancement of energy infrastructure requires continuous research in construction technologies, new materials, innovative equipment, and theoretical approaches. Improvements in these areas are crucial to enhance infrastructure performance and extend its service life. Given the diverse challenges faced by energy infrastructure—such as complex geological conditions, high operational loads, and a dynamic service environment—developing innovative materials, construction methods, and analytical tools is imperative. These advancements are vital to bridge the gap between current technological capabilities and the growing demands of modern energy systems, thereby ensuring their reliability, efficiency, and adaptability in a rapidly changing world.

This Special Issue encompasses a collection of research and review articles that address various aspects of energy infrastructure development. The topics covered include the study of failure mechanisms in transmission tower–line systems, advanced techniques for rapid installation of gas pipelines, and cutting-edge numerical simulation methods to optimize the components of energy infrastructure. By presenting these advancements, this Special Issue aims to promote knowledge sharing and foster progress in the field, with a focus on enhancing reliability, optimizing performance, and promoting sustainability in energy infrastructure construction. The insights gathered here are intended to guide future research and drive technological innovation, ultimately contributing to a more resilient and efficient energy sector. We are grateful to all authors for their valuable contributions and to readers for their engagement with this dynamic area of study.

Jianyong Han and Li Tian
Editors

energies

MDPI

Editorial

Review of Construction Technology of Advanced Energy Infrastructure

Jianyong Han [1,2,*] and Li Tian [3]

1 School of Civil Engineering, Shandong Jianzhu University, Jinan 250101, China
2 Key Laboratory of Building Structural Retrofitting and Underground Space Engineering, Shandong Jianzhu University, Ministry of Education, Jinan 250101, China
3 School of Civil Engineering, Shandong University, Jinan 250061, China; tianli@sdu.edu.cn
* Correspondence: hanlwb@163.com

Abstract: Energy is crucial to the development of human civilization. Energy infrastructure, such as oil and gas pipelines, power generation systems, and storage facilities, provide core support for the exploitation and utilization of various types of energy. Thus, energy infrastructure is vital to the economic sustainable development of a country. This paper provides the motivations and a brief introduction to the Special Issue entitled "Frontiers in Construction Technology of Advanced Energy Infrastructure", which aims to present advanced technologies and theories for energy infrastructure. The primary challenges in the current construction technology of energy infrastructure are described. Furthermore, the focus of current research in this field addressed in this Special Issue is presented. A comparison of the articles included or considered for inclusion in this Special Issue with other available literature on this topic is performed, which proves the prospects and relevance of this Special Issue. Finally, perspectives on future directions of energy utilization and energy infrastructure construction are provided.

Keywords: energy infrastructure; energy utilization; construction technologies; storage facilities; emerging marine industries; technological challenges

check for updates

Citation: Han, J.; Tian, L. Review of Construction Technology of Advanced Energy Infrastructure. *Energies* **2024**, *17*, 4157. https://doi.org/10.3390/en17164157

Received: 24 August 2023
Revised: 21 February 2024
Accepted: 6 June 2024
Published: 21 August 2024

1. Introduction

Energy infrastructure pertains to the comprehensive systems, structures, and facilities explicitly designed to produce, transmit, and distribute energy [1–3]. Serving as a cornerstone of modern societies, energy infrastructure powers homes, industries, and businesses [4,5]. It manifests in a variety of forms, depending on the specific energy source and other elements of its value chain, for instance, traditional energy infrastructure like gas pipelines, power generation systems for both fossil and renewable fuels, electrical transmission lines which are called lifelines, metering and distribution systems, intelligent systems, and modern electric and electronic systems [6,7]. It also includes storage facilities and different power control systems [8,9]. To expand further, present-day energy infrastructure takes the form of industrial systems related either directly or indirectly, and multiple interconnected sectors, like the transportation industry (charging units for electric vehicles, petrol station) and agricultural industry (farm machinery, drying systems) [10,11]. At a macro-level, energy infrastructure signifies a complex, multi-faceted system integral to the seamless functioning of modern societies. It serves as a pivotal engine propelling global economic growth and competitiveness [12–14].

Thus, energy development and utilization systems are fundamental cornerstones for the overall development of economies across the globe. Energy infrastructure is essential for effective resource usage, the normal function of a city, and waste energy disposal, but also indirectly affects economic and social development [15,16]. Over the past few decades, the focus of research pertaining to energy infrastructure has undeniably varied, carving a novel path in response to societal needs and technological advancements [17,18].

Bernd Resch et al. [19] analyzed the shortcomings of previous approaches to using GIS in renewable energy-related projects, extracted different challenges from these previous efforts, and finally defined a set of core future research pathways for GIS-based energy infrastructure planning, with a focus on the use of renewable energy sources. Tim Unterluggauer et al. [20] reviewed different methods and technologies for the planning of electric vehicle charging infrastructure and proposed various models and algorithms to address the location selection, scale determination, and charging demand prediction of charging stations. Yinan Liu et al. [21] gave an update on typical and innovative carbon capture and storage (CCS) technologies combined with solar energy, outlining thermodynamic processes, including chemical reactions, operating conditions, and efficiency indicators of the five typical technologies, in order to clarify the current level of development. From conventional coal-powered systems to renewable energy alternatives, from conventional oil storage to underground oil storage caverns, and from gas stations to electric vehicle charging devices, these research endeavors adopt different design philosophies and construction technologies yet share one unchanging objective [22–25]. They aim to architect and preserve efficient, reliable, and sustainable energy infrastructure systems, systems that exist symbiotically with the ecosystem and lead the charge in reducing our carbon footprint [26–29].

However, the complex terrain of a service environment, coupled with diverse and random loads, constitutes the primary obstacles preventing infrastructure engineers and researchers from materializing this objective [30,31]. While hybrid solar–wind–biomass energy systems have great potential for powering sustainable cities in the future, the complexity of their planning and design still limits their widespread implementation. By studying Vancouver, the researchers assessed the impact of economies of scale on hybrid renewable energy systems and found that the average cost of electricity (COE) of systems at different scales was slightly different [32–34]. Abdullahi Mohamed Samatar et al. [35] discussed the current situation and potential of solar energy utilization in Somalia, which, despite its abundant solar potential, still faces challenges such as a lack of infrastructure, high initial costs, and lack of energy awareness. The unpredictability of load demand, variability in energy sources, and intricacies of system design and maintenance all pose considerable challenges in achieving this goal [36,37]. Infrastructure engineers and researchers, therefore, must leverage technological advances, strategic planning, and innovative thinking to create more sustainable and efficient energy infrastructures [38,39].

The construction of energy infrastructure requires sufficient knowledge of geotechnical engineering, structural engineering, and the physical properties of energy materials, as well as the mechanical properties of the construction materials [40]. To enhance the service performance and service life of energy infrastructure, innovations in construction materials, structures, foundations, and construction measures are necessary [41–43]. Zaghloul et al. [44] conducted a study to determine the effect of fiber arrangement on the mechanical properties of glass-fiber-reinforced polymeric materials (GFRP). This paper focuses on the shrinkage factor, the tensile fatigue behavior of pultruded-glass-fiber-reinforced polyester, and its life estimation model. Cawas Phiroze Nazir [45] discussed the technical and economic feasibility of offshore hydropower plants using hydropower to develop marine energy reserves as renewable energy sources. Since the constantly charged stream provides the energy the plant uses to generate electricity, and the water is not used up in the process, it is a clean and renewable source of energy, the plant does not use fuel, and there is no uncertainty about transportation, storage, treatment costs, or fuel pricing, reducing the dependence on conventional fuels in other countries and being more environmentally friendly than traditional hydroelectric power plants [46,47]. Feng Shan et al. [48] introduced the performance evaluation and application of photovoltaic collectors and systems, analyzed and discussed the performance of photovoltaic hot-air collectors and photovoltaic hot water collectors, and introduced the application of photovoltaic thermal systems, such as building-integrated photovoltaic thermal systems, concentrated photovoltaic thermal systems, and photovoltaic heat pump systems. Jasem Alazemi et al. [49] discussed that hydrogen stations are one of the most important parts of the power distribution infrastructure needed to support the operation of hydrogen fuel cell electric vehicles and hydrogen combustion engine vehicles. Up to this point, notwithstanding significant strides made in improving construction methodologies, developing sophisticated construction materials, and optimizing construction theories, a noticeable disconnect still exists [50–53]. Such a gap shifts between the specific requirements of sophisticated energy systems and the real-world conditions currently characterizing energy infrastructure. This indicates that developing advanced methods for effective and efficient technologies, equipment, and theoretical concepts to improve the design, construction, and maintenance of energy infrastructure remains a persistent challenge [54–57].

This Special Issue seeks to showcase and deliberate on recent studies, innovative methodologies, case studies, and review articles that describe the current status and notable advances within this research area. Submissions focusing on confronting current challenges associated with physical experiments, computational models, and theoretical analyses pertinent to the construction technology of energy infrastructure are welcomed. When feasible, such submissions should incorporate pertinent field observations from practice. Papers submitted on new and emerging topics within this discipline are also encouraged. Contributions in the form of theoretical papers are certainly welcomed, with a particular emphasis placed on submissions of practice-oriented papers and those based on computational methods. This Special Issue seeks contributions spanning a broad range of topics that are related, but not limited, to the following:

- Stability analysis of oil and gas pipelines/tunnels in multi-field coupling environments;
- Multi-hazards on energy infrastructure and lifeline structures;
- Reinforcement analysis of infrastructure and lifeline structures;
- Construction technology of renewable energy infrastructure;
- Hazard analysis of earthquakes and strong winds;
- Monitoring, spatial–temporal prediction modeling, and early warning of facility failures using advanced measurement systems;
- Sustainable and innovative materials for energy infrastructure;
- Construction technology and durability of nuclear waste treatment facilities under extreme conditions;
- New computational and experimental methods investigating the foundations of advanced energy infrastructures.

2. Outlook of This Special Issue

This Special Issue of *Energies* was successfully organized with the unwavering support from the editorial team of the journal and the MDPI publishing team. The processing time for the articles averaged approximately 39 days. The Guest Editor is very much grateful to all reviewers for reviewing and revising the manuscripts. Devoting their valuable time to reviewing papers is essential for upholding the voluntary peer review process and is highly commendable. Through their constructive feedback for authors and confidential reports to the editors, they have played a significant role in maintaining the high academic standard of this journal.

The response to our call for papers was excellent, with the following statistics:

- Submissions: (23);
- Publications: (10);
- Rejections: (13);

Article types: review articles (1); research articles (9).

The geographical distribution of authors (corresponding author and first author) by country for the published articles is presented in Table 1, in which it is possible to observe 20 authors from two countries. Note that it is usual for an item to be signed by more than one author and for authors to collaborate with others from different affiliations.

Table 1. Authors' geographical distribution.

Country	Authors
China	18
Korea	2

The corresponding authors and first authors of this Special Issue and their first affiliations are reflected in Table 2.

Table 2. Authors and affiliations.

Author	First Affiliation	Country	Reference
Chaozhe Zhang	Southeast University	China	[58]
Jianyong Han	School of Civil Engineering, Shandong Jianzhu University	China	[58]
Li Tian	Shandong University	China	[58]
Duc-Vu Ngo	Kunsan National University	Korea	[59]
Dong-Hyawn Kim	Kunsan National University	Korea	[59]
Hong Yu	Electric Power Research Institute, Yunnan Power Grid Co., Ltd.	China	[60]
Hua-Dong Zheng	Wuhan University of Technology	China	[60]
Shidan Chi	Shandong Electric Power Engineering Consulting Institute Co., Ltd.	China	[61]
Tao Luan	Shandong University	China	[61]
Sifeng Zhang	Shandong Jianzhu University	China	[62]
Guojian Zhang	Shandong Jianzhu University	China	[62]
Hao Liu	China University of Geosciences	China	[63]
Meixia Zhang	China University of Geosciences	China	[63]
Yunjuan Chen	Shandong Jianzhu University	China	[64]
Tianshuo Xu	China University of Geosciences	China	[65]
Peng Zhang	China University of Geosciences	China	[65]
Junkuo Li	State Grid Hebei Economic Research Institute	China	[66]
Chuncheng Liu	Northeast Electric Power University	China	[66]
Peinan Li	Donghua University	China	[67]
Xi Wang	Tongji University	China	[67]

Table 3 summarizes the research carried out by identifying the topics to which they belong, according to the proposed topics in this Special Issue. A clear dominance is observed in two research topics in this Special Issue, specifically "Pipeline" and "Transmission Tower".

Table 3. Research topics.

Research Topics	Number of Publications	References
Low-carbon pile	1	[58]
Non-destructive test	1	[59]
Transmission tower	2	[60,66]
Grate biomass boiler	1	[61]
Pipeline	3	[62,63,65]
Hydraulic coupling	1	[64]
Special-shaped shield tunnel	1	[67]

The summary of the articles published in this Special Issue is discussed in the subsequent sections of this editorial.

3. A Review of This Special Issue

A number of articles involving several subjects about the construction technology of advanced energy infrastructure have been published in this Special Issue. Their details and a brief summary of the addressed issues are provided in the following.

A Review of Efficient and Low-Carbon Pile Technologies for Extra-Thick Soft Strata by Chaozhe Zhang, Jianyong Han, Songyu Liu, Zhenglong Cao, Chen Jiang, Xuhan Diao, Guangwei Chen and Li Tian [58].

The authors discuss the challenges posed by limited bearing capacity in soft strata and the need for innovative pile technologies. They focus on composite pile technologies that combine rigid piles with flexible columns, such as the stiffened deep mixing (SDM) column, squeezed branch pile, pre-bored grouting plated nodular (PGPN) pile, precast cement pile reinforced by cemented soil with a variable section (PCCV), and carbonized composite pile (CCP). The authors also review low-carbon energy-efficient curing technologies that have been developed for geotechnical engineering. This paper provides insights into efficient and low-carbon pile technologies and offers guidance for industry practitioners.

Reliability Analysis and Life Prediction of Aging LNG Unloading Arms based on Non-Destructive Test Data by Duc-Vu Ngo, Jong-Kwon Lim and Dong-Hyawn Kim [59].

This study examines unloading arms (ULAs) that have been in service for over 30 years and analyzes their structural deterioration and integrity. The authors use a finite-element program to model the aging ULAs and update the thickness dimensions of structural members based on NDT data. A reliability assessment is then conducted using the stress distribution of the main structural components under external loads, specifically wind speed. This study also constructs a time-dependent reliability index curve to predict the failure probability of the components during their remaining lifetime. The findings suggest that the current condition of the ULA system is satisfactory for current loading conditions.

Failure Patterns of Transmission Tower-Line System Caused by Landslide Events by Hong Yu, Hao Li, Zhi-Qiang Zhang, Gui-Feng Zhang, Da-Hai Wang and Hua-Dong Zheng [60].

This article investigates the collapse patterns of transmission tower-line structures under landslides. This study aims to understand the failure mechanisms of these structures and the impact of landslides on their collapse patterns. This research uses the explicit analysis method to analyze the nonlinear dynamic response equations of the structure. Two foundation failure cases under landslides are considered, and the displacement responses and progressive collapse behavior are analyzed. This study finds that the failure patterns of the tower-line structure under landslides are influenced by the failure mode of the tower foundation and the pulling effect of cables.

Numerical Simulation of Tail Over-Fire Air Supply of a Grate Biomass Boiler by Shidan Chi, Yan Liang, Weixi Chen, Zhen Hou and Tao Luan [61].

This article explores the effects of tail burnout air's incidence angle, wind speed, and pipe diameter on the flow field distribution and movement of unburned carbon particles in a grate biomass boiler. This study uses ANSYS software to simulate these effects and obtain the influence rules of the tail burnout air parameters on combustion and particle movement. The results show that the setting of burnout air at the tail of the grate can improve air flow field organization, change particle distribution, and prolong particle residence time on the grate. Increasing the outlet velocity and pipe diameter of the tail burnout air leads to an increased movement and burnout of grate particles, resulting in improved boiler efficiency.

Seepage Performance of Fibre Bundle Drainage Pipes: Particle Flow Simulation and Laboratory Testing by Sifeng Zhang, Guozhang Ren, Guojian Zhang, Ziyin Ren, Chong Xia and Yuan Gao [62].

This article explores the seepage performance of fiber bundle drainage pipes in slope engineering projects. The authors conducted particle flow simulations and laboratory tests to investigate the influences of soil particles on the clogging of geotextile filters and drainage pipes, as well as the seepage rates. They found that higher water pressure, smaller flower hole intervals in the pipe, greater soil friction angles, and smaller pipe inclination angles are less conducive to drainage. They also discovered that installing a fiber bundle in drainage pipes can improve seepage and drainage performance under silting conditions. The authors tested five types of fiber bundles and found that plastic rope provided the best drainage effect. The arrangement of fiber bundles also affected drainage, with uneven arrangements being more effective with plastic rope and cotton rope, and uniform arrangements being more effective with nylon rope, hemp rope, and polyester rope.

Properties of lightweight Controlled Low-Strength Materials using construction waste and EPS for Oil and Gas Pipelines by Hao Liu, Kaizhi Liu, Yiqi Xiao, Peng Zhang, Meixia Zhang, Youzeng Zhu, Kaixin Liu, Tianshuo Xu and Rui Huang [63].

This study investigates the use of construction waste and expanded polystyrene (EPS) as fine aggregate to prepare lightweight controlled low-strength materials (CLSMs) for oil and gas pipelines. The EPS is modified with ethylene vinyl acetate-resin (EVA) to improve its compatibility with cementitious materials. The CLSM's fluidity, strength, and microstructure are studied through experiments and microscopic analysis. The results show that the surface modification of EPS with EVA improves its compatibility with cementitious materials and allows for a fluidity greater than 200 mm. The volume ratio of EPS to construction waste and the cement content significantly affect the material's strength development. This study suggests that the EPS content should not be too large to achieve the desired mechanical properties. Microscopic analysis reveals that an increase in EPS content leads to poor pore uniformity and a loose mesh structure, which hinders strength development.

A study on the influence of joint locations and hydraulic coupling actions on rock masses' failure process was conducted by Yunjuan Chen, Tao Gao, Fuqiang Yin, Xiaozhi Liu, and Jun Wang [64].

This article explores the distribution of joints and fissures in rock masses and their impact on the stability of surrounding rock in underground engineering construction. The authors compare and analyze two simulation programs, Discontinuous Deformation Analysis for Rock Failure (DDARF) and Rock Failure Process Analysis (RFPA), based on laboratory tests. They then use RFPA to analyze the failure process, stress state, acoustic emission characteristics, and energy dissipation laws of rock masses with different joint locations under hydraulic coupling conditions. The results show that the location of joints affects the failure process, with a large tensile stress region generated on both ends of the original joint and the damage effect angle increasing gradually from the middle joint to the marginal joint. The water pressure in cracks promotes the generation of tensile stress, and its effect on crack failure depends on the proximity of the joint to the middle part of the specimen.

Key techniques for gas pipelines rapid jacking and lying: A case study in project of 'Jingshihan' gas pipeline in China by Tianshuo Xu, Le Wang, Peng Zhang, Yuheng Zhou, Kaixin Liu, Xin Feng, Yongchun Qi and Cong Zeng [65].

This article discusses the key techniques for the rapid jacking and laying of pipelines, using the case study of the 'Jingshihan' gas pipelines in China. The authors highlight the need to improve the construction process of oil and gas pipelines due to the rapid growth of the industry. They introduce the direct pipe laying method, which combines the advantages of horizontal directional drilling and pipe jacking construction. The method is suitable for pipeline crossings in different strata and offers advantages such as less construction land, high speed, and reversibility. The authors investigate the application status of the direct pipe laying method in the 'Jingshihan' gas pipeline project and discuss various aspects of the construction process, including the working well, equipment selection, guiding control technology, supporting equipment installation, and drag reduction measures. They also analyze the factors influencing thrust force and trajectory deviation.

Collapse Mechanism of Transmission Tower Subjected to Strong Wind Load and Dynamic Response of Tower-line System by Junkuo Li, Fan Gao, Lihuan Wang, Yaning Ren, Chuncheng Liu, Aiquan Yang, Zhao Yan, Tao Jiang and Chengbo Li [66].

This article investigates the collapse mechanism and failure modes of transmission towers under strong wind loads. The authors use the finite element method (FEM) to analyze the ultimate bearing capacity of a typical 220 kV transmission tower. The results show that the collapse of the tower under strong wind loads is usually due to the buckling of the leg members. The authors also establish a finite element model to analyze the dynamic response of the tower-line system under fluctuating wind. The results show that the wind-induced coefficients designed by current codes accurately assess the collapse displacement of the transmission tower, but the load-carrying capacity of the tower in the plastic phase can be overestimated by static calculation results.

A Study of the Segment Assembly Error and Quality Control Standard of Special-Shaped Shield Tunnels by Peinan Li, Zeyu Dai, Xi Wang, Jun Liu, Yi Rui, Xiaojun Li, Jie Fan and Peixin Chen [67].

This study focuses on the segment assembly error and quality control standard of specially shaped shield tunnels. The researchers investigated the formation mechanism and control measures of lining segment assembly defects in a quasi-rectangular shield tunnel project. They developed a simulation program to calculate the error in segment assembly by quantifying the manufacturing error and positioning error. They also established a finite element model to analyze the mechanical behavior of key blocks and proposed a control standard for the key block assembly error. The results showed that the assembly quality can be improved by assembling the LZ block first and applying error control measures.

4. Conclusions

The contributions in this Special Issue discussed a wide range of subjects relevant to the construction technology and applications of advanced energy infrastructure. Construction technologies are crucial for the update and development of energy infrastructure. Research on advanced construction technology and energy infrastructure theory is therefore very important for the energy industry. Even though the articles reported very interesting applications and relative techniques developed thus far, there are still many gaps to fill to satisfy the rapidly increasing demand for energy utilization.

Combined with the ever-increasing global demand for energy, it has become essential to find new solutions, based on the background of the complex service conditions of energy infrastructure and the development of technology of renewable energy resources. The expansion of new energy sources, particularly in the realm of electric vehicles, has underscored the significance of robust charging infrastructure. Simultaneously, the impact of conventional power facilities on the environment remains a pivotal concern. Traditional electricity production and transmission, often reliant on fossil fuels, have contributed significantly to air pollution and greenhouse gas emissions. However, the steady advancement of renewable energy sources, including wind and solar power, has provided a means to mitigate environmental degradation. By enhancing power infrastructure and embracing greener energy production methods, substantial progress can be made in minimizing adverse environmental effects.

Addressing the security of energy infrastructure also holds paramount importance, which includes two primary issues: structural safety and network security. Regardless of whether it is the collapse of transmission towers or hydroelectric dam failures, incessant issues regarding the structural stability of energy infrastructure bring vast damages to the energy industry and national economy. To address these structural safety issues, further advancements in the technology and theory of infrastructure construction should be enhanced, and they should propel the development of associated construction standards for energy infrastructure. On the other hand, the evolution of interconnected systems, from charging stations to intelligent grids, accentuates the need for a holistic security approach encompassing both physical and cyber safeguards. Governments and industry stakeholders must collaborate to establish robust standards, protocols, and contingency plans to counter potential threats. This comprehensive strategy entails safeguarding not only physical assets but also bolstering cybersecurity measures to thwart cyberattacks and prevent data breaches.

Funding: This research work is supported by the Key Research and Development Program of GansuProvince, China (Grant No. 22YF7FH224) and the Doctoral Research Fund of Shandong JianzhuUniversity (Grant No. X19080Z).

Data Availability Statement: Data are contained within the article.

Acknowledgments: The Guest Editor would also like to thank the authors for submitting their excellent contributions to this Special Issue. Thanks is also extended to the reviewers for evaluating the manuscripts and providing helpful suggestions.

Conflicts of Interest: No potential conflicts of interest were reported by the authors.

References

1. Bridge, G.; Özkaynak, B.; Turhan, E. Energy infrastructure and the fate of the nation: Introduction to special issue. *Energy Res. Soc. Sci.* **2018**, *41*, 1–11.
2. Kumar, N.M.; D'adamo, I.; Hait, S.; Priya, A.; Kichou, S.; Gastaldi, M. Editorial: Sustainable planning and lifecycle thinking of energy infrastructure. *Front. Energy Res.* **2023**, *11*, 1196826.
3. Xie, D.; Tian, Y.L.; Wang, G.Q.; Li, T.J.; Zhong, D.Y. Study on the pathway and strategy of carbon neutralization in Qinghai province. *J. Basic Sci. Eng.* **2022**, *30*, 1331–1345.
4. Jones, C.F. Building More Just Energy Infrastructure: Lessons from the Past. *Sci. Cult.* **2013**, *22*, 157–163.
5. Pandey, V. Energy infrastructure for sustainable development. In *Affordable and Clean Energy*; Springer: Cham, Switzerland, 2020; pp. 1–13.
6. Weng, L.X.; Zhao, J.F.; Lin, R.; Wu, B.; Chen, F.B. Investigation of distribution characteristics of typhoon in Fujian coastal area and its influence on transmission lines. *J. Chang. Univ. Sci. Technol. (Nat. Sci.)* **2020**, *17*, 95–101.
7. Zhuo, G.P.; Nazir, S.; Khan, H.U.; Mukhtar, N. An efficient multifeature model for improving the performance of critical energy infrasture. *J. Adv. Transp.* **2021**, *2021*, 8411379.
8. Wang, H.X.; Zhang, B.; Yu, X.; Xu, N.X.; Ye, J.H. Long-Term Stability and Deformation Behaviour of Anhydrite Mine-Out for Crude Oil Storage. *Rock Mechnics Rock Eng.* **2020**, *53*, 1719–1735.
9. Zhang, B.; Wang, H.X.; Wang, L.; Mei, G.; Shi, L.; Xu, N.X.; Li, J.Y. Large-scale field test on abandoned deep anhydrite mine-out for reuse as crude oil storage—A case study. *Eng. Geol.* **2020**, *267*, 105477.
10. Alhazmi, Y.A.; Mostafa, H.A.; Salama, M.M.A. Optimal allocation for electric vehicle charging stations using Trip Success Ratio. *Int. J. Electr. Power Energy Syst.* **2017**, *91*, 101–116.
11. Bayram, I.S.; Galloway, S.; Burt, G. A probabilistic capacity planning methodology for plug-in electric vehicle charging lots with on-site energy storage systems. *J. Energy Storage* **2021**, *32*, 101730.
12. Alkhalidi, A.; Qoaider, L.; Khashman, A.; Al-Alami, A.R.; Jiryes, S. Energy and water as indicators for sustainable city site selection and design in Jordan using smart grid. *Sustain. Cities Soc.* **2018**, *37*, 125–132.
13. Köktürk, G.; Tokuç, A. Vision for wind energy with a smart grid in Izmir. *Renew. Sustain. Energy Rev.* **2017**, *73*, 332–345.
14. Mosannenzadeh, F. Smart energy city development: A story told by urban planners. *Cities* **2017**, *64*, 54–65.
15. Adom, P.K.; Amuakwa-Mensah, F.; Amuakwa-Mensah, S. Degree of financialization and energy efficiency in Sub-Saharan Africa: Do institutions matter? *Financ. Innov.* **2020**, *6*, 33. [CrossRef]
16. Zakari, A.; Musibau, H.O. Sustainable economic development in OECD countries: Does energy security matter? *Sustain. Dev.* **2023**, *32*, 1337–1353. [CrossRef]
17. Wang, J.; Han, J.Y.; Zhu, C.G.; Lu, W.Y.; Zhang, Y.; He, X.S.; Liu, K.M.; Chen, Y. Structural optimization and engineering application of concrete-filled steel tubular composite supports. *Eng. Fail. Anal.* **2024**, *159*, 108082.
18. Rees, M.T.; Wu, J.; Jenkins, N.; Abeysekera, M. Carbon constrained design of energy infrastructure for new build schemes. *Appl. Energy* **2014**, *113*, 1220–1234.
19. Resch, B. GIS-Based Planning and Modeling for Renewable Energy: Challenges and Future Research Avenues. IScharging infrastructure planning for integrated transportation aPRS. *Int. J. Geo-Inf.* **2014**, *3*, 662–692.
20. Unterluggauer, T. Electric vehicle nd power distribution networks: A review. *E Transp.* **2022**, *12*, 100163.
21. Liu, Y. Energy-saving pathway exploration of CCS integrated with solar energy: A review of innovative concepts. *Renew. Sustain. Energy Rev.* **2017**, *77*, 652–669.
22. Gallardo-Lozano, J.; Milanés-Montero, M.I.; Guerrero-Martínez, M.A.; Romero-Cadaval, E. Electric vehicle battery charger for smart grids. *Electr. Power Syst. Res.* **2012**, *90*, 18–29.
23. Kesler, M.; Kisacikoglu, M.C.; Tolbert, L.M. Vehicle-to-Grid Reactive Power Operation Using Plug-In Electric Vehicle Bidirectional Offboard Charger. *IEEE Trans. Ind. Electron.* **2014**, *61*, 6778–6784.
24. Dai, Y.F.; Zhou, Z.F. Steady seepage simulation of underground oil storage caverns based on Signorini type variational inequality formulation. *Geosci. J.* **2015**, *19*, 341–355.

25. Qiao, L.P.; Wang, Z.C.; Li, S.C.; Bi, L.P.; Xu, Z.H. Assessing containment properties of underground oil storage caverns: Methods and a case study. *Geosci. J.* **2017**, *21*, 579–593.
26. Güney, T. Renewable energy, non-renewable energy and sustainable development. *Int. J. Sustain. Dev. World Ecol.* **2019**, *26*, 389–397.
27. Barone, G.; Buonomano, A.; Giuzio, G.F.; Palombo, A. Towards zero energy infrastructure buildings: Optimal design of envelope and cooling system. *Energy* **2023**, *279*, 128039.
28. Ostergaard, P.A.; Duic, N.; Noorollahi, Y.; Kalogirou, S. Renewable energy for sustainable development. *Renew. Energy* **2022**, *199*, 1145–1152. [CrossRef]
29. Shrestha, A.; Mustafa, A.A.; Htike, M.M.; You, V.; Kakinaka, M. Evolution of energy mix in emerging countries: Modern renewable energy, traditional renewable energy, and non-renewable energy. *Renew. Energy* **2022**, *199*, 419–432. [CrossRef]
30. Han, J.Y.; Wang, J.; Cheng, C.; Zhang, C.Z.; Liang, E.B.; Wang, Z.K.; Song, J.J.; Leem, J. Mechanical response and parametric analysis of a deep excavation structure overlying an existing subway station: A case study of the Beijing subway station expansion. *Front. Earth Sci.* **2023**, *10*, 1079837.
31. Han, J.Y.; Wang, J.; Jia, D.F.; Yan, F.S.; Zhao, Y.; Bai, X.Y.; Yan, N.; Yang, G.; Liu, D. Construction technologies and mechanical effects of the pipe-jacking crossing anchor-cable group in soft stratum. *Front. Earth Sci.* **2023**, *10*, 1019801.
32. Bagheri, M. Optimal planning of hybrid renewable energy infrastructure for urban sustainability: Green Vancouver. *Renew. Sustain. Energy Rev.* **2018**, *95*, 254–264.
33. Kabeyi, M.J.B.; Olanrewaju, O.A. The levelized cost of energy and modifications for use in electricity generation planning. *Energy Rep.* **2023**, *9*, 495–534. [CrossRef]
34. Shea, R.P.; Ramgolam, Y.K. Applied levelized cost of electricity for energy technologies in a small island developing state: A case study in Mauritius. *Renew. Energy* **2019**, *132*, 1415–1424. [CrossRef]
35. Samatar, A.M. The utilization and potential of solar energy in Somalia: Current state and prospects. *Energy Strategy Rev.* **2023**, *48*, 101108.
36. Paraschiv, S. Analysis of the variability of low-carbon energy sources, nuclear technology and renewable energy sources, in meeting electricity demand. *Energy Rep.* **2023**, *9*, 276–283. [CrossRef]
37. Todd, A.C.; Optis, M.; Bodini, N.; Fields, M.J.; Perr-Sauer, J.; Lee, J.C.Y.; Simley, E.; Hammond, R. An independent analysis of bias sources and variability in wind plant pre-construction energy yield estimation methods. *Wind Energy* **2022**, *25*, 1775–1790. [CrossRef]
38. Liu, J.; Huang, Z.Y.; Fan, M.; Yang, J.H.; Xiao, J.; Wang, Y. Future energy infrastructure, energy platform and energy storage. *Nano Energy* **2022**, *104*, 107915.
39. Shally, S.K.; Gupta, P. Energy efficient resource optimization algorithm for cloud infrastructure. *J. Intell. Fuzzy Syst.* **2023**, *44*, 409–419.
40. Yu, J.L.; Jia, Q.; Hu, H.Q. Charging infrastructure construction from the perspective of new infrastructure. *Energy Rep.* **2021**, *7*, 224–229.
41. Campana, R.; Sabatini, L.; Frangipani, E. Moulds on cementitious building materials-problems, prevention and future perspectives. *Appl. Microbiol. Biotechnol.* **2020**, *104*, 509–514. [CrossRef]
42. Ma, Y.X.; Li, D.; Yang, R.T.; Zhang, S.; Arici, M.; Liu, C.Y.; Zhang, C.J. Energy and daylighting performance of a building containing an innovative glazing window with solid-solid phase change material and silica aerogel integration. *Energy Convers. Manag.* **2022**, *271*, 116341. [CrossRef]
43. Baetens, R.; Jelle, B.P.; Gustavsen, A. Phase change materials for building applications: A state-of-the-art review. *Energy Build.* **2010**, *42*, 1361–1368.
44. Mahmoud Zaghloul, M.Y.; Yousry Zaghloul, M.M.; Yousry Zaghloul, M.M. Physical analysis and statistical investigation of tensile and fatigue behaviors of glass fiber-reinforced polyester via novel fibers arrangement. *J. Compos. Mater.* **2023**, *57*, 147–166.
45. Nazir, C.P. Offshore hydroelectric plant: A techno-economic analysis of a renewable energy source. *Renew. Sustain. Energy Rev.* **2014**, *34*, 174–184.
46. Doria-Iriarte, J.J.; Doria-Elejoste, I. A new theory on ocean wave mechanics and its application in energy power generation. *Dyna* **2021**, *96*, 276–280.
47. Silva, A.R.; Pimenta, F.M.; Assireu, A.T.; Spyrides, M.H.C. Complementarity of Brazil's hydro and offshore wind power. *Renew. Sustain. Energy Rev.* **2016**, *56*, 413–427.
48. Shan, F. Performance evaluations and applications of photovoltaic–thermal collectors and systems. *Renew. Sustain. Energy Rev.* **2014**, *33*, 467–483.
49. Alazemi, J.; Andrews, J. Automotive hydrogen fuelling stations: An international review. *Renew. Sustain. Energy Rev.* **2015**, *48*, 483–499.
50. Bo, W.; Jun, W.; Bing, L.; Zonggang, X.F.W.; Tongle, J. Performance test and application of steel pipe concrete support structure for soft rock tunnel in deep wells. *Rock Mech. Eng.* **2010**, *29*, 2604–2609.
51. Chen, Y.; Okudan, G.E.; Riley, D.R. Sustainable performance criteria for construction method selection in concrete buildings. *Autom. Constr.* **2010**, *19*, 235–244.
52. Choi, Y.S.; Lim, I.; Kim, T.; Cho, H.; Kang, K.I. Case Study of the Core Structure Succeeding Method for Tall Building Construction. *J. Constr. Eng. Manag.* **2016**, *142*, 05016017.

53. Ferrada, X.; Serpell, A. Selection of Construction Methods for Construction Projects: A Knowledge Problem. *J. Constr. Eng. Manag.* **2014**, *140*, B4014002.
54. Abeysundara, U.G.Y.; Babel, S.; Gheewala, S. A matrix in life cycle perspective for selecting sustainable materials for buildings in Sri Lanka. *Build. Environ.* **2009**, *44*, 997–1004.
55. Kanniyapan, G.; Nesan, L.J.; Mohammad, I.S.; Keat, T.S.; Ponniah, V. Selection criteria of building material for optimising maintainability. *Constr. Build. Mater.* **2019**, *221*, 651–660.
56. Zhan, J.Y.; Liu, W.; Wu, F.; Li, Z.H.; Wang, C. Life cycle energy consumption and greenhouse gas emissions of urban residential buildings in Guangzhou city. *J. Clean. Prod.* **2018**, *194*, 318–326.
57. Xie, S.; Ji, Z.J.; Zhu, L.C.; Zhang, J.J.; Cao, Y.X.; Chen, J.H.; Liu, R.R.; Wang, J. Recent progress in electromagnetic wave absorption building materials. *J. Build. Eng.* **2020**, *27*, 100963.
58. Zhang, C. A Review of Efficient and Low-Carbon Pile Technologies for Extra-Thick Soft Strata. *Energies* **2023**, *16*, 2836. [CrossRef]
59. Ngo, D.; Lim, J.; Kim, D. Reliability Analysis and Life Prediction of Aging LNG Unloading Arms Based on Non-Destructive Test Data. *Energies* **2022**, *15*, 9408. [CrossRef]
60. Yu, H.; Li, H.; Zhang, Z.Q.; Zhang, G.F.; Wang, D.H.; Zheng, H.D. Failure Patterns of Transmission Tower-Line System Caused by Landslide Events. *Energies* **2022**, *15*, 7155. [CrossRef]
61. Chi, S.; Liang, Y.; Chen, W.; Hou, Z.; Luan, T. Numerical Simulation of Tail Over-Fire Air Supply of a Grate Biomass Boiler. *Energies* **2022**, *15*, 7664. [CrossRef]
62. Zhang, S.; Ren, G.; Zhang, G.; Ren, Z.; Xia, C.; Gao, Y. Seepage Performance of Fibre Bundle Drainage Pipes: Particle Flow Simulation and Laboratory Testing. *Energies* **2022**, *15*, 7270. [CrossRef]
63. Liu, H.; Liu, K.; Xiao, Y.; Zhang, P.; Zhang, M.; Zhu, Y.; Liu, K.; Xu, T.; Huang, R. Properties of Lightweight Controlled Low-Strength Materials Using Construction Waste and EPS for Oil and Gas Pipelines. *Energies* **2022**, *15*, 4301.
64. Chen, Y.; Gao, T.; Yin, F.; Liu, X.; Wang, J. Study on Influence of Joint Locations and Hydraulic Coupling Actions on Rock Masses' Failure Process. *Energies* **2022**, *15*, 4024. [CrossRef]
65. Xu, T.; Wang, L.; Zhang, P.; Zhou, Y.; Liu, K.; Feng, X.; Qi, Y.; Zeng, C. Key Techniques for Rapid Jacking and Laying of Pipelines: A Case Study on 'Jingshihan' Gas Pipelines in China. *Energies* **2022**, *15*, 2918. [CrossRef]
66. Li, J.; Gao, F.; Wang, L.; Ren, Y.; Liu, C.; Yang, A.; Yan, Z.; Jiang, T.; Li, C. Collapse Mechanism of Transmission Tower Subjected to Strong Wind Load and Dynamic Response of Tower-Line System. *Energies* **2022**, *15*, 3925. [CrossRef]
67. Li, P.; Dai, Z.; Wang, X.; Liu, J.; Rui, Y.; Li, X.; Fan, J.; Chen, P. A Study of the Segment Assembly Error and Quality Control Standard of Special-Shaped Shield Tunnels. *Energies* **2022**, *15*, 2512. [CrossRef]

energies

Review

A Review of Efficient and Low-Carbon Pile Technologies for Extra-Thick Soft Strata

Chaozhe Zhang [1], Jianyong Han [2,*], Songyu Liu [1], Zhenglong Cao [3], Chen Jiang [4], Xuhan Diao [3], Guangwei Chen [4] and Li Tian [5,*]

1 Institute of Geotechnical Engineering, School of Transportation, Southeast University, Nanjing 211189, China; chaozhe_zhang@163.com (C.Z.)
2 School of Civil Engineering, Shandong Jianzhu University, Jinan 250101, China
3 Shandong Provincial Airport Management Group Co., Ltd., Jinan 250107, China
4 Jiangsu Provincial Transportation Engineering Construction Bureau, Nanjing 210004, China
5 School of Civil Engineering, Shandong University, Jinan 250061, China
* Correspondence: hanlwb@163.com (J.H.); tianli@sdu.edu.cn (L.T.)

Abstract: With the development of urban underground space and increased infrastructure functions, both the scale of engineering construction and engineering difficulties have increased globally. In the construction of structures in soft strata, especially in coastal areas, the limited bearing capacity of the foundations poses a significant challenge. The composite pile technologies employing an organic combination of the rigid pile andthe flexible column can enable efficient soft ground treatment. In light of prominent global environmental issues, low-carbon energy-saving curing technologies have been rapidly developed for application in geotechnical engineering. This paper discusses progress in research on the mechanical properties of the efficient and low-carbon pile technologies, including the stiffened deep mixing (SDM) column, squeezed branch pile, pre-bored grouting plated nodular (PGPN) pile, precast cement pile reinforced by cemented soil with a variable section (PCCV), and carbonized composite pile (CCP). In addition, it reviews the technical characteristics and recent progress of feasible low-carbon energy-efficient curing technologies. The paper also proposes future directions for theoretical research and technological development of low-carbon pile technologies. The key contribution of this review is to provide insights into efficient and low-carbon pile technologies. In addition, the findings from the study of the pile technologies used in extra-thick soft strata also provide industry practitioners with a comprehensive guide regarding the specific applications and mechanical performance of the pile technologies, which can serve as a stepping stone to facilitate the technological development of the underground space industry.

Keywords: underground space; pile foundation; precast cement pile; deep mixing column; industrial solid waste; reactive magnesia

Citation: Zhang, C.; Han, J.; Liu, S.; Cao, Z.; Jiang, C.; Diao, X.; Chen, G.; Tian, L. A Review of Efficient and Low-Carbon Pile Technologies for Extra-Thick Soft Strata. *Energies* **2023**, *16*, 2836. https://doi.org/10.3390/en16062836

Academic Editor: Krzysztof Skrzypkowski

Received: 19 February 2023
Revised: 14 March 2023
Accepted: 16 March 2023
Published: 18 March 2023

1. Introduction

With the rapid development of the global economy and urbanization, the construction of infrastructure such as high-speed railways and expressways, and underground space exploitation, have undergone rapid change [1–4]. Both the construction scale and degree of engineering difficulty are at an unprecedentedly high level internationally. Moreover, deep soft and weak soils are employed widely in coastal and river areas; their low strength, large deformation, poor stability, and seismic performance pose significant challenges to various engineering construction projects [5,6] and are also the main cause of various engineering accidents [7–9]. The existing technical specifications for the ground treatment primarily apply to soft soil foundations with a depth less than 20 m. Rigid pile composite foundations and pile foundations are mainly used for deeper soft soil strata, which are also the primary form of the foundation used for high-rise buildings, highways, and railway roadbeds, as well as bridge foundations. The main types of rigid piles and pile foundations are precast concrete (PC) piles, pipe piles, and bored piles [10], which play an important role in infrastructure modernization and construction. However, conventional rigid piles are often expensive, and the seismic performance of PC piles does not adequately meet design requirements [11,12], thus limiting their application.

To overcome these problems, some novel types of composite piles have been developed in recent years, such as stiffened deep mixing (SDM) columns [13] and pre-bored grouting planted nodular (PGPN) piles [14]. These piles consist of an organic combination of a flexible column and rigid pile, i.e., inserting the PC pile with different cross-sections into the deep mixing (DM) column. This combination can exploit the advantages of high bearing capacity of concrete piles, compensate for the low strength of the DM column, and improve the strength of the surrounding soil [15,16], significantly improving the performance of precast piles and expanding their application range. Thus, it has good potential for widespread application [17,18].

However, in recent years, there have been breakthroughs in both the pile types and materials used in mixing column technology. Two-way variable-section mixing pile technology, and powder and wet spraying integrated construction technology, have been popularized [19]. T-shaped deep mixing (TDM) columns [20,21] and squeezed branch piles as the variable-section piles are the choices for layered soft ground [22,23]. Furthermore, since the production process of conventional Portland cement consumes large amounts of resources and energy, and emits a large amount of harmful gases including CO_2, dust, soot, and SO_2, technologies employing Portland cement as a curing material can no longer satisfy the requirements of modern low-carbon environmental protection and sustainable development [24]. For this reason, new low-carbon energy-saving curing agents, such as blast furnace slag, steel slag, fly ash [25,26], and reactive magnesia [27], which enable green and environmental protection, have been rapidly developed. In this paper, we review a series of novel pile technologies, with a focus on presenting the interface characteristics and vertical and horizontal bearing behaviors of the SDM column under the building loads, the bearing mechanism and failure mode of the SDM column-supported embankment, and the status of research into the squeezed branch pile and PGPN pile. Based on the studies of feasible low-carbon energy-efficient curing agents, a basis for the further development of variable-section composite pile technology and a carbonized composite pile (CCP) is proposed.

2. Current Status of Research on the SDM Column

The use of the SDM column as a composite pile originated at Hebei Cangzhou Machinery Construction Co., Ltd., Cangzhou, China, in 1994. The column is formed by inserting the PC or pre-stressed high-intensity concrete (PHC) piles, termed the inner core or cored pile, into the DM column, termed the outer core, after the construction of the DM column [28]. It is worth noting that the PC pile improved with cement-treated soil (PCCS) [29,30] is another term for the SDM column; however, despite some similarities in the configuration, they are nevertheless based on different ideas. The goal of the PCCS is to improve the bearing capacity of the PC pile with cemented soils. The design concept of the SDM column is to reinforce the DM column with reinforcement (i.e., concrete or an H-type steel pile). For the sake of consistency, the term "SDM column" has been adopted in this study. In 1998, Tianjin University and Hebei University of Technology conducted a series of model experiments and field tests on SDM columns and initially derived the bearing characteristics, design, and construction method. They found that the bearing capacity of the SDM columns can increase by 4.4–5.4 times and 1.36–1.58 times compared with that of the DM columns and cast-in-place piles with the same diameter and length, respectively [31]. Due to their high efficiency, SDM columns have been widely used for pile foundations under rigid loads, and in subgrade and foundation pit support projects, in China and Thailand. The Technical Specification for Strength Composite Piles (JGJ/T 327-2014) was drawn up on the basis of previous research. SDM columns are divided into three types of piles according to the relationship between the length of inner and outer core: short core, equal core, and long core [30], as shown in Figure 1. To date, the inner–outer core interface behaviors [32,33], vertical bearing characteristics [34,35], and horizontal bearing characteristics of the SDM column under rigid loads have been extensively studied [36–38]. Furthermore, preliminary investigations into the bearing mechanism and failure form of the SDM column under embankment loads have been performed. In the analysis of the bearing characteristics of this composite pile, the core length ratio (L_{pc}/L_{dm}) and core diameter ratio (D_{pc}/D_{dm}) are often involved.

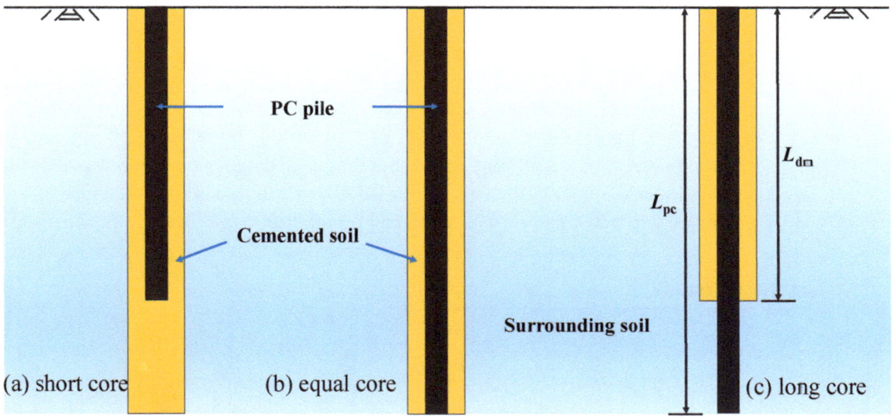

Figure 1. Schematic diagram of the SDM column.

2.1. Interface Properties

The SDM column has two interfaces, i.e., the inner–outer core interface and the pile–soil interface, whose characteristics are the main factors controlling the bearing and deformation characteristics of this composite pile. The driving inner core can compress and expand the cemented soil and surrounding soil to make the pile–soil interface rougher and tighter, increasing the interface friction resistance of the SDM column [18]. Using field vertical tensile capacity tests, Jamsawang et al. [39] illustrated that the interfacial shear strength can ensure that inner and outer cores do not misalign and slip. In addition, the interface damage mode changed from plastic damage to brittle damage with the increase in the unconfined compressive strength of the cemented soil. Test results indicated that the cored pile and cemented soil have good synergistic working characteristics, and ultimate interface resistance varies approximately linearly with the cemented soil strength [32]. Table 1 summarizes the ratio of friction resistance at the inner–outer core interface and the cemented soil strength in existing references. It can be seen that the range of the interfacial strength of the inner–outer cores of the SDM column differs greatly in the relevant Chinese codes.

Table 1. The ratio of friction resistance at the inner–outer core interface and the cemented-soil strength.

Reference	Ratio
Zou et al. [40]	0.133
Wu et al. [41]	0.194
Li [33]	0.188
Technical Specification for Concrete Core Mixing Pile (YB2007) [13]	0.19
Technical Specification for Strength Composite Piles (JGJ/T 327-2014) [42]	0.04–0.08
Technical Specification for Pile Foundation of Pipe Pile Embedded in Cement Soil (JGJ/T 330-2014) [43]	0.16

2.2. Vertical Bearing Characteristics of the SDM Column

In recent years, many scholars have studied the vertical bearing characteristics of the SDM column using indoor model tests, field tests, and numerical simulations. Liu et al. [44] studied the load transfer mechanism of the SDM column through full-scale model tests, with the results indicating that the deformation of the inner and outer cores is approximately coordinated under a vertical load. Furthermore, the axial force ratio of inner and outer cores in the same section is approximately equivalent to the ratio of their elastic modulus. Wang et al. [45] found that the SDM column exhibited the working characteristics of a friction pile, and its vertical load was borne mainly by the lateral frictional force of the pile. Zhou et al. [46,47] concluded that inner and outer cores can be entirely taken into account in analyses of the vertical bearing characteristics of the SDM column. Wonglert et al. [48] and Voottipruex et al. [49] studied the bearing behavior and failure mode of the SDM column by performing model tests and numerical simulations. It was found that when the core length ratio exceeded the optimum value, the failure of the SDM column was mainly attributed to the plastic damage of the surrounding soil, i.e., the pile body sank entirely, and the various failure nodes were summarized based on the numerical results, as shown in Figure 2.

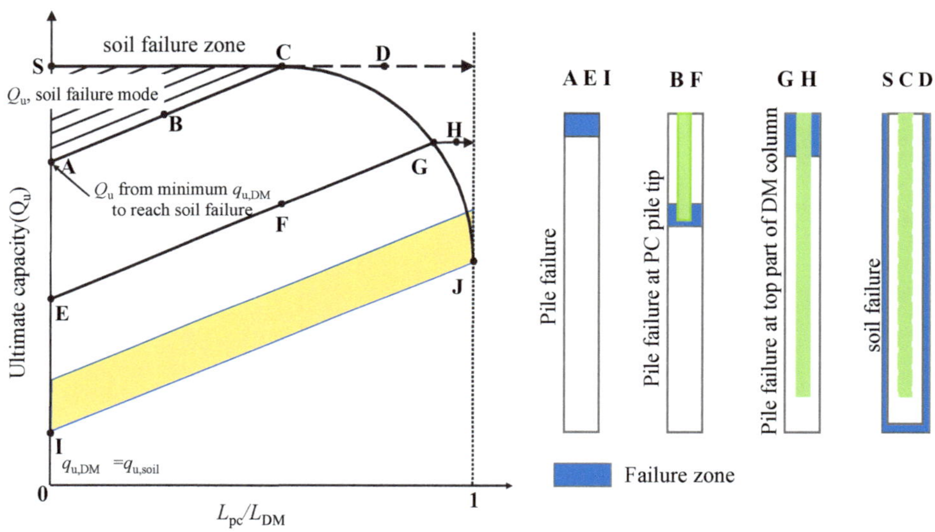

Figure 2. The relationship between the core length ratio and ultimate capacity, as well as the associated failure modes [48].

Regarding theoretical analysis, Gu et al. [50] derived a theoretical method for calculating the axial stress and interface shear stress of the inner core of the SDM column based on the principle of composite mechanics and Mindlin's solution. However, this method was limited to elastic analyses. Based on the load transfer method, Ren et al. [51] proposed a simplified calculation method for the load transfer of the SDM column, considering inner–outer core and pile–soil interface friction, whilst ignoring the continuity of the surrounding soil. In addition, the corresponding calculation process involved complicated iterative trial calculations. Wang et al. [35] developed a simplified approach for the axial response analysis of SDM column reinforcement embedded in non-homogeneous soil considering elastic non-slippage and elastoplastic slippage at the cemented soil–soil interface, and the increase in the strength and stiffness of subsoil with depth, as shown in Figure 3. Based on the deformation model of the surrounding soil, cemented soil, and the composite pile, Ye et al. [15] proposed an analytical solution for calculating the stress ratio between the inner core, outer core, and the surrounding soil in the SDM column composite foundation under a rigid load. The predicted results indicate that the pile–soil stress ratio gradually increases with the diameter of the inner core and the length of DM column. According to the model test results, Li [33] classified the failure modes of three modes of SDM columns with a short core, equal core, and long core. Among these, SDM columns with a short core include four modes, i.e., overall pile sinking, cross-sectional failure of the composite section, core pile piercing, and cross-sectional failure of the non-composite section. SDM columns with an equal core and long core will result in three failure modes, namely, overall pile sinking, cross-sectional damage of the composite section, and core pile piercing.

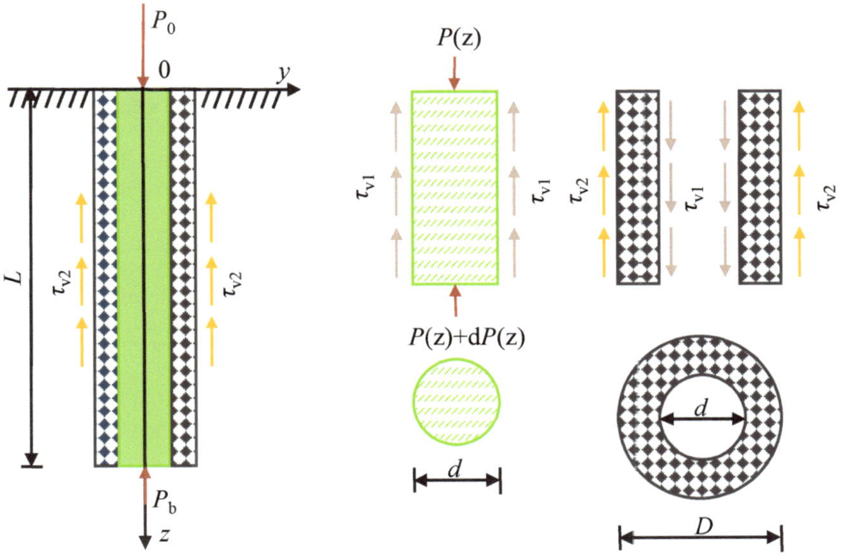

Figure 3. Force diagram of the SDM column under rigid loads [35].

2.3. Horizontal Bearing Characteristics of the SDM Column

In projects such as high embankments, ports, pits, and slope supports, pile foundations are subjected to both vertical and horizontal loads. Similarly, research on horizontal load-bearing characteristics of the SDM column has been conducted. Based on field tests and numerical simulation results, Voottipruex et al. [49] found that the horizontal ultimate bearing capacity of the SDM column is approximately 10 times that of a DM column with the same parameters. Liu [52] concluded that the bearing characteristics of the SDM column under a horizontal load are similar to those of the PC pile. However, the presence of the outer core can effectively reduce the maximum bending moment and shear force of the inner core, as well as enhance the pile head stiffness. Wang et al. [53] reported that the horizontal ultimate bearing capacity of this composite pile formed by bored piles constructed in jet-grouted piles can be increased by 64–69% compared with bored piles with the same parameters. He et al. [54,55] systematically evaluated the effect of using jet-grouted pile reinforcement on the horizontal force performance of bored piles in a soft clay. The results showed that the jet-grouted reinforcement could significantly improve the pile head stiffness, and reduce the cumulative displacement of the bored piles under static loads and one-way and two-way cyclic loads. Based on the comparative analysis of the horizontal bearing behaviors of the SDM column and PHC pile using numerical simulations, Du et al. [36] concluded that the horizontal displacement and bending moment of the SDM column were significantly lower than those of the PHC pile under the same horizontal load.

Concerning the theoretical calculation of the horizontal bearing capacity of composite piles, Huang et al. [56] proposed the calculation method of the proportional coefficient, m, of the horizontal resistance coefficient of the SDM column. The results showed that flexural stiffness of the horizontal bearing capacity calculated by the m method should be taken as the algebraic sum of the inner core and outer core, where the m value of the SDM column is closer to that of the bored pile recommended by the Technical Code for Building Pile Foundations (JGJ 94-2008) [57]. By considering the contribution of the DM column, Li et al. [58] derived the formula for calculating the horizontal bearing capacity of the SDM column using the modified m approach, and the range of m values and influencing factors were discussed.

2.4. Status of Research on the SDM Column-Supported Embankment

To further improve the application of the SDM column-supported embankment reinforcement on soft soil, researchers have studied the load transfer mechanism, consolidation characteristics, embankment stability, and settlement calculation of the SDM column composite foundation under embankment loads. To examine the load-bearing characteristics, Zhang et al. [59] performed a series of centrifuge model tests and numerical analysis to study the pile–soil load transfer mechanism and consolidation characteristics of the SDM column-supported embankments with a slab on soft soil, as shown in Figure 4. The results showed that the slab above the gravel cushion effectively reduced differential settlement, accelerated subsoil consolidation, and improved the load-bearing ratio of the SDM column. When the core length ratio falls within the range of 0.5 to 0.8, it ensures high load transfer efficiency between the pile and soil, while causing stress concentrations in the slab and inner core under load. Ye et al. [60] found that the soil arching heights of the SDM column-supported embankment are 1.7 and 1.3 times the net spacing of inner cores according to the soil stress and deformation control conditions, respectively. Thus, it is proposed that the embankment height should be 1.5 times the soil arching height, and the area replacement rate of inner cores should be within the range of 13–25% during engineering design after conducting a multi-parameter analysis. Regarding the calculation of embankment settlement, Ye et al. [61] proposed a method of settlement calculation for SDM column-supported embankments over soft soil considering the core pile piercing the outer core. The total settlement comprised three parts: soil settlement within the length of the inner core, soil settlement from the inner core tip to the tip of the SDM column, and soil settlement below the bottom of the SDM column. A case analysis revealed that the results of the proposed calculation model are in better agreement with the embankment settlement determined from measurements.

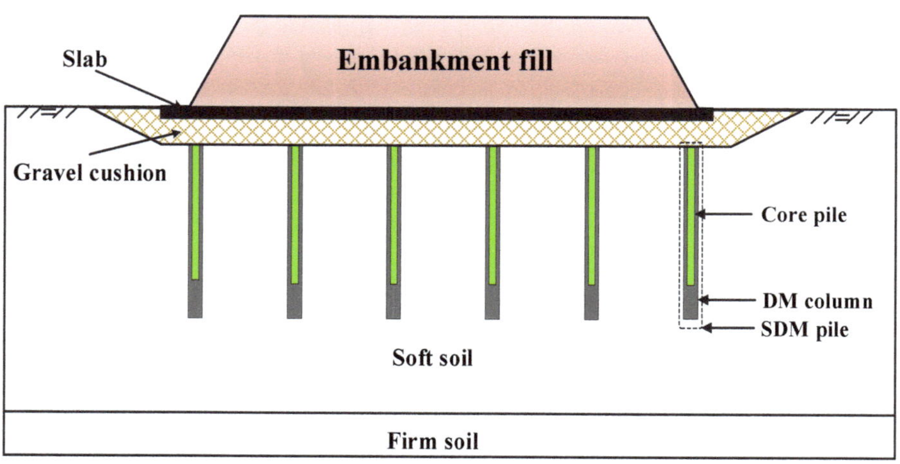

Figure 4. Schematic diagram of SDM column-supported embankment on soft soil [59].

The consolidation of soft soil is a key issue affecting the settlement of the embankment. Based on the assumption of a constant stress concentration ratio during the consolidation,

Zhang et al. [62] proposed an analytical model of consolidation for the SDM column-supported embankment without considering pile–soil interaction, as shown in Figure 5. Compared to a pure DM column, the existence of the PC pile in the DM column significantly reduced the maximum excess pore water pressure induced by embankment loads, which was attributed to the core pile playing a significant role in load shear due to its high modulus. Yang et al. [63] derived the consolidation equations for SDM columns (floating long core, penetrated long core, and short core) composite foundations. Subsequently, the analytical solutions for the excess pore water pressure and the overall average consolidation degree of the SDM column composite foundation for the surrounding soil and underlying stratum were established. The results indicate that the consolidation rate increases with the penetration ratio of the pile and the compression modulus of the underlying stratum for the floating long-core SDM column composite foundation. However, the length and stiffness of the outer core do not affect the consolidation rate. The consolidation rate increases with an increasing compression modulus for the SDM column composite foundations with a long core [64]. Regarding the SDM column composite foundation with a short core, the consolidation rate increases with the replacement ratio of this composite pile, the core length ratio, and the cemented soil compression modulus under a variable load [65]. However, the aforementioned solutions do not consider the pile pierced into the gravel cushion and the additional stress that varies with time and depth. Zhang et al. [30] developed an analytical solution for the consolidation of the composite ground reinforced by SDM columns based on a modified equal strain assumption.

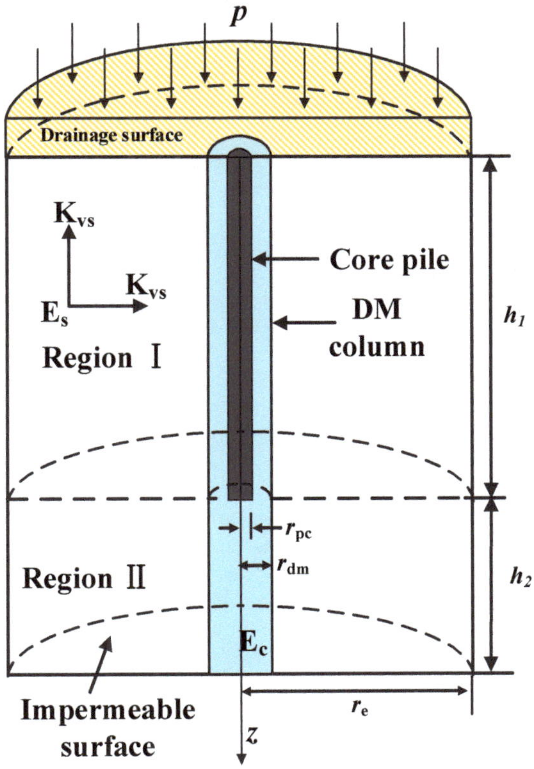

Figure 5. Analytical consolidation model of the SDM column-supported embankment [62].

The main function of the SDM column is to transfer the embankment loads to the subsoils, thus reducing settlement through the arching effect. The SDM column also plays a crucial role of increasing the slope stability against horizontal or sliding forces under embankment loads. Zhang et al. [66] conducted scale-down 1g (normal gravity) model tests to illustrate the failure mechanism of the SDM column-supported embankment. The results show that the SDM columns under the embankment firstly failed by compression with local bulging of the inner core bottom, followed by the sequence bending failures of the columns below the embankment slope subjected to the horizontal force of the soil. The sliding surface of the foundation was not completely through the pile damage location. In addition, the sequence and modes of the columns' failure, as well as the load transfer obtained from the indoor tests, were verified by the numerical analysis considering the strain softening of the SDM column. It was concluded that the progressive failure mechanism of the SDM column-supported embankment is due to the strain softening of the material and the uneven distribution of additional stresses [67]. In addition, due to the lack of a technical code for the calculation of the safety factor of the embankment supported by SDM columns, Zhang et al. [66] adopted the semi-rigid column method based on residual strength in the currently used code. This yielded a close safety factor for the tested results, although its feasibility still needs further study.

Combined with existing findings, the SDM column is currently widely employed in industrial and civil construction engineering. Nonetheless, its applications in the fields of highways and railways have been limited because the force form of the SDM column composite foundation under flexible loads is different from that under rigid loads. The existing research mainly concentrates on the load transfer mechanism, the settlement calculation, and the consolidation of the SDM column-supported embankment with the rigid slab. However, there have been few systematic studies of stability calculations, consolidation characteristics, and the seismic performance of this type of column-supported embankment.

3. Status of Research on Variable-Section Pile Technology

Most recent ground treatment and pile foundation engineering has involved the use of the piles with equal sections. However, the conventional equal-section piles often fail to consider the pile tip resistance and lateral friction resistance, resulting in certain limitations for deep and weak soil treatment; in addition, piled foundation engineering requires high bearing capacity and strict settlement control. Piles with a variable section, such as squeezed branch piles and PGPN piles, have been improved compared to traditional piles by enhancement of the frictional properties of the pile–soil interface and the expanded head at the pile tip [68]. This enhancement leads to an improved pile-end bearing performance, allowing for the full mobilization of the bearing capacity of the surrounding soil and effectively increasing the bearing capacity of the single pile, as well as meeting the requirements of complex projects [69,70]. In particular, the stratum often appears to be layered due to deposition, indicating a stratified distribution of soft soil and medium compressible soil. If the conventional equal-section piles are adopted to reinforce this kind of layered soft ground, the design leads to a high replacement rate of piles in medium compressible soil layers, which leads to resource waste and increases the cost of the projects. For such a situation, bi-directional mixing technology using automatic expanding blades (see Figure 6) enables the realization of the variable-section mixing pile. This variable-section mixing pile (shown in Figure 7) optimizes the vertical stiffness for layered foundations, forming a layered composite foundation in which the replacement rate varies with the nature of the soil layer, and realizing a cost-effective treatment of layered foundations [71]. To date, three kinds of piles with variable sections, namely, the squeezed branch pile, PGPN pile, and variable-section deep mixing (VDM) column, have been widely studied and applied in coastal regions because of their superior engineering properties.

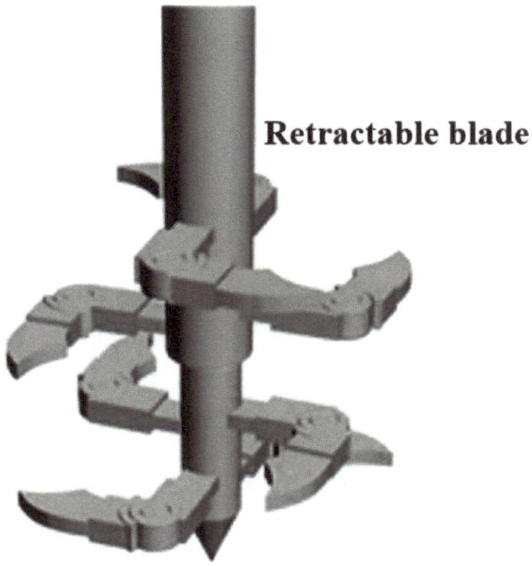

Figure 6. Schematic of the automatic retractable mixing blade.

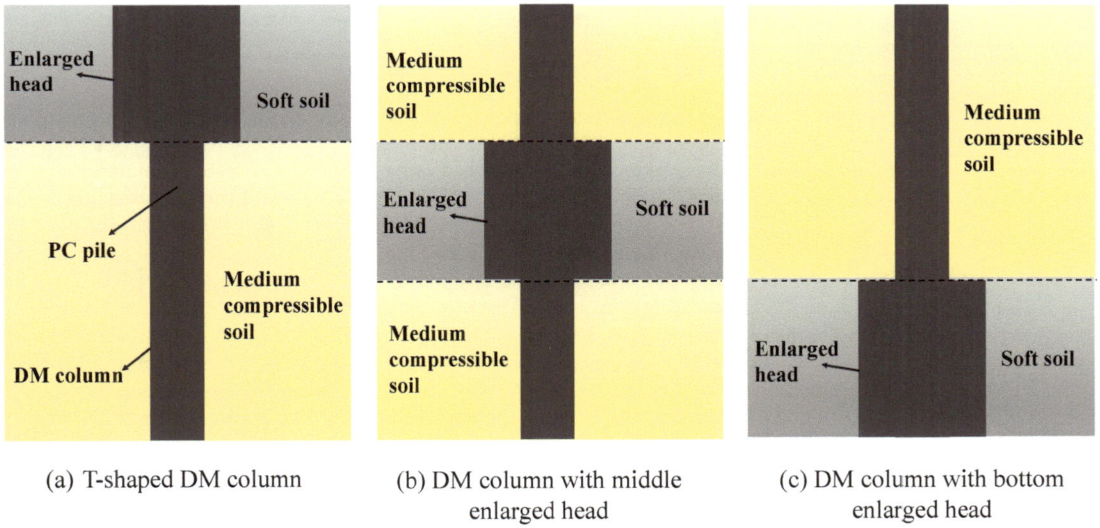

Figure 7. Schematic diagram of VDM columns reinforcing layer-forming soft ground.

3.1. Squeezed Branch Pile Technology

The squeezed branch pile (shown in Figure 8) is a novel variable-section pile that features multiple enlarged parts at specific locations along the bored pile shaft, designed according to the distribution of soil strata. Compared with conventional bored piles, it can significantly improve the vertical bearing capacity and pull-out resistance [72]. Since the 1990s, this pile type has been widely used in foundation engineering in China. Subsequently, the compressive, pullout, horizontal, and seismic performances of the squeezed branch piles have been extensively studied.

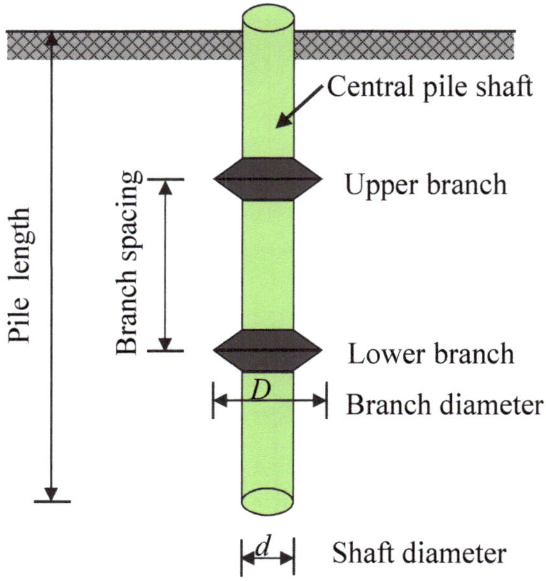

Figure 8. Schematic diagram of the squeezed branch pile.

The effect of compressive behaviors on the bearing mechanism [73,74], the calculation of the bearing capacity [75,76], and the optimization of parameters [77–80] of the squeezed branch pile have been extensively studied using laboratory model tests, numerical analyses, and field tests. Yang et al. [81,82] found that the pile lateral friction resistance and the branch resistance have an obvious time sequence effect, and the friction resistance occurs before the plate. Zhang et al. [83] indicated that the displacement field of both the branch and the pile tip had a radial shape, and to reduce the negative effect of soil displacement on the frictional resistance at the interface under the branches, it was proposed that the branch spacing exceeds three times the diameter of the branch. Concerning the theoretical calculation of bearing capacity and settlement, Li et al. [84] proposed a simplified nonlinear method for calculating the bearing capacity and settlement of the squeezed branch pile using a hyperbolic model and applying a segmental displacement coordinated iterative algorithm, which was verified by the measured results in field tests. Gao et al. [85] proposed a method for calculating the horizontal bearing capacity of the squeezed branch pile; its results were in good agreement with the numerical results. When the squeezed branch pile is used to support foundation pits, its maximum bending moment and axial force are reduced by 20–60% compared with the traditional cantilever support, and the overturning moment provided by the branches can effectively improve the bending resistance of the squeezed branch pile [86].

Unlike ordinary uplift piles, whose bearing capacity is contributed by the lateral frictional resistance and the pile's self-weight, the presence of the branches can effectively provide the tip load, thus effectively increasing its extracted load capacity [73]. The load transfer process of the squeezed branch pile is gradually exerted from the lateral frictional resistance of the pile shaft to each branch. During the damage process, the plastic zone at the branch initially appears and then gradually develops into the damage surface as the uplift load increases, caused by the pile–soil sliding [87]. By fitting the load-deformation curves of 22 squeezed branch piles obtained from the field tests using the modified hyperbolic model, Jiang et al. [88] found that the calculated uplift loads of the fitted curves were in agreement with the measured values. In addition, the tensile strength provided by the unit pile length of the squeezed branch pile was 7.5% higher than that of the straight pile in clayey strata [89].

3.2. Pre-Bored Grouting Planted Nodular (PGPN) Pile

Since its emergence, the pre-bored grouting planted nodular (PGPN) pile has attracted significant attention from scholars at home and abroad. This pile is formed by inserting a precast nodular pile and cemented soil using the static drilling and rooting method, as shown in Figure 9. Numerous studies of its compressive [47,90] and pullout-bearing characteristics [68,69] have been carried out. Kiya et al. [91] analyzed the effect of the size and shape of the enlarged base on the bearing performance of the PGPN pile by conducting experimental tests, and they concluded that increasing the diameter and height of the enlarged head could significantly improve the bearing capacity of the PGPN pile. Karkee et al. [92] proposed a reliability-based model for calculating the ultimate bearing capacity of PGPN piles with 95% confidence based on the field loading tests of PGPN piles with a pile top displacement of 13 mm. Hirai [93] studied the compressive bearing characteristics of a piled raft with PGPN piles using a Winkler model approach. The theoretical relationship between the pile-end bearing capacity and the displacement was established.

In addition, the vertical stiffness coefficients in non-homogeneous soil have been determined. Professor Gong's group systematically studied the compressive, pullout, and horizontal bearing characteristics of PGPN piles in soft soil, which included the load transfer mechanism [94–96], the calculation method of settlement and load carrying capacity [97–99], the influence of key parameters [100–103], and the longitudinal dynamic properties [104–108]. Field static loading tests revealed that the bearing capacity of PGPN piles exceeds that of ordinary bored piles, whose displacement is controlled by the precast pile, and the lateral friction resistance is 1.05–1.10 times that of bored piles [94]. To address the problem of the low strength of the bearing layer, sand filling can effectively improve the soil properties around the PGPN pile and the frictional behavior of the pile–soil interface; the pile shaft resistance coefficient can then reach values of 1.15–1.40 [100]. Based on the results of laboratory model tests, Zhou et al. [95] verified that the pile tip displacement of the PGPN pile can be calculated using the theoretical method of the conventional pile.

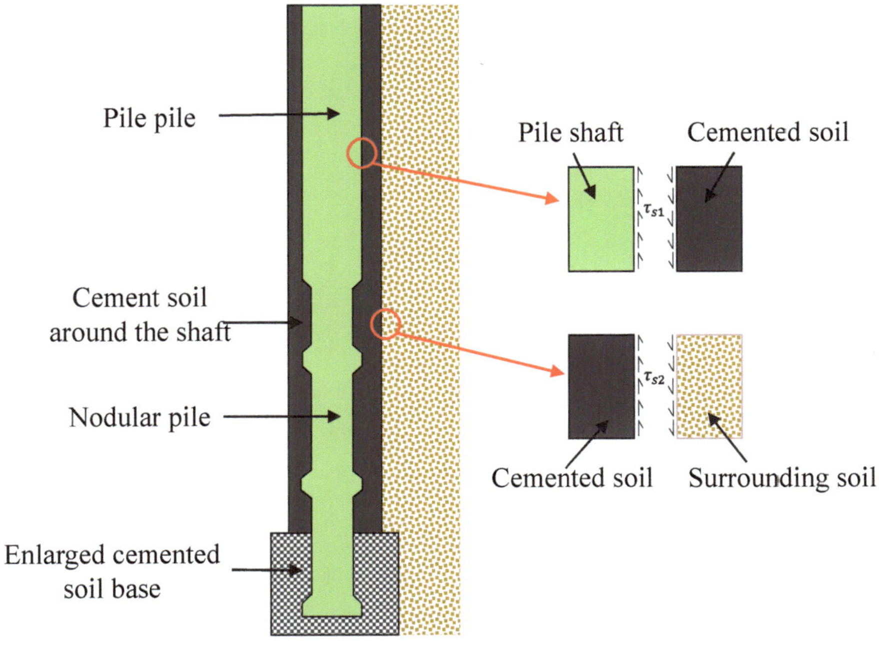

Figure 9. Schematic diagram of the PGPN pile [109].

Numerical analysis shows that the compressive characteristics of the PGPN pile are better than the pullout characteristics. The cemented soil in the non-expanded section of a PGPN pile only transforms the shear stress, but does not bear the load, and the proportion of the total tip resistance of the precast pile is smaller than that borne by the enlarged base of the cemented soil [110]. The total lateral friction reduction factor (λ) was defined as the ratio of the shaft resistance between the uplift pile and the compressive pile, and it was found that λ does not vary with the enlarged base diameter [111].

3.3. Variable-Section Composite Pile Technology

The construction technology, mechanism, and design method of the VDM column have been developed and have matured. However, due to the limited strength of cemented soil, it is difficult for the VDM column to meet the requirements of high bearing capacity and strict settlement control for deep soft soil ground treatment. The idea of a novel composite pile, termed the precast cement pile reinforced by cemented soil with a variable section (PCCV), which is formed by inserting a PC pile into a VDM column, can effectively solve the problem of the insufficient bearing capacity of the VDM column [112], as shown in Figure 10. PCCVs can be divided into three types of piles having top, middle, and bottom enlarged heads according to the location of the variable section. Among these, the replacement rate of the DM column varies with the mechanical properties of the soil layer, which can ensure that the PCCV has a high bearing capacity and can reduce resource use and project costs, reflecting the concepts of efficient energy saving and environmental ground treatment.

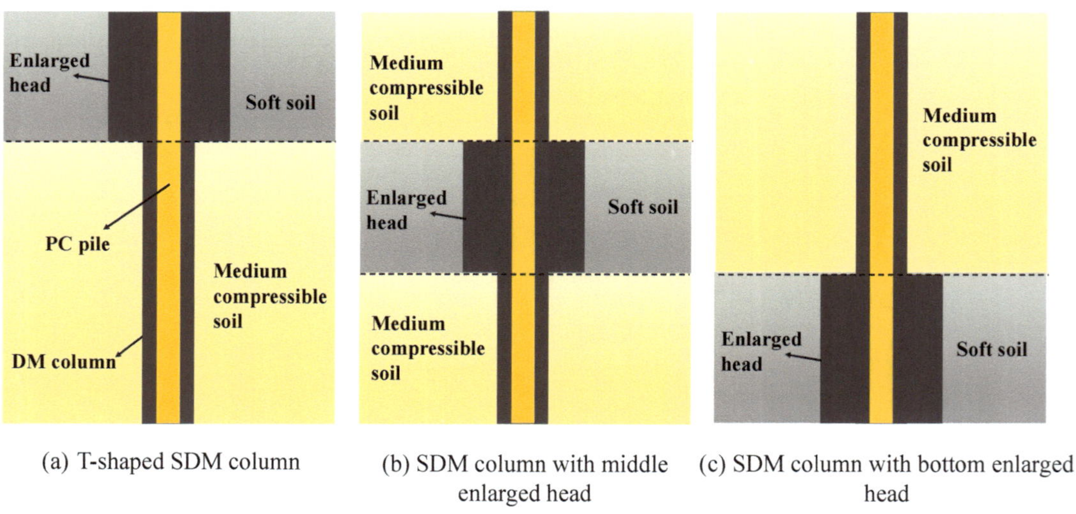

Figure 10. Schematic diagram of the PCCV.

4. Status of Research on Low-Carbon Energy-Saving Reinforcement Technologies

Traditionally, Portland cement has been used as the main curing material for ground improvement and piling works. However, cement-based curing agents cannot be adapted to the needs of low-carbon green engineering construction, with the main problems being: ① Significant consumption of resources and energy: the calcination temperature during cement production is about 1450 °C, which requires a large amount of energy [113,114]. In China, for example, each ton of cement clinker produced requires an average consumption of 0.95 tons of limestone, 0.13 tons of clay, and 0.11 tons of standard coal, and it is an irreversible process [115,116]. China's limestone resources can last for about 35–40 years. ② Serious CO_2 emissions and air pollution: the production of 1 ton of cement emits about 0.85 tons of CO_2, and the cement industry is one of the main sources of anthropogenic

CO_2 emissions. ③ Cement production also emits large amounts of dust, soot, and toxic gases, such as sulfur dioxide, nitrogen oxides, and fluorine gas, causing significant environmental pollution. ④ Cement-cured soils react more slowly, generally requiring 28 days of maintenance before they can be carried [117,118].

For this reason, in recent years, studies concerning new low-carbon and energy-saving curing agents have become a focus of research in the field of engineering. New curing materials, such as reused industrial waste and magnesium oxide cement, have been studied and applied.

4.1. Status of Research on Industrial Solid Waste Curing Technology

With rapid modern industrialization, the generation of a large amount of industrial waste residue can result in environmental pollution and resource waste if the waste is not effectively treated in a timely manner. In recent years, the use of alkali-activated binding materials, as new environmentally friendly curing agents, has attracted extensive attention worldwide [119]. Alkali-activated binding materials use an alkaline activator to activate industrial by-products/waste with cementing components, resulting in common industrial by-products such as blast furnace slag, fly ash, phosphate slag, steel slag, and silicon powder. Among these, blast furnace slag and fly ash have been more widely researched and are considered mature binding materials due to their more stable and excellent performance [120,121].

(1) Blast furnace slag

In the 1940s, researchers began to focus on alkali-activated binding materials, studying the binding properties of blast furnace slag after activation of KOH and NaOH [122,123]. However, to increase the activity of blast furnace slag, which is low at room temperature, the slag is water quenched and finely ground to obtain ground granulated blast furnace slag (GGBS), which has been widely used due to its desirable properties [120,121]. The current activators for GGBS include NaOH, Na_2CO_3, Na_2SO_4, lime, and MgO. Yi et al. [27] found that the activation effect of Na_2CO_3 on a GGBS–soil admixture was very limited, while NaOH had the best activation effect on a GGBS–soil admixture, with the highest strength at different ages among the aforementioned four activators. Furthermore, lime has been used as an activator of GGBS for soft soil subgrade improvement, non-sintered bricks, and clay stabilization [124–127]. In addition, the micro-mechanism [128,129], strength characteristics, and durability [130,131] of lime-activated GGBS–soil admixtures have been systematically studied. The results show that lime-activated GGBS can effectively reduce the swelling characteristic of the soil, because of its higher unconfined compressive strength and better durability, which can better mitigate the adverse effects of flooding on the embankment than cement-activated GGBS soil.

As part of the continuous research into the GGBS activated by alkaline materials, Liu et al. presented a series of studies of soft soil treatment with active MgO as the activation material of GGBS, investigating the action mechanism [25,26,132], microstructure, strength properties [119], acid resistance, and durability [133–135]. The results indicate that the effect of MgO-activated GGBS is significantly better than that of lime-activated GGBS, and that the strength of GGBS-MgO reinforced soil can exceed that of Portland cemented soil with the same dosage. Micromechanical analysis revealed that calcium silicate hydrate (CSH) is the main hydration product of both MgO–GGBS and lime–GGBS; the latter can effectively encapsulate and connect the soil particles, whereas the former can produce more hydrotalcite to fill the pores, thus explaining the main reason for the superior strength of MgO–GGBS over GGBS–lime and Portland cement (PC).

(2) Fly ash

Fly ash is a common component of cement and concrete and is widely used in projects such as railway bridges, highway pavements, and dams. Research has revealed that mixing various activators such as lime, cement, and MgO with fly ash can be used as fly-ash-based cementitious materials. At the end of the 1990s, preliminary investigations into the strength properties, reaction mechanism, and microstructure of cement and fly-ash-solidified clay were performed [136–138]. Nalbantoğlu et al. [139] found that fly ash can effectively reduce the swelling of highly plastic clay. Kolias et al. [140] performed a microscopic analysis of high-calcium fly ash and cement-solidified silty clay, and found that cement can provide an alkaline environment for the reaction, and can thus promote the formation of fly ash into calcium silicate hydrate. Among the continuous research into fly-ash-solidified soft soil technology, extensive research on the performance of different types of activators to excite fly-ash-cured soft soil has been carried out. The activations studied in more detail include NaOH, Na_2CO_3, and $Na_2SiO_3{\cdot}9H_2O$, as well as active MgO. It has been shown that the first three of these activators can effectively excite fly ash–sludge admixtures by improving their unconfined compressive strength, and that NaOH and $Na_2SiO_3{\cdot}H_2O$ are more effective than Na_2CO_3. Furthermore, the whole process model of the sludge improvement induced by the intrinsic chemical reaction of alkali-activated fly ash has been established.

Concerning MgO-activated fly-ash-cured soft soils, Professor Wang's team carried out a series of studies of the microscopic mechanism [141,142], strength properties [143–145], and durability [143–145] of stabilized dredged sludge. X-ray diffraction analysis and scanning electron microscopy indicated that MgO-activated fly ash could generate cementation products, $Mg(OH)_2$ and magnesium silicate hydrate (M-S-H), which transformed the intragranular pores into inter-particle pores and promoted the microstructure of soft soils to be denser and more integral. The whole process model of stabilized silt induced by the intrinsic chemical reaction of reactive MgO-fly ash was established. In addition, the results of the durability tests showed that the reactive MgO-fly ash stabilization method could effectively enhance the freeze–thaw resistance and water stability of the sludge. In addition, after 10 freeze–thaw cycles, the surface of the specimens did not show significant cracks.

4.2. Research Status of Magnesium Oxide Curing Technology

In 2001, based on the concept of environmental protection, Harrison [146,147] was the first to invent a new curing material by mixing active magnesium oxide (MgO) and Portland cement. Compared with conventional PC, the MgO mixture, which has a lower calcination temperature, can absorb a large amount of CO_2 gas during the hydration process, and the hydrated magnesium carbonates generated by the reaction have better curing strength, durability, carbonation resistance, and salt resistance, as shown in Figure 11. Moreover, the hydrated magnesium carbonates can re-generate MgO after calcination, which realizes the recycling of resources and effectively reduces the emission of greenhouse gases. Figure 11 shows the schematic diagram of the cycle of MgO and CO_2.

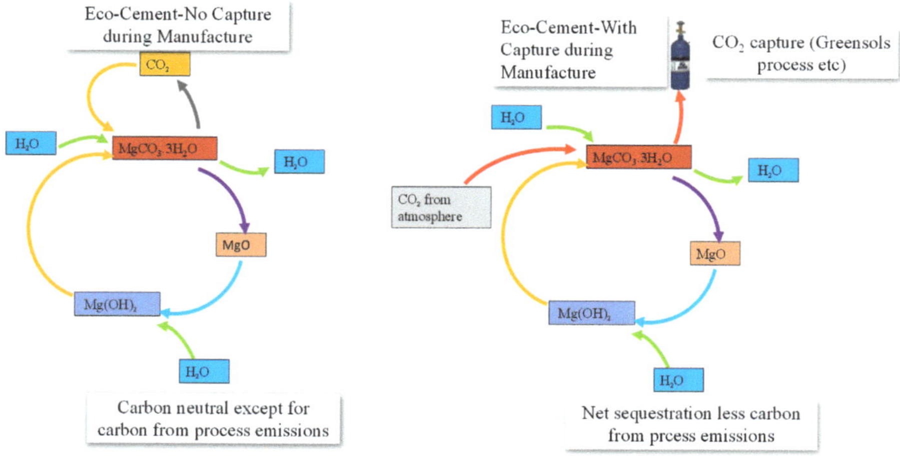

Figure 11. Schematic diagram of MgO and CO_2 cycles [146].

Subsequently, Professor Al-Tabbaa's group at the University of Cambridge conducted a detailed study of reactive MgO cements, including of their hydration properties [148,149], microstructure [150], carbonation properties [151,152], and applications of building blocks [153]. It was shown that the hydration process of activated MgO–PC blended cement proceeds independently to produce $Mg(OH)_2$ and hydrated magnesium silicate. In addition, the former has low cementitious strength and the involvement of activated MgO does not contribute to the mechanical properties of the cement. However, the strength of blocks using independent activated MgO as a binding agent was 2–3 times that of PC blocks under the forced carbonation with a CO_2 concentration of 20%, which produces hydrated magnesium carbonate with high cementitious strength due to the carbonation of MgO. Based on the results of acid resistance and corrosion resistance tests, it was found that the carbonized blocks were more resistant to sulfuric acid than PC blocks; however, they were less resistant to hydrochloric acid erosion than PC blocks [154,155].

On the basis of the successful application of magnesium oxide in building masonry, Professor Liu first proposed the soft soil ground treatment reinforced by active MgO under forced carbonation. The micro-mechanisms, freeze–thaw cycles [116,156–158], permeability [159,160], resistivity [113,161,162], and acid resistance [163] of different types of soft soils, including sandy, silt, and clayey soils treated with active MgO, were investigated. Laboratory experiments, microscopic tests, and model tests were used to systematically analyze the effects of crucial parameters, such as MgO dosage, carbonization time, initial moisture content, MgO activity index, CO_2 pressure, and compaction degree, on the performance of carbonized soil [164]. The critical dosage of MgO (20–25%), the critical carbonation time (8–10 h), the prediction equation of the carbonation strength, and the prediction model of the resistivity of carbonated soil were obtained. Based on the above research results and in response to the problems with respect to field application, Professor Liu et al. [165–167] proposed deep MgO mixing (DMM) column technology using active MgO as an alternative to PC under CO_2 gas at a specific pressure to carbonize the mixing soil for ground treatment.

The chemical reaction between MgO, water, and CO_2 can rapidly reduce the water content and the porosity of soil, and increase the cementation strength. The results of indoor model tests indicate that the strength of deep MgO mixing (DMM) columns decreased exponentially with the increasing initial water content, and conversely increased with the CO_2 pressure [168]. Yi et al. [169] concluded that the DMM column still has considerable unconfined compressive strength with low MgO content in soft soil. In addition, the time required for carbonation reactions is significantly less than the 28 days curing time required for conventional cement curing.

However, the complex DMM column method under forced CO_2 carbonation fails to ensure good agreement with the original approach. The CO_2 gas tends to flow upward, thus affecting the uniformity of the DMM column and resulting in a limited depth of soft ground treatment. To address these problems, Liu et al. [170] first proposed the carbonized composite pile (CCP), which is typically formed by inserting a pervious concrete pile into a DMM column, after the construction of the DMM column, as shown in Figure 12. The advantage of this method is that the CO_2 gas can enter and react with the soil–MgO mixture through the pervious concrete pile, where the uniformly distributed gas outlet channels can promote a more uniform carbonization effect of the DMM column. Wang [171] proved that the carbonation strength using this carbonization method can reach that obtained from triaxial compression tests under constant conditions.

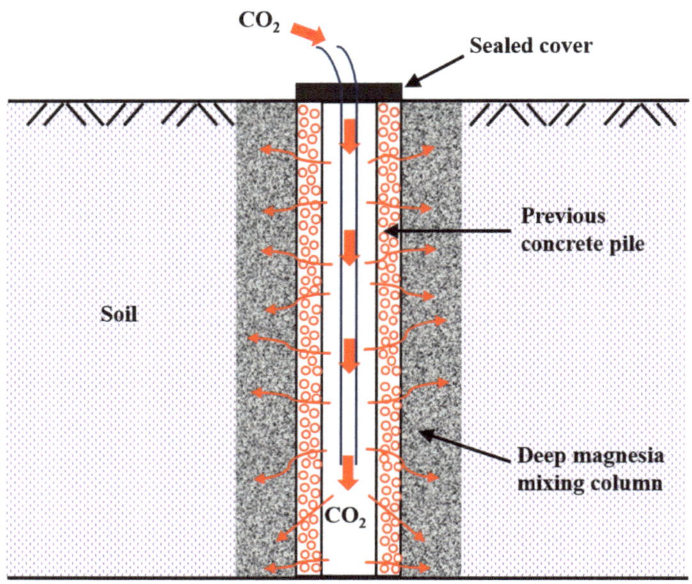

Figure 12. Schematic diagram of the carbonized composite pile.

The research described above reveals that reducing the amount of cement or using low-carbon materials to replace cement is the current goal of the engineering community globally, and significant progress has been made in recent years. In particular, the reinforcement mechanism and effect of industrial waste utilization and MgO carbonization technologies have been clarified, and these technologies have been successfully applied in engineering practice. It is still necessary to identify methods for further enhancing the combination of low-carbon curing materials and engineering applications, study their mechanism of action and effects, and establish practical engineering design and analysis methods. In addition, it is necessary to better understand the requirements of low-carbon sustainable development strategies. Table 2 summarizes the characteristics of various efficient and low-carbon pile technologies, including the depth, cross-section type, and technical maturity, which can provide some suggestions for engineering practice.

Table 2. The characteristics of various efficient and low-carbon pile technologies.

<table>
<tr><th>Pile Type</th><th>Depth</th><th>Cross-Section Type</th><th>Technical Maturity</th></tr>
<tr><td>CCP</td><td>Undetermined</td><td rowspan="3">Constant section</td><td>Novel</td></tr>
<tr><td>DM column</td><td rowspan="4"><20 m</td><td>Mature</td></tr>
<tr><td>DMM column</td><td>Further development</td></tr>
<tr><td>TDM column</td><td rowspan="2">Variable section</td><td>Mature</td></tr>
<tr><td>VDM column</td><td>Further development</td></tr>
<tr><td>PC pile</td><td rowspan="2">Penetrated soft soil layers</td><td rowspan="2">Constant section</td><td>Mature</td></tr>
<tr><td>PHC pile</td><td>Mature</td></tr>
<tr><td>PCCV</td><td rowspan="3">Controlled by length of DM column</td><td rowspan="2">Variable section</td><td>Novel</td></tr>
<tr><td>PGPN</td><td>Mature</td></tr>
<tr><td>SDM column</td><td>Constant section</td><td>Mature</td></tr>
</table>

5. Conclusions

This paper reviews and summarizes the current research status and practical applications of efficient and low-carbon pile technologies in deep soft soil strata, including the SDM column, the squeezed branch piles, the PGPN pile, the PCCV piles, and the CCP, as well as several low-carbon and energy-saving curing agents. Thus, this review may be helpful for overcoming the technical bottleneck of deep and weak soil foundation treatment, and promoting low-carbon environmental protection and sustainable development of foundation treatment technology. The conclusions that can be drawn are as follows:

(1) The interfacial mechanical properties, and horizontal and vertical bearing properties, of the SDM column, as well as the bearing mechanism and failure mode of the SDM column-supported embankment, have been systematically analyzed. Studies of stability calculations, consolidation characteristics, and the seismic performance of this column-supported embankment need to be conducted in further research.

(2) Squeezed branch piles, PGPN piles, and PCCV piles, which are variable-section piles, have significant advantages for the layered strata because they adopt an optimal replacement ratio of branches or enlarged heads to effectively improve the bearing capacity. Compared to traditional piles, their enhancements have been achieved by increasing the frictional properties of the pile–soil interface and the expanded head to fully mobilize the bearing capacity of the surrounding soil.

(3) The low-carbon curing agents of pile materials widely used for soft soil stabilization can provide a solution for novel piles, whose application in engineering is insufficient. In addition, determining how to further enhance the combination of low-carbon curing agents and existing pile technologies, studying their mechanism of action and effects, and establishing practical engineering design and analysis methods remain to be addressed.

Author Contributions: Conceptualization, C.Z., S.L. and J.H.; methodology, J.H. and C.J.; investigation, C.Z. and J.H.; resources, Z.C., C.J., X.D. and G.C.; writing—original draft preparation, C.Z.; writing—review and editing, J.H. and L.T.; visualization, S.L. and Z.C.; supervision, S.L. and L.T.; funding acquisition, S.L., Z.C., C.J. and G.C. All authors have read and agreed to the published version of the manuscript.

Funding: This research work is supported by the National Natural Science Foundation of China (No. 52078129 & 42277146),the Transportation Science and Technology Project of Jiangsu Province of China, the Introduction Program of High-Level Innovation and Entrepreneurship Talents in Jiangsu Province (No. JSSCBS20210385) and the Natural science research project of colleges and universities of Jiangsu Province (No. 21KJB170019).

Data Availability Statement: Not applicable.

Acknowledgments: All of these financial supports are gratefully acknowledged. The authors also wish to thank Southeast University for providing the access to the software and other facilities.

Conflicts of Interest: No potential conflict of interest was reported by the authors.

Nomenclature

CCP	Carbonized composite pile
CSH	Calcium silicate hydrate
DM	Deep mixing
DMM	Deep MgO mixing
D_{pc}/D_{dm}	Core diameter ratio
GGBS	Ground granulated blast furnace slag
L_{pc}/L_{dm}	core length ratio
m	Proportional coefficient of the horizontal resistance coefficient
PC	Precast concrete
PCCV	Precast cement pile reinforced by cemented soil with variable section

PGPN	Pre-bored grouting plated nodular
PHC	Pre-stressed high-strength concrete
SDM	Stiffened deep mixing
TDM	T-shaped deep mixing
VDM	Variable-section deep mixing

References

1. Lai, F.; Yang, D.; Liu, S.; Zhang, H.; Cheng, Y. Towards an improved analytical framework to estimate active earth pressure in narrow c–ϕ soils behind rotating walls about the base. *Comput. Geotech.* **2022**, *141*, 104544. [CrossRef]
2. Wang, X.; Shen, L.; Shi, S. Evaluation of underground space perception: A user-perspective investigation. *Tunn. Undergr. Space Tech.* **2023**, *131*, 104822. [CrossRef]
3. Wang, X.; Li, S.; Xu, Z.; Li, X.; Lin, P.; Lin, C. An interval risk assessment method and management of water inflow and inrush in course of karst tunnel excavation. *Tunn. Undergr. Space Tech.* **2019**, *92*, 103033. [CrossRef]
4. Wu, W.; Zhang, Y. A Review of Pile Foundations in Viscoelastic Medium: Dynamic Analysis and Wave Propagation Modeling. *Energies* **2022**, *15*, 9432. [CrossRef]
5. Liu, H. Study on green ground improvement technique. *China Civ. Eng. J.* **2018**, *51*, 121–128.
6. Liu, S.; Zhou, J.; Zhang, D. State of the art of the ground improvement technology in China. *China Civ. Eng. J.* **2020**, *53*, 94–110.
7. Zheng, G.; Long, X.; Xie, Y. State-of-the-art techniques for ground improvement in China. *China Civ. Eng. J.* **2012**, *45*, 127–146.
8. Behnood, A. Soil and clay stabilization with calcium- and non-calcium-based additives: A state-of-the-art review of challenges, approaches and techniques. *Transp. Geotech.* **2018**, *17*, 14–32. [CrossRef]
9. Gomes Correia, A.; Winter, M.G.; Puppala, A.J. A review of sustainable approaches in transport infrastructure geotechnics. *Transp. Geotech.* **2016**, *7*, 21–28. [CrossRef]
10. *GB/T 50783-2012*; Techniacl Code for Composite Foundation. Ministry of Housing and Urban-Rural Development of the People's Republic of China, China Planning Press: Beijing, China, 2012.
11. Dai, X.; Zheng, G.; Zhang, N. Seismic performance of PHC pipe piles of high-rise building in layered soft soils. *China Civ. Eng. J.* **2019**, *52*, 248–256.
12. Zheng, G.; Zhang, N.; Ba, Z. Seismic performance of PHC pipe piles in layered soft soils. *Chin. J. Geotech. Eng.* **2013**, *35*, 506–510.
13. *JGJ/T 327-2014*; Techinal Specification for Strength Composite Piles. Ministry of Housing and Urban-Rural Development of the People's Republic of China, China Architecture & Building Press: Beijing, China, 2014.
14. Xiao, H.; Luo, Q.; Deng, J. Test study of reinforcing railway subgrades by using concrete compacted base-enlarged piles and geogrids. *Rock Soil Mech.* **2008**, *29*, 2157–2162.
15. Ye, G.; Cai, Y.; Zhang, Z. Research on calculation of pile-soil stress ratio for composite foundation reinforced by stiffened deep mixed piles. *Rock Soil Mech.* **2016**, *37*, 672–678.
16. Li, J.; Zhang, Y.; Deng, Y. Load transfer mechanism of composite pile composed of jet-mixing cement and phc pile with core concrete. *Chin. J. Rock Mech. Eng.* **2014**, *33*, 3068–3076.
17. Qian, Y.; Xu, Z.; Deng, Y. Engineering application and test analysis of strength composite piles. *Chin. J. Geotech. Eng.* **2013**, *35*, 998–1001.
18. Li, J.; Deng, Y.; Song, G. Analysis of load-bearing mechanism of composite foundation of plain concrete reinforced ce-ment-soil mixing piles. *Rock Soil Mech.* **2009**, *30*, 181–185.
19. Liu, S. *Innovative Deep Mixing Method–Theory and Technology*; Southeast University Press: Nanjing, China, 2014.
20. Liu, S.; Du, Y.; Yi, Y.; Puppala, A.J. Field Investigations on Performance of T-Shaped Deep Mixed Soil Cement Column–Supported Embankments over Soft Ground. *J. Geotech. Geoenviron.* **2012**, *138*, 718–727. [CrossRef]
21. Yi, Y.; Liu, S.; Puppala, A.J. Laboratory modelling of T-shaped soil–cement column for soft ground treatment under embankment. *Géotechnique* **2015**, *66*, 85–89. [CrossRef]
22. Li, L.; Cheng, X. Load Transfer Method for Squeezed and Branch Piles Considering Cavity Esspansion Theory. *China J. Highw. Transp.* **2018**, *31*, 20–29.
23. Lu, C.; Jia, Y.; Zhou, L. Project performance study of model piles with branches and plates under repeated load in silty clay. *Rock Soil Mech.* **2008**, *29*, 431–436.
24. Li, M.; Cai, G.; Wang, Q.; Liu, S.; He, H.; Liu, X.; Shi, W. The state of the art of carbonation technology in geotechnical engineering: A comprehensive review. *Renew. Sustain. Energy Rev.* **2023**, *171*, 112986. [CrossRef]
25. Yi, Y.; Liska, M.; Jin, F.; Al-Tabbaa, A. Mechanism of reactive magnesia–ground granulated blastfurnace slag (GGBS) soil stabilization. *Can. Geotech. J.* **2016**, *53*, 773–782. [CrossRef]
26. Yi, Y.; Liska, M.; Al-Tabbaa, A. Properties and microstructure of GGBS–magnesia pastes. *Adv. Cem. Res.* **2014**, *26*, 114–122. [CrossRef]
27. Yi, Y.; Gu, L.; Liu, S.; Jin, F. Magnesia reactivity on activating efficacy for ground granulated blastfurnace slag for soft clay stabilisation. *Appl. Clay Sci.* **2016**, *126*, 57–62. [CrossRef]
28. Zhu, S.; Chen, C.; Cai, H.; Mao, F. Analytical modeling for the load-transfer behavior of stiffened deep cement mixing (SDCM) pile with rigid cap in layer soils. *Comput. Geotech.* **2022**, *144*, 104618. [CrossRef]

29. Zhang, C.; Zhang, D.; Liu, S.; Han, J.; Jiang, C.; Zhao, Y. Analytical Solution for Consolidation Behaviors of Combined Composite Foundation Reinforced with Penetrated PCCSs and Floating DM Columns. *Geofluids* **2023**, *2023*, 7371850. [CrossRef]
30. Zhang, C.; Liu, S.; Zhang, D.; Lai, F.; Lu, T.; Liu, Y. A modified equal-strain solution for consolidation behavior of composite foundation reinforced by precast concrete piles improved with cement-treated soil. *Comput. Geotech.* **2022**, *150*, 104905. [CrossRef]
31. Zhou, M.; Li, Z.; Han, Y.; Ni, P.; Wang, Y. Experimental Study on the Vertical Bearing Capacity of Stiffened Deep Cement Mixing Piles. *Int. J. Geomech.* **2022**, *22*, 04022043. [CrossRef]
32. Zhou, J.; Yu, J.; Gong, X.; El Naggar, M.H.; Zhang, R. The effect of cemented soil strength on the frictional capacity of precast concrete pile–cemented soil interface. *Acta Geotech.* **2020**, *15*, 3271–3282. [CrossRef]
33. Li, L. *Study on the Bearing Capacity of Stiffened DCM Pile*; Southeast University: Nanjing, China, 2016.
34. Han, Y.; Cheng, J.; Zhou, M.; Ni, P.; Wang, Y. Experimental Study of Compaction and Expansion Effects Caused by Penetration of Core Pile During Construction of SDCM Pile. *Int. J. Geomech.* **2022**, *22*, 04022041. [CrossRef]
35. Wang, A.; Zhang, D.; Deng, Y. A Simplified Approach for Axial Response of Single Precast Concrete Piles in Cement-Treated Soil. *Int. J. Civ. Eng.* **2018**, *16*, 1491–1501. [CrossRef]
36. Du, G.; Wang, A.; Li, L.; Zhang, D. Calculation Approach for Lateral Bearing Capacity of Single Precast Concrete Piles with Improved Soil Surrounds. *Adv. Civ. Eng.* **2018**, *2018*, 5127927. [CrossRef]
37. Li, H.; Liu, S.; Tong, L. Field investigation of the performance of composite foundations reinforced by DCM-bored piles under lateral loads. *Constr. Build. Mater.* **2018**, *170*, 690–697. [CrossRef]
38. Wang, A.; Zhang, D.; Deng, Y. Lateral response of single piles in cement-improved soil: Numerical and theoretical investigation. *Comput. Geotech.* **2018**, *102*, 164–178. [CrossRef]
39. Jamsawang, P.; Bergado, D.T.; Voottipruex, P. Field behaviour of stiffened deep cement mixing piles. *Proc. Inst. Civ. Eng.-Ground Improv.* **2011**, *164*, 33–49. [CrossRef]
40. Zhou, Z.; Lin, F. Design and construction reason of reinforced concrete cement-soil composite pile with strong core–New pile type used in soft soil foundation. *Geotech. Eng.* **2000**, *12*, 19–21.
41. Wu, M.; Zhao, X. Bearing Behaviors of Stiffened Deep Cement Mixed Pile. *Trans. Tianjin Univ.* **2006**, *12*, 209–214.
42. *DBJ 53/T-19-2007*; Technical Specification for Concrete Core Mixing Pile. Construction Department of Yunnan Province, Yunnan Science and Technology Press: Kunming, China, 2007.
43. *JGJ/T 330-2014*; Technical Specification for Pile Foundation of Pipe Pile Embedded in Cement Soil. Ministry of Housing and Urban-Rural Development of the People's Republic of China, China Architecture & Building Press: Beijing, China, 2014.
44. Liu, H.; Ren, L.; Zheng, H. Full-scale model test on load transfer mechanism for jet grouting soil-cement-pile strengthened pile. *Rock Soil Mech.* **2010**, *31*, 1395–1401.
45. Wang, C.; Xu, Y.; Dong, P. Plate Load Tests of Composite Foundation Reinforced by Concrete-Cored DCM Pile. *Geotech. Geol. Eng.* **2014**, *32*, 85–96. [CrossRef]
46. Zhou, J.-J.; Wang, K.-H.; Gong, X.-N.; Zhang, R.-H. Bearing capacity and load transfer mechanism of a static drill rooted nodular pile in soft soil areas. *Zhejiang Univ. Sci. A* **2013**, *14*, 705–719.
47. Zhou, J.; Gong, X.; Wang, K.; Zhang, R.; Yan, T. A Model Test on the Behavior of a Static Drill Rooted Nodular Pile Under Compression. *Mar. Georesources Geotechnol.* **2015**, *34*, 293–301. [CrossRef]
48. Wonglert, A.; Jongpradist, P. Impact of reinforced core on performance and failure behavior of stiffened deep cement mixing piles. *Comput. Geotech.* **2015**, *69*, 93–104. [CrossRef]
49. Voottipruex, P.; Suksawat, T.; Bergado, D.T.; Jamsawang, P. Numerical simulations and parametric study of SDCM and DCM piles under full scale axial and lateral loads. *Comput. Geotech.* **2011**, *38*, 318–329. [CrossRef]
50. Gu, S.; Shi, J.; Wang, C. Theoretical study of core pile load transfer regularity of reinforced mixing pile. *Rock Soil Mech.* **2011**, *32*, 2473–2478.
51. Ren, L.; Liu, X.; Wang, G. Simplified calculation and analysis of load transfer behavior for single JPP. *Chin. J. Rock Mech. Eng.* **2010**, *29*, 1279–1287.
52. Liu, B. Experimental Study on the Rein Forced Mixing Pile Respectively Subjected to a Vertical Load or a Lateral load. Master's Thesis, Tianjin University, Tianjin, China, 2006.
53. Wang, L.; He, B.; Guo, Z.; Li, L. Field Tests of the Lateral Monotonic and Cyclic Performance of Jet-Grouting-Reinforced Cast-in-Place Piles. *J. Geotech. Geoenviron.* **2015**, *141*, 06015001. [CrossRef]
54. He, B.; Wang, L.; Hong, Y. Field testing of one-way and two-way cyclic lateral responses of single and jet-grouting reinforced piles in soft clay. *Acta Geotech.* **2017**, *12*, 1021–1034. [CrossRef]
55. He, B.; Wang, L.; Hong, Y. Capacity and failure mechanism of laterally loaded jet-grouting reinforced piles: Field and numerical investigation. *Sci. China Technol. Sci.* **2016**, *59*, 763–776. [CrossRef]
56. Huang, X.; Yue, J.; Li, L. Calculation method for horizontal resistance coefficient of foundation soil with composite piles. *Chin. J. Geotech. Eng.* **2011**, *33*, 192–196.
57. *JGJ 94-2008*; Technical Code for Building Pile Foundations. Ministry of Housing and Urban-Rural Development of the People's Republic of China, China Architecture & Building Press: Beijing, China, 2008.
58. Li, L.; Liu, S.; Zhang, D.; Deng, Y. Bearing Capacity Calculations of Strength Composite Piles. *Chin. J. Undergr. Space Eng.* **2015**, *11*, 43–47.

59. Zhang, Z.; Ye, G.; Cai, Y.; Zhang, Z. Centrifugal and numerical modeling of stiffened deep mixed column-supported embankment with slab over soft clay. *Can. Geotech. J.* **2019**, *56*, 1418–1432. [CrossRef]
60. Ye, G.; Cai, Y.; Zhang, Z. Numerical study on load transfer effect of Stiffened Deep Mixed column-supported embankment over soft soil. *KSCE J. Civ. Eng.* **2017**, *21*, 703–714. [CrossRef]
61. Ye, G.; Rao, F.; Zhang, Z.; Wang, M. Calculation Method for Settlement of Stiffened Deep Mixed Column-Supported Embankment over Soft Clay. In Proceedings of the GeoShanghai International Conference, Singapore, 27–30 May 2018.
62. Zhang, Z.; Rao, F.; Ye, G. Analytical modeling on consolidation of stiffened deep mixed column-reinforced soft soil under embankment. *Int. J. Numer. Anal. Methods Geomech.* **2019**, *44*, 137–158. [CrossRef]
63. Yang, T.; Dai, J.; Wang, H. Analytical solutions for consolidation of a composite ground with floating stiffened deep cement mixing columns with long core piles. *Chin. J. Geotech. Eng.* **2019**, *42*, 246–252.
64. Yang, T.; Dai, J.; Wang, H. Analytical solution for consolidation of a compoaite foundation with t-shaped stiffened deep cement mixing piles under rigid foundtion. *Ind. Constr.* **2018**, *48*, 105–110.
65. Yang, T.; Tang, F. Consolidation analysis of composite ground with stiffened deep cement mixing piles under time-dependent loading. *J. Build. Struct.* **2017**, *38*, 160–166.
66. Zhang, Z.; Ye, Y.; Ye, G. Model test on stability failure of short-cored stiffened deep mixed column-supported embankment. *J. Eng. Geol.* **2019**, *27*, 1063–1069.
67. Zhang, Z.; Zhang, C.; Ye, G. Progressive failure mechanism of stiffened deep mixed column-supported embankment. *Rock Soil Mech.* **2020**, *41*, 2122–2131.
68. Jia-Jin, Z.; Xiao-Nan, G.; Ri-Hong, Z.; Kui-Hua, W.; Tian-Long, Y. Shaft capacity of pre-bored grouted planted nodular pile under various overburden pressures in dense sand. *Mar. Georesources Geotechnol.* **2020**, *38*, 97–107. [CrossRef]
69. Zhou, J.; Gong, X.; Wang, K.; Zhang, R. Shaft capacity of the pre-bored grouted planted pile in dense sand. *Acta Geotech.* **2018**, *13*, 1227–1239. [CrossRef]
70. Zhou, J.; Gong, X.; Wang, K.; Zhang, R.; Yan, J. Testing and modeling the behavior of pre-bored grouting planted piles under compression and tension. *Acta Geotech.* **2017**, *12*, 1061–1075. [CrossRef]
71. Yan, C. Research on the Stability Analysis of Variable Deep Mixed Column Composite Foundation. Ph.D. Thesis, Southeast University, Nanjing, China, 2016.
72. Li, Q.; He, M.; Tang, Y. Experimental study on squeezed branch piles as retaining and protecting of foundation excavation. *Rock Soil Mech.* **2005**, *26*, 1667–1670.
73. Kong, L.; Jiang, L.; Chen, R. Response of squeezed branch piles under inclined uplift loads. *Chin. J. Appl. Mech.* **2013**, *30*, 228–233.
74. Lu, C.; Wang, C.; Meng, F. Model tests on piles with branches and plates in unsaturated silt under cyclic loads. *Chin. J. Geotech. Eng.* **2007**, *29*, 603–607.
75. Lu, C.; Wang, Z.; Meng, F. Experimental research on model pile with plates and branches under repeated loading in different soils. *Chin. J. Rock Mech. Eng.* **2007**, *1*, 3141–3147.
76. Chen, L.; Chang, D.; Li, G. Finite element anaylsis of bearing capacity of a single DX pile. *Eng. Mech.* **2002**, *19*, 67–72.
77. Wang, Y.; Xu, L.; Li, B. Finite element numerical study on the axial bearing behaviors and factors of squeezed branch pile. *China Civ. Eng. J.* **2015**, *48*, 158–162.
78. Li, L.; Li, X. Influence of squeezed branch pile bearing capacity on different numbers and positions of enlarged parts. *J. Shandong Univ. (Eng. Sci.)* **2016**, *46*, 88–94.
79. Kou, H.; Zhang, M.; Bai, X. Numerical Simulation Analysis of a New Pressured Multi-Branch Bored Pile. *Chin. J. Undergr. Space Eng.* **2012**, *8*, 602–607.
80. Lu, C.; Chen, Z.; Wang, J. Experimental study on bearing capacity behavior of disk pile with different shape of disk. *J. Zhejiang Univ. Technol.* **2015**, *43*, 279–282.
81. Chen, F.; Wu, K.; He, S. Field tests on load transfer performances of squeezed branch piles. *Chin. J. Geotech. Eng.* **2013**, *35*, 990–993.
82. Yang, J.; Pu, C. Distributed detection and bearing mechanism of squeezed branch piles. *Chin. J. Geotech. Eng.* **2013**, *35*, 1232–1235.
83. Zhang, M.; Cui, W.; Xu, X. Research on soil displacement field around the squeezed branch pile under vertical load. *Chin. J. Rock Mech. Eng.* **2017**, *36*, 3569–3577.
84. Li, X.; Li, X.; Liu, B. A simplified Nonlinear Settlement Analysis Method for Axial Loaded Squeezed and Branch Piles. *Chin. J. Undergr. Space Eng.* **2018**, *14*, 904–911.
85. Gao, X.; Liu, F.; Li, Y. Calucuation and influence factor analysis on lateral bearing capacity of squeezed branch and plate pile. *Eng. Mech.* **2009**, *26*, 97–104.
86. Wang, S.; Li, Y.; Li, Z. Experimental on squeezed pan piles in single and double rows under loading for protecting founda-tion excavation. *J. China Coal Soc.* **2009**, *34*, 537–541.
87. Zhao, M.; Li, W.; Shan, Y. Study on uplift mechanism and calculation of DX pile. *Rock Soil Mech.* **2006**, *27*, 199–203.
88. Jiang, J.; Gao, G.; Liu, W. Modified Hyperbolic Model Describing Q-s Curves of Squeezed Branch Pile. *J. Basic Sci. Eng.* **2010**, *18*, 999–1009.
89. Yang, Z.; Xie, Y.; Fan, X. A study on the relationship between first disk depth and pullout resistance of uplift branch pile. *J. Railw. Sci. Eng.* **2019**, *16*, 2442–2450.
90. Zhou, J.; Gong, X.; Zhang, R. Model tests comparing the behavior of pre-bored grouted planted piles and a wished-in-place concrete pile in dense sand. *Soils Found.* **2019**, *59*, 84–96. [CrossRef]

91. Kiya, Y.; Kato, Y.; Kuwabara, F. Model tests on Vertical Bearing Performance of Enlarged Base Bulb of Buried Piles. *J. Struct. Constr. Eng.* **2007**, *615*, 137–144. [CrossRef] [PubMed]
92. Karkee, M.B.; Horiguchi, T.; Kishida, H. Limit state formulation for the vertical resistance of bored PHC nodular piles based on field load test results. In Proceedings of the Eleventh Asian Regional Conference on Soil Mechanics and Geotechnical Engineering, Seoul, Republic of Korea, 16–20 August 1999.
93. Hiroyoshi, H. Analysis of piled raft with nodular pile subjected to vertical load using a Winkler model approach. *Int. J. Numer. Anal. Methods Geomech.* **2016**, *40*, 1863–1889.
94. Zhou, J.; Wang, K.; Long, X. Bearing capacity and load transfer mechanism of static drill rooted nodular piles. *Rock Soil Mech.* **2014**, *35*, 1367–1376.
95. Zhou, J.; Long, X.; Wang, K. Model test load transfer mechanism of a static drill rooted nodular pile. *J. Zhejiang Univ. (Eng. Sci.)* **2015**, *49*, 531–537, 546.
96. Zhou, J.; Long, X.; Wang, K. Performance of static drill rooted nodular piles under cpmpression. *J. Zhejiang Univ. (Eng. Sci.)* **2014**, *48*, 835–842.
97. Zhou, J.; Wang, K.; Long, X. A test on base bearing capacity of static drill rooted nodular pile. *Rock Soil Mech.* **2016**, *37*, 2603–2609.
98. Zhou, J.; Long, X.; Wang, K. Application of static drill rooted precast nodular pile in soft soil foundation and calculation for bearing capacity. *Chin. J. Rock Mech. Eng.* **2014**, *33*, 4359–4366.
99. Zhou, J.; Long, X.; Wang, K. A simplified approach to calculating settlement of a single pre-bored grouting planted nodular pile in layered soils. *Rock Soil Mech.* **2017**, *38*, 109–116.
100. Zhou, J.; Long, X.; Yan, T. Behavior of sand filled nodular piles under compression in soft soil areas. *Rock Soil Mech.* **2018**, *39*, 3425–3432.
101. Zhou, J.; Long, X.; Wang, K. Behavior of the static drill rooted nodular piles under tension. *Chin. J. Geotech. Eng.* **2015**, *37*, 570–576.
102. Zhou, J.; Wang, K.; Long, X. Numerical simulation on behavior of static drill rooted pile under tension. *J. Zhejiang Univ. (Eng. Sci.)* **2015**, *49*, 2135–2141.
103. Zhou, J.; Zhang, H.; Huang, C. Field Behavior of Pre-Stressed High-Strength Concrete Nodular Piles and PHC Pipe Piles Under Compression in Soft Soil Areas. *J. Tianjin Univ. (Sci. Technol.)* **2019**, *52*, 9–15.
104. Wu, J.; Wang, K.; Xiao, S. Longitudinal vibration characteristics of static drill rooted tubular piles with variable section. *Chin. J. Rock Mech. Eng.* **2018**, *37*, 1030–1040.
105. Wang, K.; Xiao, S.; Gao, L. Vertical dynamic response of a static drill rooted nodular pile. *J. Viberation Shock* **2019**, *38*, 49–56, 86.
106. Wang, K.; Li, Z.; Lu, S. Vertical vibration characteristics of static drill rooted nodular pile and its applications. *J. Zhejiang Univ. (Eng. Sci.)* **2015**, *49*, 522–530.
107. Li, Z.; Wang, K.; Gao, L. Longitudinal dynamic impedance of static drill rooted nodular pile considering vertical wave effect of soil. *Chin. J. Rock Mech. Eng.* **2017**, *36*, 1791–1802.
108. Zhou, J.J.; Gong, X.N.; Wang, K.H.; Ri-Hong, Z.; Guo-Lin, X. Effect of cemented soil properties on the behavior of pre-bored grouted planted nodular piles under compression. *J. Zhejiang Univ. Sci. A* **2018**, *19*, 534–543. [CrossRef]
109. Zhou, J.J.; Gong, X.N.; Wang, K.H.; Zhang, R.H.; Yan, J.J. A simplified nonlinear calculation method to describe the settlement of pre-bored grouting planted nodular piles. *J. Zhejiang Univ. Sci. A* **2017**, *18*, 895–909. [CrossRef]
110. Gong, X.; Xie, C.; Shao, J. Analysis of Bearing Characteristics of Static Drill Rooted Nodular Piles Under Tesion and Compression. *Adv. Eng. Sci.* **2018**, *50*, 102–109.
111. Gong, X.; Xie, C.; Zhou, J. A Comparative Study on the Static Drill Rooted Nodular Piles Under Tension and Compression. *J. Shanghai Jiao Tong Univ.* **2018**, *52*, 1467–1474.
112. Liu, S.; Zhang, C.; Lai, F.; Cheng, Y. A Construction Method of the Composite Pile with Variable-Section. CN111364454A, 3 July 2020.
113. Cai, G.H.; Du, Y.J.; Liu, S.Y.; Singh, D.N. Physical properties, electrical resistivity, and strength characteristics of carbonated silty soil admixed with reactive magnesia. *Can. Geotech. J.* **2015**, *52*, 1699–1713. [CrossRef]
114. Cai, G.; Liu, S.; Du, Y.; Cao, J. Influences of Activity Index on Mechanical and Microstructural Characteristics of Carbonated Reactive Magnesia-Admixed Silty Soil. *J. Mater. Civ. Eng.* **2016**, *29*, 04016285. [CrossRef]
115. Cai, G.; Liu, S.; Du, Y.; Zhang, D.; Zheng, X. Strength and deformation characteristics of carbonated reactive magnesia treated silt soil. *J. Cent. South Univ.* **2015**, *22*, 1859–1868. [CrossRef]
116. Liu, S.Y.; Zheng, X.; Cai, G.H. Effects of Drying-Wetting Cycles on Durability of Carbonated Reactive Magnesia-Admixed Clayey Soil. *J. Mater. Civ. Eng.* **2019**, *31*, 04019276.
117. Yi, Y.; Lu, K.; Liu, S.; Al-Tabbaa, A. Property changes of reactive magnesia–stabilized soil subjected to forced carbonation. *Can. Geotech. J.* **2016**, *53*, 314–325. [CrossRef]
118. Yi, Y.; Liska, M.; Unluer, C.; Al-Tabbaa, A. Carbonating magnesia for soil stabilization. *Can. Geotech. J.* **2013**, *50*, 899–905. [CrossRef]
119. Yi, Y.; Liska, M.; Al-Tabbaa, A. Properties of Two Model Soils Stabilized with Different Blends and Contents of GGBS, MgO, Lime, and PC. *J. Mater. Civ. Eng.* **2014**, *26*, 267–274. [CrossRef]
120. Haha, M.; Lothenbach, B.; Le, S. Influence of slag chemistry on the hydration of alkali-activated blast-furnace slag—Part I: Effect of MgO. *Cem. Concr. Res.* **2011**, *41*, 955–963. [CrossRef]

121. Higgins, D. *Soil Stabilisation with Ground Granulated Blastfurnace Slag*; UK Cementitious Slag Makers Association (CSMA): London, UK, 2005; pp. 1–15.
122. Hughes, P.; Glendinning, S. Deep dry mix ground improvement of a soft peaty clay using blast furnace slag and red gypsum. *Q. J. Eng. Geol. Hydrogeol.* **2004**, *37*, 205–216. [CrossRef]
123. Nidzam, R.; Kinuthia, J. Sustainable soil stabilisation with blastfurnace slag—A review. *Proc. Inst. Civ. Eng.-Constr. Mater.* **2010**, *163*, 157–165. [CrossRef]
124. Oti, J.; Kinuthia, J.; Bai, J. Using slag for unfired-clay masonry-bricks. *Proc. Inst. Civ. Eng.-Constr. Mater.* **2008**, *161*, 147–155. [CrossRef]
125. Oti, J.; Kinuthia, J.; Bai, J. Developing unfired stabilised building materials in the UK. *Proc. Inst. Civ. Eng.-Eng. Sustain.* **2008**, *161*, 211–218. [CrossRef]
126. Oti, J.; Kinuthia, J.; Bai, J. Engineering properties of unfired clay masonry bricks. *Eng. Geol.* **2009**, *107*, 130–139. [CrossRef]
127. James, R.; Kamruzzaman, A.H.M.; Haque, A.; Willkinson, A. Behaviour of lime–slag-treated clay. *Proc. Inst. Civ. Eng.-Ground Improv.* **2008**, *161*, 207–216. [CrossRef]
128. Wild, S.; Kinuthia, J.; Jones, G.; Higgins, D. Suppression of swelling associated with ettringite formation in lime stabilized sulphate bearing clay soils by partial substitution of lime with ground granulated blastfurnace slag (GGBS). *Eng. Geol.* **1999**, *51*, 257–277. [CrossRef]
129. Tasong, W.A.; Wild, S.; Tilley, R.J. Mechanisms by which ground granulated blastfurnace slag prevents sulphate attack of lime-stabilised kaolinite. *Cem. Concr. Res.* **1999**, *29*, 975–982. [CrossRef]
130. Obuzor, G.N.; Kinuthia, J.M.; Robinson, R.B. Enhancing the durability of flooded low-capacity soils by utilizing lime-activated ground granulated blastfurnace slag (GGBS). *Eng. Geol.* **2011**, *123*, 179–186. [CrossRef]
131. Obuzor, G.N.; Kinuthia, J.M.; Robinson, R.B. Utilisation of lime activated GGBS to reduce the deleterious effect of flooding on stabilised road structural materials: A laboratory simulation. *Eng. Geol.* **2011**, *122*, 334–338. [CrossRef]
132. Yi, Y.; Qin, X.; Zhuang, Y. Utilization of GGBS in stabilization of soft soils and its mechanism. *Chin. J. Geotech. Eng.* **2013**, *35*, 829–833.
133. Yu, B.; Du, Y.; Liu, C. Study of durability of reactive magnesia-activated ground granulatedblast-furnace slag stabilized soil attacked by sulfate sodium solution. *Rock Soil Mech.* **2015**, *36*, 64–72.
134. Wu, H.; Bo, Y.; Du, Y. Acid neutralization capacity, strength properties and micro-mechanism of Pb-contaminated soils stabilized by alkali-activated GGBS. *Chin. J. Geotech. Eng.* **2019**, *41*, 137–140.
135. YI, Y.; LI, C.; LIU, S.; AL-TABBAA, A. Resistance of MgO–GGBS and CS–GGBS stabilised marine soft clays to sodium sulfate attack. *Géotechnique* **2014**, *64*, 673–679. [CrossRef]
136. Takhelmayum, G.; Savitha, A.; Krishna, G. Laboratory study on soil stabilization using fly ash mixtures. *Int. J. Eng. Sci. Innov. Technol. (IJESIT)* **2013**, *2*, 477–482.
137. Hu, Z.; Jia, Z.; Gao, L.; Yuan, Z. The effects of sulfate on the strength of lime-fly ash stabilized soil. *Electron. J. Geotech. Eng* **2016**, *21*, 3669–3676.
138. Zentar, R.; Wang, D.; Abriak, N.E.; Benzerzour, M.; Chen, W. Utilization of siliceous–aluminous fly ash and cement for solidification of marine sediments. *Constr. Build. Mater.* **2012**, *35*, 856–863. [CrossRef]
139. Nalbantoğlu, Z. Effectiveness of Class C fly ash as an expansive soil stabilizer. *Constr. Build. Mater.* **2004**, *18*, 377–381. [CrossRef]
140. Kolias, S.; Kasselouri-Rigopoulou, V.; Karahalios, A. Stabilisation of clayey soils with high calcium fly ash and cement. *Cem. Concr. Compos.* **2005**, *27*, 301–313. [CrossRef]
141. Wang, D.; Xiao, J.; Li, L. Micro-mechanism of durability evolution of sludge dredged from East Lake, Wuhan based on carbonation-solidification technique. *Rock Soil Mech.* **2019**, *40*, 3045–3053.
142. Wang, D.; Wang, H.; Wang, R. Micro-mechanisms of dredged sludge solidified with reactive MgO-fly ash. *Chin. J. Rock Mech. Eng.* **2019**, *38*, 3717–3725.
143. Wang, D.; Wang, H.; Xiao, J. Strength and micromechanism of reactive MgO-activated fly ash as alternative soft soil stabilizer. *J. China Univ. Min. Technol.* **2018**, *47*, 879–884.
144. Wang, D. Research on strength and durability of sediments solidified with high volume fly ash. *Rock Soil Mech.* **2012**, *33*, 3659–3664.
145. Wang, D.; Gao, X.; Du, Y. Experimental Investigation on Shear Properties of Reactive MgO-fly Ash Stabilized Loess. *J. Disaster Prev. Mitig. Eng.* **2018**, *38*, 822–829.
146. Harrison, J.; FCPA, B. New cements based on the addition of reactive magnesia to Portland cement with or without added pozzolan. In Proceedings of the 21st Biennial Conference of the Concrete Institute of Australia 2003: Concrete in the Third Millenium, Brisbane, Australia, 17–19 July 2003.
147. Harrison, J. Reactive Magnesium Oxide Cement. WO/055049, 2 August 2001.
148. Vandeperre, L.J.; Liska, M.; Al-Tabbaa, A. Hydration and Mechanical Properties of Magnesia, Pulverized Fuel Ash, and Portland Cement Blends. *J. Mater. Civ. Eng.* **2008**, *20*, 375–383. [CrossRef]
149. Jin, F.; Al-Tabbaa, A. Thermogravimetric study on the hydration of reactive magnesia and silica mixture at room temperature. *Thermochim. Acta* **2013**, *566*, 162–168. [CrossRef]
150. Vandeperre, L.; Liska, M.; Al-Tabbaa, A. Microstructures of reactive magnesia cement blends. *Cem. Concr. Compos.* **2008**, *30*, 706–714. [CrossRef]

151. Unluer, C.; Al-Tabbaa, A. Impact of hydrated magnesium carbonate additives on the carbonation of reactive MgO cements. *Cem. Concr. Res.* **2013**, *54*, 87–97. [CrossRef]
152. Vandeperre, L.J.; Al-Tabbaa, A. Accelerated carbonation of reactive MgO cements. *Adv. Cem. Res.* **2007**, *19*, 67–79. [CrossRef]
153. Jin, F.; Gu, K.; Al-Tabbaa, A. Strength and drying shrinkage of reactive MgO modified alkali-activated slag paste. *Constr. Build. Mater.* **2014**, *51*, 395–404. [CrossRef]
154. Liska, M.; Al-Tabbaa, A. Performance of magnesia cements in porous blocks in acid and magnesium environments. *Adv. Cem. Res.* **2012**, *24*, 221–232. [CrossRef]
155. Liska, M.; Al-Tabbaa, A. Ultra-green construction: Reactive magnesia masonry products. *Proc. Inst. Civ. Eng.-Waste Resour. Manag.* **2009**, *162*, 185–196. [CrossRef]
156. Cwirzen, A.; Habermehl-Cwirzen, K. Effects of Reactive Magnesia on Microstructure and Frost Durability of Portland Cement–Based Binders. *J. Mater. Civ. Eng.* **2013**, *25*, 1941–1950. [CrossRef]
157. Cai, G.-H.; Liu, S.-Y.; Zheng, X. Influence of drying-wetting cycles on engineering properties of carbonated silt admixed with reactive MgO. *Constr. Build. Mater.* **2019**, *204*, 84–93. [CrossRef]
158. Zheng, X.; Liu, S.; Cai, G. Experimental study on freeze-thaw properties of carbonated reactive MgO-stabilized soils. *J. Southeaxt Univ. (Nat. Sci. Ed.)* **2015**, *45*, 595–600.
159. Wang, L.; Liu, S.; Cai, G. Permeability properties of carbonated reactive MgO-stabilized soils. *Chin. J. Geotech. Eng.* **2018**, *40*, 953–959.
160. Cai, G.; Lu, H.; Liu, S. Moisture-Heat Migration Laws and Permeability of Compacted Clay Under Temperature Gradient. *J. Northeast. Univ. (Nat. Sci.)* **2017**, *38*, 874–879.
161. Cai, G.; Liu, S.; Du, G.; Wang, L. Effect of MgO Activity Index on Physicochemical, Electrical and Mechanical Properties of Carbonated MgO-admixed Silt. *KSCE J. Civ. Eng.* **2019**, *23*, 3807–3817. [CrossRef]
162. Cai, G.; Liu, S.; Cao, Q. Influence of Initial Water Content on Strengh and Electrical Resistivity of MgO-carbonated Silt. *China J. Highw. Transp.* **2017**, *30*, 18–26.
163. Liu, S.; Zheng, X.; Cai, G. Study of resistance to sulfate attack of carbonated reactive MgO-stabilized soils. *Rock Soil Mech.* **2016**, *37*, 3057–3064.
164. Cai, G. Experimental and Application Studies on Soft Soil Carbonated and Stabilized by Reactive Magnesia. Ph.D. Thesis, Southeast University, Nanjing, China, 2017.
165. Cai, G.; Yi, Y.; Liu, S. The Invention Relates to a Treatment System and a Method for Strengthening Soft Soil Foundation by Heat from Industrial Waste Gas. CN103147434A, 10 April 2013.
166. Liu, S.; Cai, G.; Yi, Y. The Utility Model Relates to a Treatment System for Foundation Reinforcement and a Carbonized Pile Forming Method. CN103981854A, 13 August 2014.
167. Yi, Y.; Martin, L. The Invention Relates to a Carbonization Curing Method of Soil and an Apparatus Thereof. CN102102360A, 22 June 2011.
168. Liu, S.; Cai, G.; Du, G. Model tests on carbonated reactive MgO mixing piles. *Chin. J. Geotech. Eng.* **2017**, *39*, 136–139.
169. Yi, Y. New Technology and Theory of Agitating Pile Series based on Sustainable Development. Ph.D. Thesis, Southeast University, Nanjing, China, 2013.
170. Liu, S.; Cai, G.; Wang, L. The Invention Relates to a Carbonized Mixing Pile-Breathable Pipe Pile Composite Foundation and a Construction Method Thereof. CN106869120B, 31 July 2018.
171. Wang, L. Experiment and Technology Research of Carbonized Composite Pile based on Permeable Pipe Pile. Master's Thesis, Southeast University, Nanjing, China, 2019.

energies

Article

Failure Patterns of Transmission Tower-Line System Caused by Landslide Events

Hong Yu [1], Hao Li [1], Zhi-Qiang Zhang [2], Gui-Feng Zhang [2], Da-Hai Wang [3,4] and Hua-Dong Zheng [3,4,*]

1 Electric Power Research Institute, Yunnan Power Grid Co., Ltd., Kunming 650233, China
2 Electric Power Research Institute, China Southern Power Grid Co., Ltd., Guangzhou 510663, China
3 School of Civil Engineering and Architecture, Wuhan University of Technology, Wuhan 430074, China
4 Sanya Science and Education Innovation Park, Wuhan University of Technology, Sanya 572024, China
* Correspondence: zhenghd@whut.edu.cn

Abstract: A transmission tower-line system consists of many towers and cables, which is a long-span structure. Due to topographical restrictions, the structure inevitably passes through wide areas prone to landslide hazards. The landslide causes the failure of the tower foundation and threatens the safety of the power transmission system. Therefore, the aim of this study was to investigate the collapse patterns of a transmission tower-line structure under landslides. The explicit analysis method was used to deal with the nonlinear dynamic response equations of the structure. To prove the effect of the landslides on the collapse patterns of the transmission tower-line structure, two foundation failure cases (Case 1 and Case 2) under landslides were considered. For each case, the displacement responses at tower tops and the progressive collapse behavior were analyzed, which illustrate that the failure patterns of the tower-line structure under landslides are dominated by two factors, i.e., the failure mode of the tower foundation and the pulling effect of cables. The overall tilting and hinge formation in the middle of the tower body are the main collapse patterns of the structure under Case 1 and Case 2, respectively. In addition, because of the pulling of cables, the collapse of one tower induced by the landslides always triggers the local failure of neighboring tower heads.

Keywords: collapse analysis; tower-line system; landslide; collapse pattern; explicit dynamic analysis

check for updates

Citation: Yu, H.; Li, H.; Zhang, Z.-Q.; Zhang, G.-F.; Wang, D.-H.; Zheng, H.-D. Failure Patterns of Transmission Tower-Line System Caused by Landslide Events. *Energies* **2022**, *15*, 7155. https://doi.org/10.3390/en15197155

Academic Editors: Jianyong Han and Li Tian

Received: 19 August 2022
Accepted: 23 September 2022
Published: 28 September 2022

Publisher's Note: MDPI stays neutral with regard to jurisdictional claims in published maps and institutional affiliations.

1. Introduction

Due to the increasing demand for electricity, the malfunctions of the power grid can produce adverse effects on people's livelihoods and the national economy. For example, the massive outage in the state of Texas in the United States [1] has caused a more than $200 billion economic loss [2] and 70 deaths [3], and the severe electricity shortage in China has caused financial and operational crises for many companies [4]. The tower-line structure is a carrier for the transmission of electric energy, which belongs to the main part of the power grids. For such a long-span structure, the towers are inevitably constructed on high-altitude mountainous areas, where the geohazard, especially the landslide, usually occurs at the steep slopes and ridges of the mountains [5]. Landslides can probably lead to the tilting and even collapse of transmission towers, as shown in Figure 1. A recent report [6] by China Electric Power Research indicates that there are more than 1500 transmission towers that are in the potential threat of landslides. Figure 2 illustrates the field pictures of the landslide located at Tower N18 in the Dan-Gong transmission line in China. Therefore, the investigation into the failure mechanisms of transmission tower-line structures under landslides is significant.

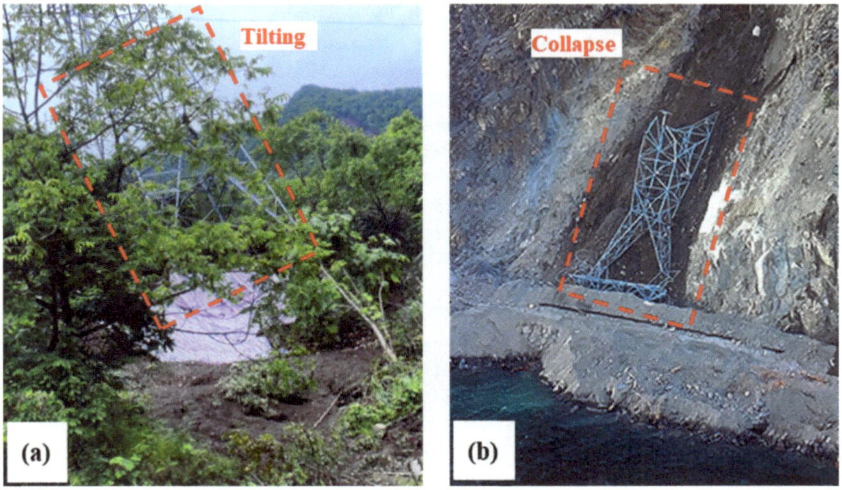

Figure 1. The grid accidents induced by landslides: (**a**) tilting; (**b**) collapse.

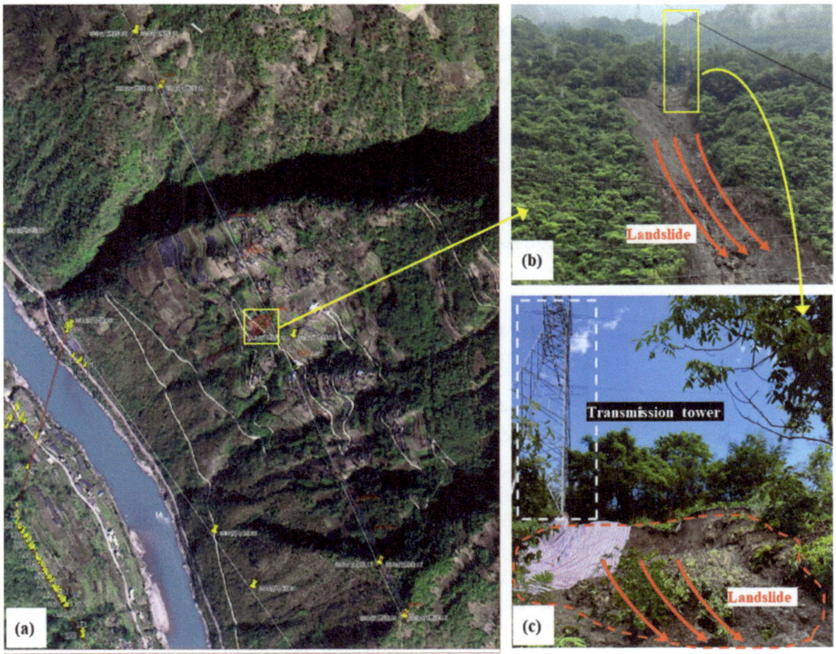

Figure 2. The landslide located at tower N18 in the Dan-Gong transmission line: (**a**) general view of the landslide, with the yellow rectangle locating the position of the tower; (**b**) landslide at the foundation of the tower; (**c**) landslide at one leg of the tower (photographed by authors).

Landslide events can be induced by earthquakes and rainfall and they are a major geomorphological process that influences landscape evolution in mountainous areas [7]. The mechanism of the development of the landslide has been studied by many researchers. Some studies [8–11] have investigated the effect of various influential factors, such as the slope angle, slope height, geological structure, and rock type, on the landslide spatial distribution. In addition, the landslide travel distance has also been evaluated by the theoretical model [12], empirical model [13,14], and numerical model [15]. Because the essential input parameters in the theoretical and numerical models are difficult to be measured during the landslide movement [16], the two models need to be greatly simplified through many assumptions. The empirical model follows the rules that are calibrated based on the experimental and field-measured data. It can describe the landslide using simple mathematical formulas and therefore it is more popular than the other two above-mentioned models.

The transmission tower-line system consists of many towers, which are connected by a series of cables. For this type of structure, the progressive collapse behavior usually occurs [17,18]. Therefore, some studies have been conducted to investigate the nonlinear dynamic response of transmission tower-line systems under extreme environmental conditions [19–23]. Zheng and Fan [19] investigated the progressive collapse mechanisms of a tower-line system subjected to hurricanes through fully coupled dynamic simulations, showing that the collapsed transmission tower would lead to the cascade collapse of the neighboring towers. Kassem et al. [20] also conducted a comparative study on the failure patterns of a transmission tower under far-field and near-field earthquakes. Li et al. [21] investigated the collapse mechanism and failure modes of the transmission tower under strong wind load, showing that the wind-induced coefficients in the current design code are not accurate enough. Fu et al. [22] comprehensively evaluated the failure modes of a transmission tower under combined wind and rain loads and they provided a critical collapse surface to quickly and conveniently judge the structural failure states at different wind intensities. However, the above-mentioned studies are mainly on the failure patterns of tower-line structures under hurricanes, earthquakes, and so on, whereas the relevant research for the structures subjected to landslides is scarce. Though one study [24] has tried to identify the failure modes of transmission tower foundations under landslides, the consequent collapse patterns of the tower-line structures have not been investigated yet.

In this research, we selected a segment of the transmission tower-line system over the Yangtze River as the study object. The nonlinear dynamic response and the collapse process of the structure under landslides were simulated. To deal with the discontinuous problems involved in the collapse simulations, the MATLAB code based on the explicit computation method was adopted. Two failure cases under landslides were considered. For each case, the structural displacement response and the collapse scenarios at the specified moments were analyzed and the corresponding collapse mechanisms of the transmission tower-line system under landslides were then identified.

2. The Yanzi Landslide

As shown in Figure 3, the Yanzi landslide is located in the power transmission zone (Tower 2) from Wuxia to Xilingxia. The overall terrain is low in the east and north and high in the west and south, where the volume of the landslide body is about 60×10^4 m^3 with an area of 60×10^4 m^2 and an average thickness of 10 m [24]. The landslide body looks like a long tongue, which has a slope of about 25–40 degrees. The material composition of the landslide is characterized by the silty clay mixed with gravel soil. The slipping zone is the contact surface between the overlying deposits and the bedrock, as shown in Figure 4. Since 2002, the Yanzi landslide has shown many creep-related slipping events due to rainfall and many cracks have presented on the surface of the slope. Then, the G209 National Highway was constructed through the slope and a 100 mm crack formed between the transmission tower feet. More information about the Yanzi landslide can be found in Ref. [24].

Figure 3. River-crossing transmission tower-line system (Google Maps, 17 August 2022).

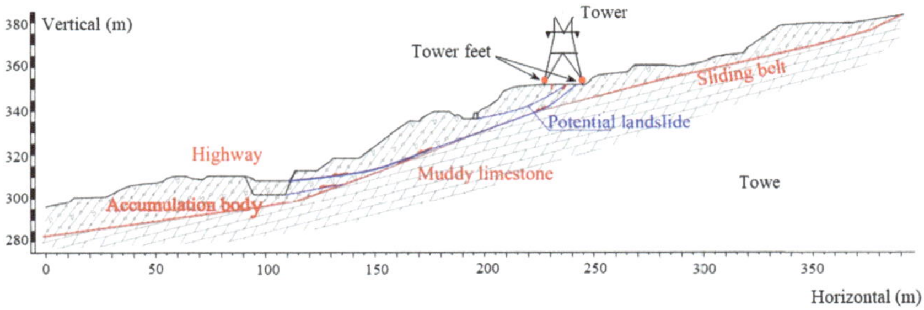

Figure 4. Geological section of Yanzi landslide.

As shown in Figure 3, the prototype of the transmission tower-line system is a segment of a 500 kV transmission line, which involves three lattice transmission towers with the same structural design and four spans of cables. The tower height is 122 m and the foot-to-foot distance is 20.24 m, as shown in Figure 5, which illustrates the top and front views of the structure. The main legs and diagonal members are steel tubes and the dimensions are also given in Figure 5. The material of the tower members is Q235 steel and the material constants are listed in Table 1. The cables can be classified into conductors and ground wires, whose span length is 800 m. The type of conductors is JL/G1A-400/35 with a diameter of 26.82 mm, a mass per unit length of 1.35 kg/m, and an elastic modulus of 65 GPa. The type of ground wires is JLB40-150 and the corresponding values are 15.75 mm, 0.7 kg/m, and 103.6 GPa, respectively.

Table 1. Material constants of tower members.

Parameter	Value
Yield stress (MPa)	235
Elastic modulus (GPa)	206
Mass density (kg/m^3)	7800
Poisson's ratio	0.3

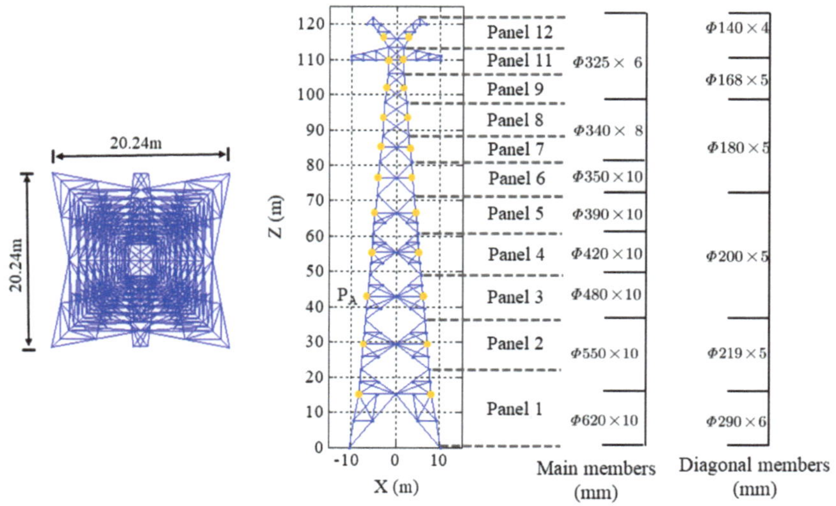

Figure 5. Top and front views of Tower 2.

3. Wind Load Formation

When the wind travels through the transmission tower-line structure, drag forces on the transmission towers are produced. For convenience in calculating the wind loads, the tower was divided into 12 panels, as illustrated in Figure 5. The drag forces, F_x (X-direction) and F_y (Y-direction), can be expressed as

$$F_x = 0.5\rho U^2 \cos^2 \alpha C_{dx} A_x \tag{1}$$

$$F_y = 0.5\rho U^2 \sin^2 \alpha C_{dy} A_y \tag{2}$$

where U is the wind speed; α is the angle between wind and panels; C_{dx} and C_{dy} are the drag coefficients of tower members [25]; A_x and A_y are the projected areas of tower members; ρ is the air density.

The drag forces on cables, F_{xl}, can be calculated using

$$F_{xl} = 0.5\rho U^2 \cos^2 \beta C_{dl} A_{xl} \tag{3}$$

where β is the angle between wind speed and cables; C_{dl} is the drag coefficient of cables, whose value is 1.0 [19]; A_{xl} is the projected area of cables.

The wind speed U is the summation of the turbulent component u and the mean component $\overline{u}_{\text{m}}$.

$$U(x, y, z, t) = u(x, y, z, t) + \overline{u}_{\text{m}}(z) \tag{4}$$

The mean wind component $\overline{u}_{\text{m}}$ varies with the height z and it can be expressed using a normal wind profile.

$$\overline{u}_{\text{m}}(z) = u_{\text{ref}}\left(\frac{z}{z_{\text{ref}}}\right)^{\eta} \tag{5}$$

where u_{ref} is the wind speed at the reference height z_{ref}; η is the ground roughness index, taken as 0.14 [26].

For a wind field point, the turbulent wind speed u can be simulated using the Kaimal spectrum [27]. However, to consider the turbulent wind speed correlations between Point i and Point j in a wind field, the cross-spectrum, $S_{ij}(f)$, (Equation (6)) should be used, which can be obtained based on the Kaimal spectrum and correlation function Coh_{ij} (Equation (7)).

$$S_{ij}(f) = \sqrt{S_{ii}(f) S_{jj}(f)} Coh_{ij}(f), \tag{6}$$

$$Coh_{ij}(f) = \exp\left\{-12\left[(f\,r/u_{\text{ref}})^2 + (0.12r/L_c)^2\right]^{0.5}\right\} \tag{7}$$

where f is the wind speed frequency; $S_{ii}(f)$ is the Kaimal spectrum at Point i; $S_{jj}(f)$ is the Kaimal spectrum at Point j; r is the distance between the two points; L_c is the correlation coefficient between the two points [27]. If there are n field points, the cross-spectrum matrix, $\mathbf{S}(f)$, can be yielded.

$$\mathbf{S}(f) = \left[S_{ij}(f)\right]_{n\times n}, \tag{8}$$

Then, the turbulent wind speed time history at Point i, which considers the influences from other field points, can be written as

$$u_i(t) = \sum_{j=1}^{i}\sum_{k=1}^{N_f}\left|H_{ij}(2\pi f_k)\right|\sqrt{4\pi\,\Delta f} \times \cos\left[2\pi f_k t - \theta_{ij}(2\pi f_k) + \varphi_{jk}\right] \tag{9}$$

where N_f is the number of sampling points; φ_{jk} is the phase angle that satisfies the uniform distribution; H_{ij} is the lower triangular matrix that can be computed via the Cholesky decomposition of $\mathbf{S}(f)$ [28]; $\theta_{ij}(2\pi f_k)= \tan^{-1}\left[\frac{\text{Im}H_{ij}(2\pi f_k)}{\text{Re}H_{ij}(2\pi f_k)}\right]$.

4. Dynamic Response Analysis Method of the Tower-Line Structure

Figure 6 provides the discretization model of the transmission tower-line structure, which is represented by a number of mass particles and member elements. The total mass of the structure is modeled by mass particles. The member element describes the topological relationship between mass particles. The element model developed by the researchers in [29] was adopted to simulate tower members, which can account for the yielding in tension, the buckling in compression, and the critical force degradation due to the Bauschinger effect [30]. For the cable members, the $\varepsilon - \sigma$ relationship is shown in Figure 7. It can be found that when the cables are tensioned, the material behaves in a bilinear way. The slope of stage OA is E_1, and the slope of the yielding stage AB is E_Y. The value of E_Y/E_1 is 0.05 in this study [31]. The cable member element in Figure 7 assumes that the cables cannot experience any compression actions, which means that the internal forces become zero when the cables are compressed.

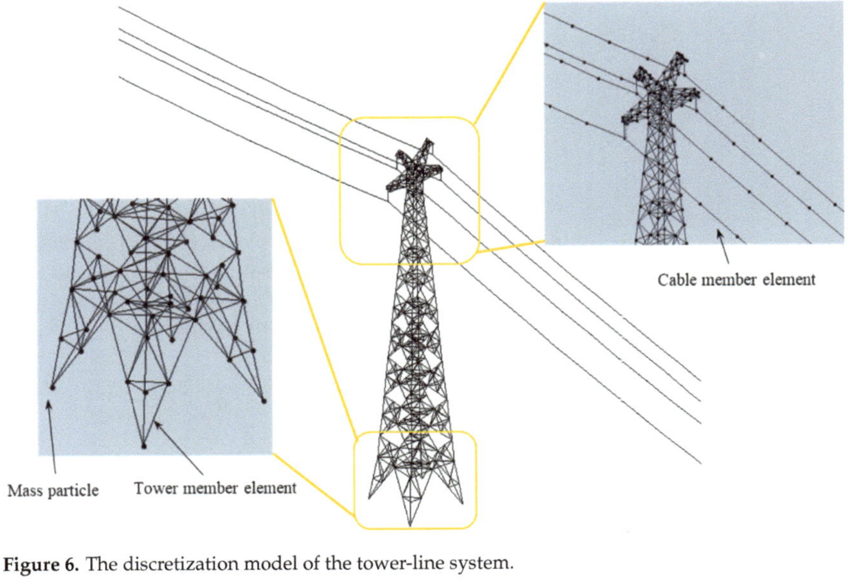

Figure 6. The discretization model of the tower-line system.

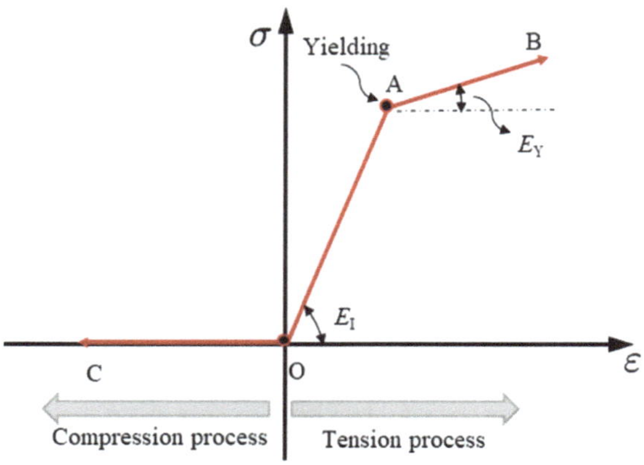

Figure 7. The $\varepsilon - \sigma$ relationship of cables in this study (OA: Elastic stage; AB: Yielding stage).

The dynamic response of structures is actually a function of time and space. The motion response of mass particles can be regarded as a series of sequential motions in small time steps [32]. The motion of mass particles in a small time step is called a path unit, which can be calculated based on Newton's second law [33]. For the mass particle i, the expression can be expressed as

$$\begin{bmatrix} m_i & 0 & 0 \\ 0 & m_i & 0 \\ 0 & 0 & m_i \end{bmatrix} \begin{bmatrix} \ddot{d}_x \\ \ddot{d}_y \\ \ddot{d}_z \end{bmatrix} = \begin{bmatrix} F_{x,\text{int}} \\ F_{y,\text{int}} \\ F_{z,\text{int}} \end{bmatrix} + \begin{bmatrix} F_{x,\text{ext}} \\ F_{y,\text{ext}} \\ F_{z,\text{ext}} \end{bmatrix} + \begin{bmatrix} F_{x,\text{damp}} \\ F_{x,\text{damp}} \\ F_{x,\text{damp}} \end{bmatrix} \tag{10}$$

where m_i is the mass of the particle i, which is the half sum of member elements connected to the particle; $\mathbf{d} = \begin{bmatrix} d_x & d_y & d_z \end{bmatrix}^{\mathrm{T}}$ and $\ddot{\mathbf{d}} = \begin{bmatrix} \ddot{d}_x & \ddot{d}_y & \ddot{d}_z \end{bmatrix}^{\mathrm{T}}$ are the vectors of displacement and acceleration of the particle i; $\begin{bmatrix} F_{x,\text{int}} & F_{y,\text{int}} & F_{z,\text{int}} \end{bmatrix}^{\mathrm{T}}$ is the vector of internal force in member elements, which can be obtained through the model of member elements; $\begin{bmatrix} F_{x,\text{ext}} & F_{y,\text{ext}} & F_{z,\text{ext}} \end{bmatrix}^{\mathrm{T}}$ is the vector of external force exerted on the particle i, which depends on the wind speed (Equations (1)–(3)) and landslide conditions used in this study. Finally, $\begin{bmatrix} F_{x,\text{damp}} & F_{y,\text{damp}} & F_{z,\text{damp}} \end{bmatrix}^{\mathrm{T}}$ is the vector of damping force, which can be further written as

$$\begin{bmatrix} F_{x,\text{damp}} \\ F_{x,\text{damp}} \\ F_{x,\text{damp}} \end{bmatrix} = -\mu \begin{bmatrix} m_i & 0 & 0 \\ 0 & m_i & 0 \\ 0 & 0 & m_i \end{bmatrix} \begin{bmatrix} \dot{d}_x \\ \dot{d}_y \\ \dot{d}_z \end{bmatrix} \tag{11}$$

where μ is the damping coefficient, which is the same as the definition in Ref. [34]; $\dot{\mathbf{d}} = \begin{bmatrix} \dot{d}_x & \dot{d}_y & \dot{d}_z \end{bmatrix}^{\mathrm{T}}$ is the velocity vector of the particle i.

To avoid the complex iteration calculations and the formation of a global stiffness matrix, Equation (10) can be solved by an explicit central difference method [35]. For this, the acceleration and velocity can be written as

$$\ddot{\mathbf{d}} = \frac{1}{\Delta t^2}\left(\ddot{\mathbf{d}}_{n+1} - 2\ddot{\mathbf{d}}_n + \ddot{\mathbf{d}}_{n-1}\right) \tag{12}$$

$$\dot{\mathbf{d}} = \frac{1}{2\Delta t}(\mathbf{d}_{n+1} - \mathbf{d}_{n-1}) \tag{13}$$

where $\mathbf{d}_{n+1}$, $\mathbf{d}_n$, and $\mathbf{d}_{n-1}$ are displacement vectors of a particle at steps n + 1, n, and n-1, respectively; Δt is the time step, taken as 10^{-5} s [36]. By substituting Equations (11)–(13) into Equation (10), the displacement response of the structure can be obtained.

Because the time step in a path unit is pretty small, the deformation and internal force increment of member elements in the path unit are linear and small [37]. The material properties of elements, the number and mass of particles, and the topological relationship can be regarded to be not changed in the path unit, but they can be changed between path units. Therefore, the total deformation of member elements could be nonlinear and very large throughout the whole analysis process, and even the discontinuous issues of member fractures can also be dealt with between path units. The above-mentioned calculation method has been programmed using MATLAB [38], which was already validated by comparison with other numerical models and was found to be accurate with an error within 4.04% [33].

5. Collapse Analysis of the Tower-Line System under Wind and Landslide Action

5.1. Numerical Model of the Tower-Line Structure

As illustrated in Figure 3, the investigated segment includes three transmission towers connected by four spans of cables. MATLAB was used to create the numerical model of the transmission tower-line structure, as shown in Figure 8. The members in the numerical model are connected by hinges. The feet of Towers 1 and 3 (P1–P4 and P9–P12) are assumed to be fixed. However, the boundary condition at the feet of Tower 2 (P5–P8) varies with the landslide cases defined in Section 5.2. The boundary conditions at P13–P16 and P17–P20 are modeled using end springs and their stiffness is defined by the elastic characteristic of the transmission towers [39]. According to previous studies [40,41], the damping of the steel tower is 2% and the damping of cables is 20%. For each simulation, the wind loads are always applied to the structure, whereas the landslide actions are exerted on the structure over 5–15 s.

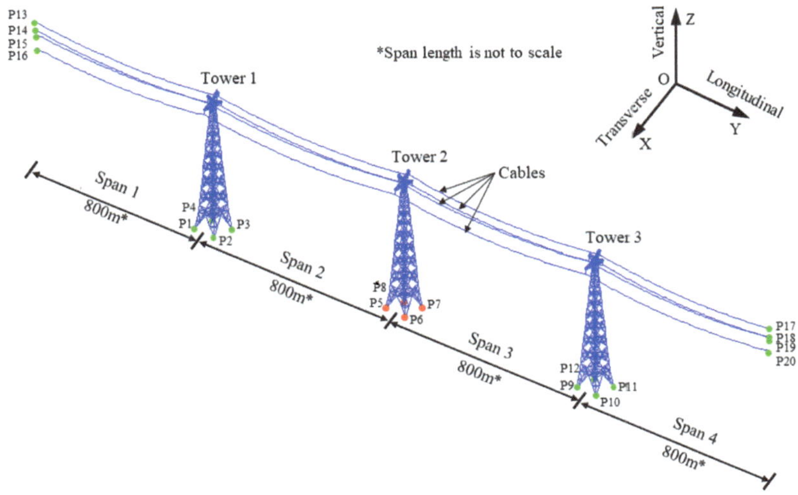

Figure 8. Numerical model of the structure.

5.2. Load Cases

The landslide would lead to the foundation failure of transmission towers. In Figure 8, the P5-P8 are feet of Tower 2. We investigated two failure cases of the tower foundation on the landslide, as illustrated in Figure 9. For Case 1, P7 and P8 of Tower 2 are located in the sliding zone (Figure 9a); for Case 2, P6, P7, and P8 of Tower 2 are located in the sliding zone (Figure 9b). We assumed that part of the tower foundation is on a potential landslide. Once the slide occurs, the part of the tower foundation that sits on the potential landslide could lose its bearing capacity and the corresponding tower feet could lose their suddenly supports [24].

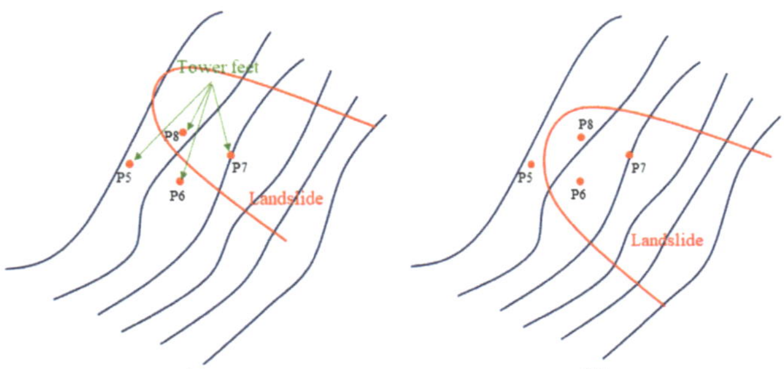

(a) (b)

Figure 9. Two failure cases of towers on the landslide. (**a**) Case 1; (**b**) Case 2.

5.3. Progressive Collapse under Case 1

For Case 1, the displacement time histories at the tops of Towers 2 and 3 are shown in Figure 10. The collapse scenarios for the transmission tower-line system at the special moments are presented in Figures 11–14. The displacement response of Tower 1 is not presented here since the response curves show similar features to those of Tower 3. The coordinate system of OXYZ is defined in Figure 8. In Figure 10, it can be found that the displacements in the *X*-, *Y*-, and *Z*-direction are very small at 5 s because there are only wind loads and dead weight applied to the structure for 1–5 s. After that, the feet of Tower 2 at P7 and P8 start losing their supports due to the landslide (Figure 9a) and the tower body gradually tilts in the *X*-direction, as shown in Figure 11. Accordingly, the displacement at the tower top also sees an increment in the *X*- and *Z*-direction, as illustrated in Figure 10a. It also can be observed in Figure 11 that the cables between transmission towers are also lifted by wind loads, which produce insignificant influences on the response of towers. As the tilt level of the tower body increases, the slack cables in the initial state (Figure 8) have straightened at t = 12 s, as shown in Figure 12. Due to the pulling effect by cables, the neighboring Towers 1 and 3 show the bending shape in the *Y*-direction at t = 14 s (Figure 13). The bending shape at the head of Towers 1 and 3 becomes more obvious at t = 14.3 s (Figure 14). This conclusion can also be drawn from the increment of the *Y*-displacement in Figure 10b. Then, the displacement at the top of Tower 3 shows a sharp decline in three directions and the duration from the start of the bending to the end of the total failure (Figure 10b) at the tower head is less than 1 s. This indicates that the collapse process of the neighboring transmission towers would be greatly accelerated by the pulling of the cables.

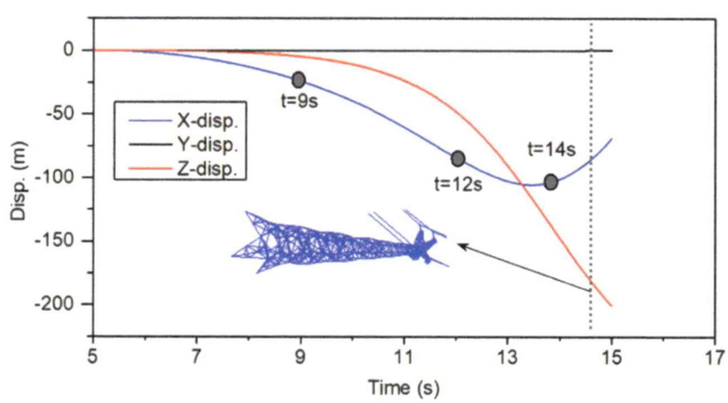

(a)

Figure 10. *Cont.*

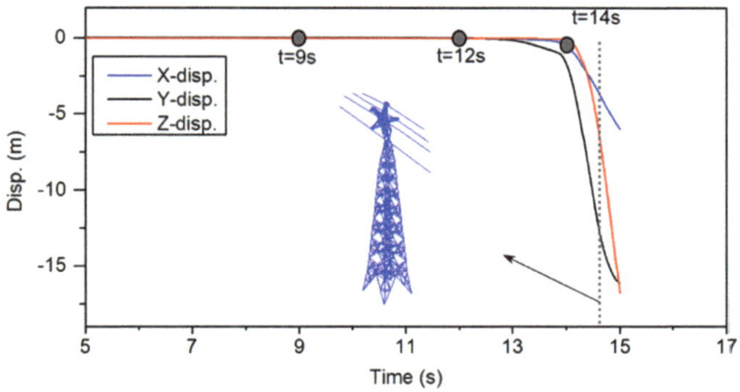

(b)

Figure 10. Displacement response at tower tops under Case 1. (**a**) Tower 2; (**b**) Tower 3.

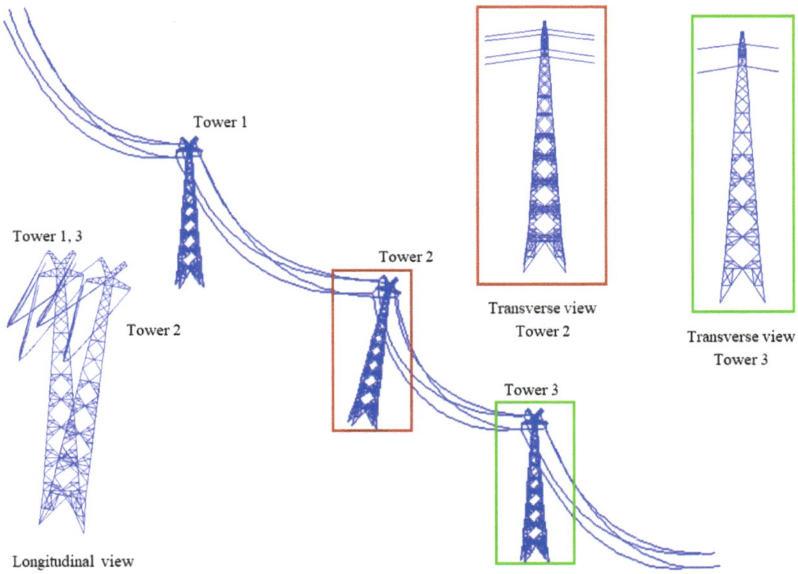

Figure 11. Behavior of the transmission tower-line structure at t = 9 s under Case 1.

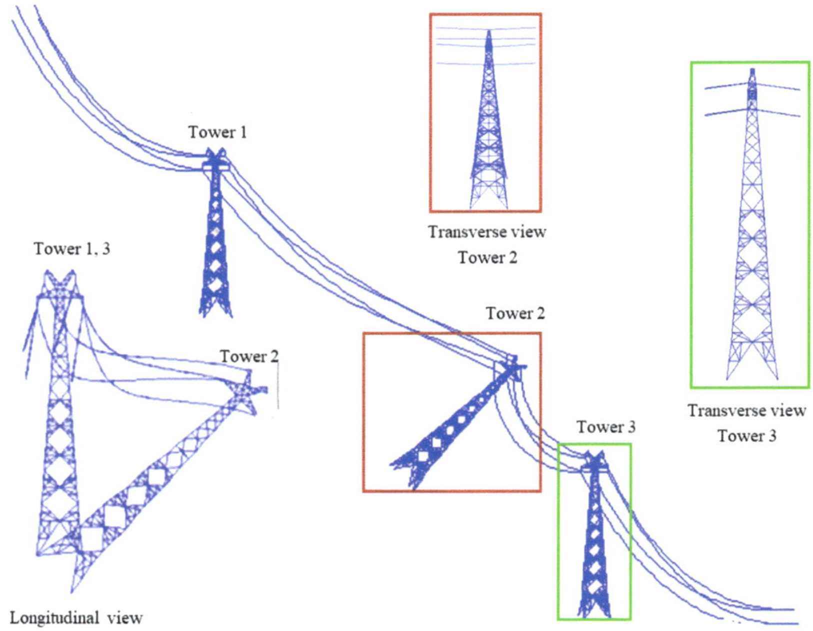

Figure 12. Behavior of the transmission tower-line structure at t = 12 s under Case 1.

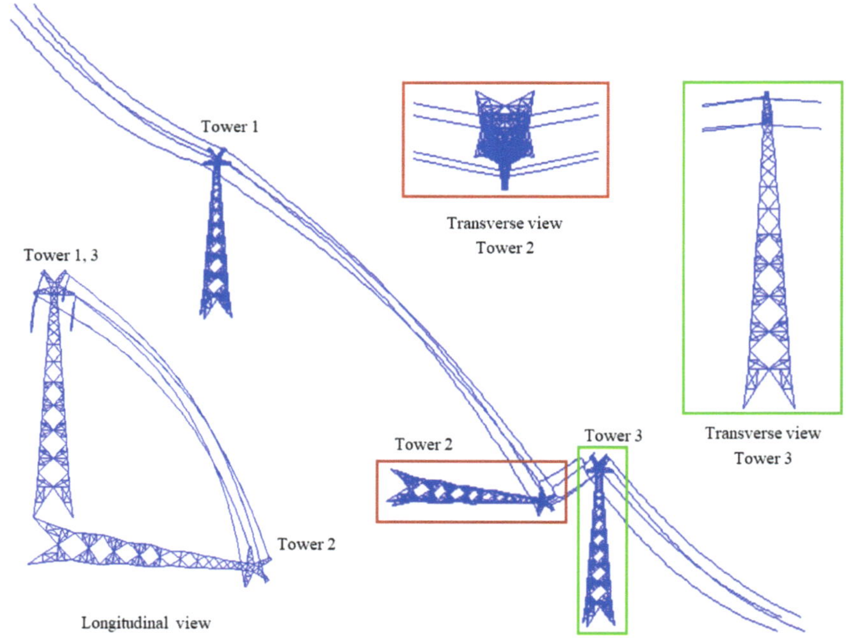

Figure 13. Behavior of the transmission tower-line structure at t = 14 s under Case 1.

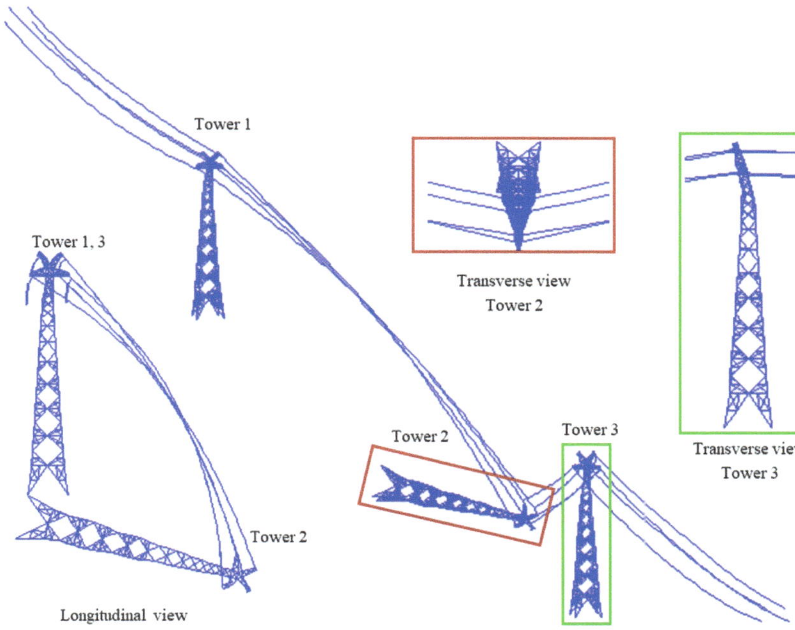

Figure 14. Behavior of the transmission tower-line structure at t = 14.3 s under Case 1.

5.4. Progressive Collapse Simulation under Case 2

For Case 2, the displacement time histories at the tops of Towers 2 and 3 are shown in Figure 15. The collapse scenarios for the transmission tower-line system at the special moments are presented in Figures 16–18. Similarly, the displacement response in the first 5 s is not obvious and therefore the curves are not plotted in Figure 15. Case 2 (Figure 9b) has one less support, namely P6, than Case 1 (Figure 9a), which means the feet of Tower 2 at P6, P7, and P8 would lose their support due to the landslide under Case 2. As illustrated in Figure 15a, the displacement response at the top of Tower 2 not only shows variations in the *X*- and *Z*-direction similar to these in Case 1 (Figure 10a), but also shows obvious changes in the *Y*-direction. At t = 9 s, it can be observed that the collapse pattern under Case 2 (Figure 16) is greatly different from that in Case 1, which presents great deformation along the tower body along with the tilting behavior in Case 1 (Figure 11). This can be explained by the fact that the cable tensions on both sides of Tower 2 are no longer equal when the displacement in the *Y*-direction occurs. As unbalanced tensions become larger and larger with the increment of the Y-displacement, a hinge forms in the middle of the tower at t = 13 s, as shown in Figure 17. After that, the head of Tower 2 drops greatly due to its own gravity and the displacement response curve in the *Z*-direction also sees an obvious decline (Figure 15a). In addition, as illustrated in Figure 18, a local failure also occurs in the upper part of the neighboring Tower 3, which is also due to the pulling of cables.

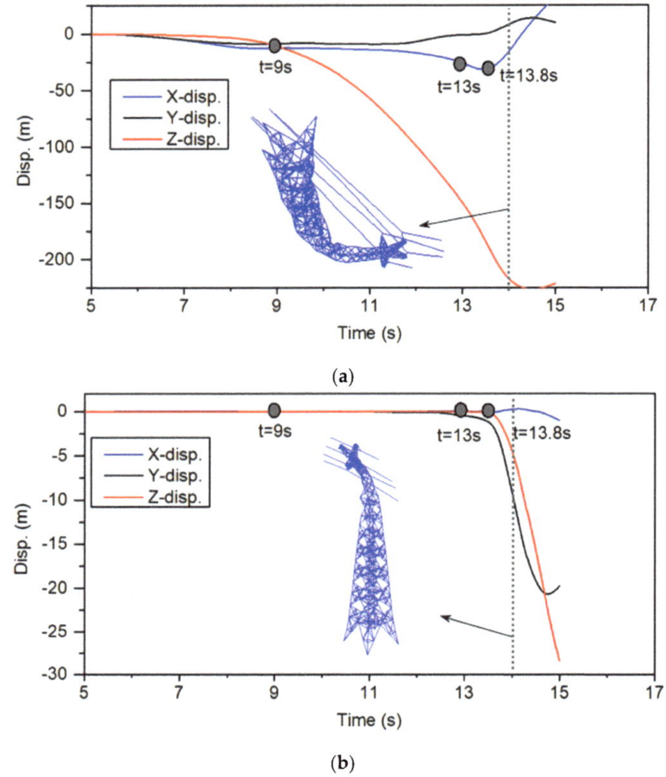

Figure 15. Displacement response at tower tops under Case 2. (**a**) Tower 2; (**b**) Tower 3.

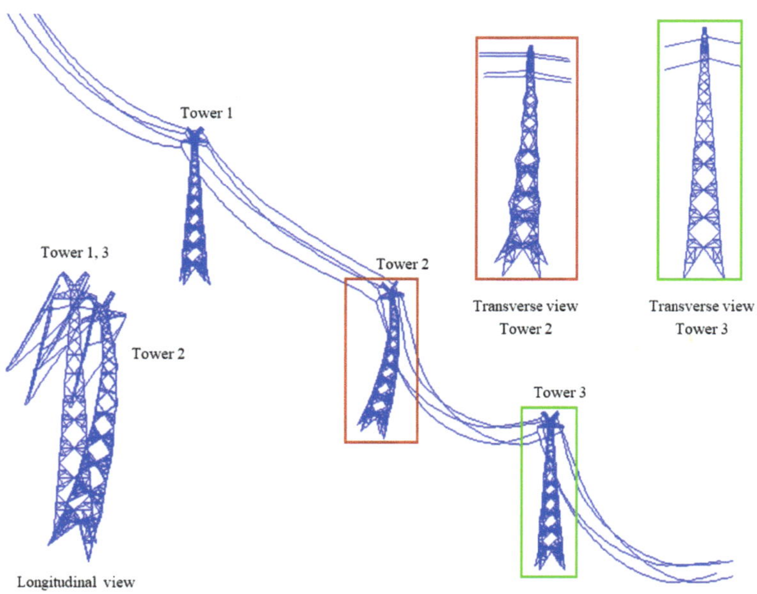

Figure 16. Behavior of the transmission tower-line structure at t = 9 s under Case 2.

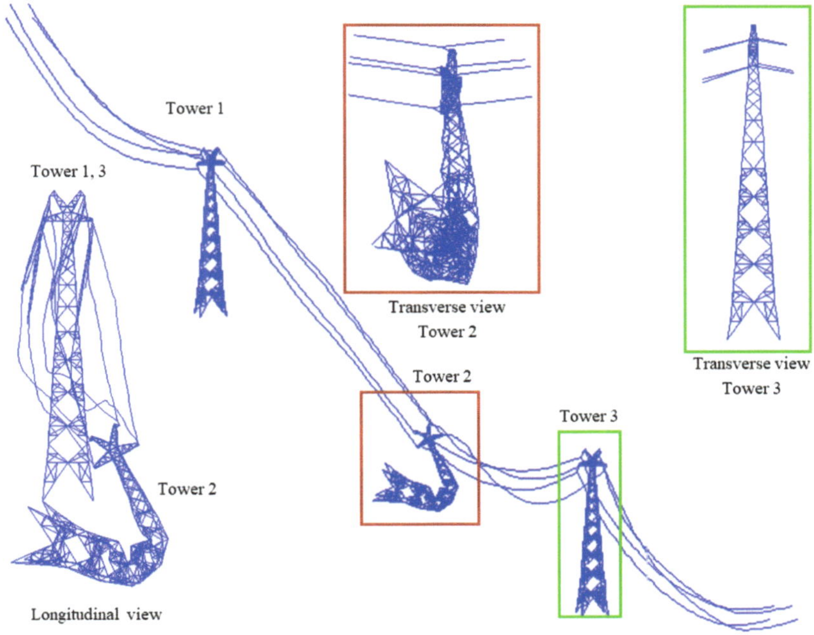

Figure 17. Behavior of the transmission tower-line structure at t = 13 s under Case 2.

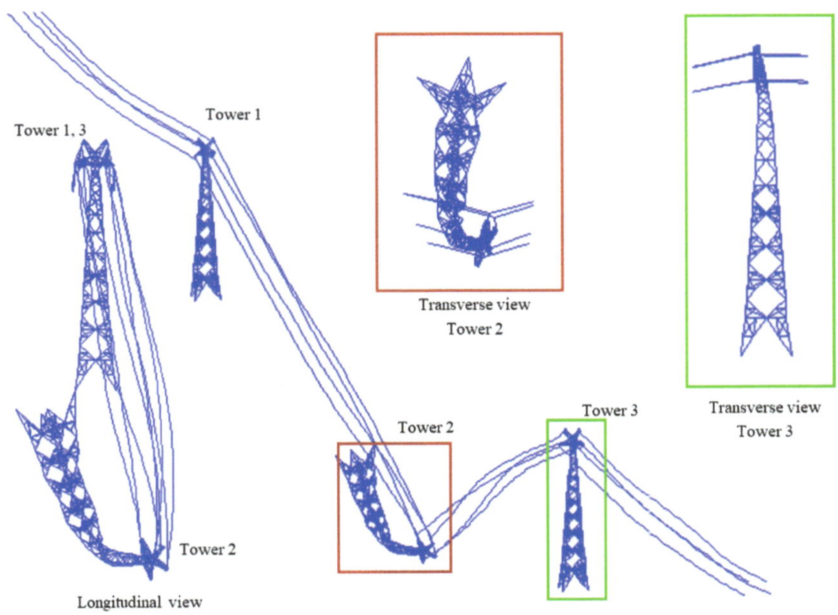

Figure 18. Behavior of the transmission tower-line structure at t = 13.8 s under Case 2.

6. Discussion

The explicit central difference method was used to solve the dynamic response equation of the structure. It does not require the formation of a global stiffness and therefore it is convenient to process the strong geometry and material nonlinearities involved in the collapse of the tower-line structure caused by landslides. It should be noted that the explicit central difference method is conditionally convergent and the time step in Equations (12) and (13) should be carefully checked. For the finite element of the tower-line system, the time step should be smaller than the critical time step, $\Delta t_c = T_n/\pi$ (T_n is the minimum natural vibration period of tower members). Based on the method, the collapse simulations of the tower-line system subjected to landslides have been performed to illustrate that the collapse patterns are greatly affected by the failure modes of the tower foundation and the pulling of cables. On one hand, the tower foundation failure modes determine the number and the location of tower feet that lose the supports under landslides, which affects the azimuthal angle as the tower body tilts under landslides. On the other hand, the contribution of the cable pulling to the structural collapse varies with the azimuthal angles. For Case 1, the tilting azimuthal angle of Tower 2 with respect to the longitudinal direction is 90 degrees and the pulling of cables at both sides of Tower 2 counteracts; for case 2, the corresponding azimuthal angle is around 135 degrees, for which the pulling of cables between Towers 2 and Tower 3 is greatly different from that between Towers 1 and Tower 2. Therefore, the collapse scenarios under Case 1 are distinguished from those under Case 2. However, the above simulations do not consider the interactions between the tower foundation and soil during landslides, which possibly affect the failure patterns of the tower-line system. Future work will include the mentioned interactions to improve the accuracy of numerical results.

7. Conclusions

In this study, we analyzed the displacement response at tower tops and the structural collapse scenarios at special moments to investigate the failure mechanism of the tower-line system under landslides. The following conclusions were obtained:

The collapse failure patterns of the transmission tower-line structure vary with the tower foundation failure modes induced by landslides. For Case 1, two tower feet lose the supports under the landslide, and the overall tilting is the main collapse pattern for the tower on the landslide. For Case 2, three tower feet lose the supports under the landslide. The displacement response in the longitudinal direction is no longer zero, which leads to the different cable tensions on both sides of the tower. A hinge forming in the middle of the tower is the main collapse pattern in this case, which is very different from the collapse pattern in Case 1.

Considering that the tower-line system includes many towers that are connected by cables, the failure of one tower would cause the cascade collapse of the neighboring towers in the transmission line due to the pulling effect of cables. For Cases 1 and 2, the neighboring towers share similar collapse patterns, which are represented by the local failure close to the tower head.

Author Contributions: Conceptualization, H.Y. and H.-D.Z.; methodology, H.L. and H.-D.Z.; software, Z.-Q.Z., G.-F.Z. and H.-D.Z.; validation, H.Y., H.L., and D.-H.W.; formal analysis, H.Y.; investigation, H.L.; resources, D.-H.W.; data curation, Z.-Q.Z.; writing—original draft preparation, H.-D.Z.; writing—review and editing, H.Y.; visualization, G.-F.Z.; supervision, D.-H.W. and H.-D.Z.; funding acquisition, D.-H.W. All authors have read and agreed to the published version of the manuscript.

Funding: This research was funded by Yunnan Power Grid Technology Project (YNKJXM20210181) and Hainan Province Science and Technology Special Fund (ZDKJ2021024).

Data Availability Statement: Not applicable.

Conflicts of Interest: The authors declare no conflict of interest.

References

1. Wu, D.; Zheng, X.; Xu, Y.; Olsen, D.; Xia, B.; Singh, C.; Xie, L. An open-source extendable model and corrective measure assessment of the 2021 texas power outage. *Adv. Appl. Energy* **2021**, *4*, 100056. [CrossRef]
2. Ivanova, I. Texas Winter Storm Costs Could Top $200 Billion—More Than Hurricanes Harvey and Ike. 2021. Available online: https://www.cbsnews.com/news/texas-winter-storm-uri-costs (accessed on 1 August 2022).

3. Tengle, J.; Renault, M. Suspected Hypothermia Deaths in Homes Mount in Texas. 2021. Available online: https://apnews.com/article/houston-hypothermia-weather-conroe-texas-8323ab5f5c1612e632f7f2e6c2c20358 (accessed on 1 August 2022).
4. Chang, J. What Happened behind the Nationwide Power Cuts? 2021. Available online: https://m.bjnews.com.cn/detail/163266456914638.html (accessed on 1 August 2022).
5. Liu, S.; Yin, K.; Zhou, C.; Gui, L.; Liang, X.; Lin, W.; Zhao, B. Susceptibility Assessment for Landslide Initiated along Power Transmission Lines. *Remote Sens.* **2021**, *13*, 5068. [CrossRef]
6. State Grid Equipment Department and China Electric Power Research Institute. *Investigation Report on Geological Hazards of Power Transmission Lines (Yearly Report), 2017–2021*; State Grid Equipment Department and China Electric Power Research Institute: Beijing, China, 2021.
7. Roering, J.J.; Kirchner, J.W.; Dietrich, W.E. Characterizing structural and lithologic controls on deep-seated landsliding: Implications for topographic relief and landscape evolution in the Oregon Coast Range, USA. *Geol. Soc. Am. Bull.* **2005**, *117*, 654–668. [CrossRef]
8. Xu, C.; Dai, F.C.; Yao, X.; Chen, J.; Tu, X.B.; Sun, Y.; Wang, Z.Y. GIS-based landslide susceptibility assessment using analytical hierarchy process in Wenchuan earthquake region. *Chin. J. Rock Mech. Eng.* **2009**, *28*, 3978–3985.
9. Chigira, M.; Wu, X.; Inokuchi, T.; Wang, G. Landslides induced by the 2008 Wenchuan earthquake, Sichuan, China. *Geomorphology* **2010**, *118*, 225–238. [CrossRef]
10. Qi, S.; Xu, Q.; Lan, H.; Zhang, B.; Liu, J. Spatial distribution analysis of landslides triggered by 2008.5. 12 Wenchuan Earthquake, China. *Eng. Geol.* **2010**, *116*, 95–108. [CrossRef]
11. Guo, D.; Hamada, M. Qualitative and quantitative analysis on landslide influential factors during Wenchuan earthquake: A case study in Wenchuan County. *Eng. Geol.* **2013**, *152*, 202–209. [CrossRef]
12. TAKAHASHI, T. Estimation of potential debris flows and their hazardous zones: Soft countermeasures for a disaster. *Nat. Disaster Sci.* **1981**, *3*, 57–89.
13. Rickenmann, D. Empirical relationships for debris flows. *Nat. Hazards* **1999**, *19*, 47–77. [CrossRef]
14. Rickenmann, D. Runout prediction methods. In *Debris-Flow Hazards and Related Phenomena*; Springer: Berlin/Heidelberg, Germany, 2005; pp. 305–324.
15. McDougall, S.D.; Hungr, O. Objectives for the development of an integrated three-dimensional continuum model for the analysis of landslide runout. In *Debris-Flow Hazards Mitigation: Mechanics, Prediction, and Assessment, Proceedings of the 3rd International DFHM Conference, Davos, Switzerland, Millpress 10–12 September 2003*; Rickenmann, D., Chen, C.L., Eds.; Ios Press Inc.: Amsterdam, The Netherlands, 2003; pp. 481–490.
16. Guo, D.; Hamada, M.; He, C.; Wang, Y.; Zou, Y. An empirical model for landslide travel distance prediction in Wenchuan earthquake area. *Landslides* **2014**, *11*, 281–291. [CrossRef]
17. Tian, L.; Bi, W.; Liu, J.; Dong, X.; Xin, A. Seismic and collapse analysis of a UHV transmission tower-line system under cross-fault ground motions. *Earthq. Struct.* **2020**, *19*, 445–457.
18. Tian, L.; Pan, H.; Qiu, C.; Ma, R.; Yu, Q. Wind-induced collapse analysis of long-span transmission tower–line system considering the member buckling effect. *Adv. Struct. Eng.* **2019**, *22*, 30–41. [CrossRef]
19. Zheng, H.D.; Fan, J. Progressive collapse analysis of a truss transmission tower-line system subjected to downburst loading. *J. Constr. Steel Res.* **2022**, *188*, 107044. [CrossRef]
20. Kassem, M.M.; Beddu, S.; Qi Min, W.; Tan, C.G.; Mohamed Nazri, F. Quantification of the Seismic Behavior of a Steel Transmission Tower Subjected to Single and Repeated Seismic Excitations Using Vulnerability Function and Collapse Margin Ratio. *Appl. Sci.* **2022**, *12*, 1984. [CrossRef]
21. Li, J.; Gao, F.; Wang, L.; Ren, Y.; Liu, C.; Yang, A.; Yan, Z.; Jiang, T.; Li, C. Collapse Mechanism of Transmission Tower Subjected to Strong Wind Load and Dynamic Response of Tower-Line System. *Energies* **2022**, *15*, 3925. [CrossRef]
22. Fu, X.; Li, H.N.; Li, G.; Dong, Z.Q. Fragility analysis of a transmission tower under combined wind and rain loads. *J. Wind. Eng. Ind. Aerodyn.* **2020**, *199*, 104098. [CrossRef]
23. Guan, R.; Xiang, C.; Jia, Z. Anti-Wind Experiments and Damage Prediction of Transmission Tower under Typhoon Conditions in Coastal Areas. *Energies* **2022**, *15*, 3372. [CrossRef]
24. Zhou, Y.; Sheng, Q.; Chen, J.; Li, N.; Fu, X.; Zhou, Y. The failure mode of transmission tower foundation on the landslide under heavy rainfall: A case study on a 500-kV transmission tower foundation on the Yanzi landslide in Badong, China. *Bull. Eng. Geol. Environ.* **2022**, *81*, 125. [CrossRef]
25. Wang, F.Y.; Xu, Y.L.; Qu, W.L. Multi-scale failure analysis of transmission towers under downburst loading. *Int. J. Struct. Stab. Dyn.* **2018**, *18*, 1850029. [CrossRef]
26. Zheng, H.D.; Zhang, B.; Wang, S.B.; Zhou, G.Z. Effects of Blade Numbers on Wind-Induced Fatigue Lives of Straight-Bladed Vertical-Axis Wind-Turbine Tower Bases. *Metals* **2022**, *12*, 321. [CrossRef]
27. International Electrotechnical Commission. *Wind Turbines-Part 1: Design Requirements. IEC 614001 Ed. 3*; International Electrotechnical Commission: Geneva, Switzerland, 2006.
28. Higham, N.J. *Analysis of the Cholesky Decomposition of a Semi-Definite Matrix*; Oxford University Press: Oxford, UK, 1990.
29. Zheng, H.D.; Fan, J.; Long, X.H. Analysis of the seismic collapse of a high-rise power transmission tower structure. *J. Constr. Steel Res.* **2017**, *134*, 180–193. [CrossRef]

30. Zheng, H.D.; Fan, J. Phenomenological hysteretic model for steel braces including inelastic postbuckling and low-cycle fatigue prediction. *J. Struct. Eng.* **2019**, *145*, 04019037. [CrossRef]
31. Yang, S.C.; Hong, H.P. Nonlinear inelastic responses of transmission tower-line system under downburst wind. *Eng. Struct.* **2016**, *123*, 490–500. [CrossRef]
32. Long, X.; Wang, W.; Fan, J. Collapse analysis of transmission tower subjected to earthquake ground motion. *Model. Simul. Eng.* **2018**, *2018*, 2687561. [CrossRef]
33. Zheng, H.D.; Fan, J. Analysis of the progressive collapse of space truss structures during earthquakes based on a physical theory hysteretic model. *Thin-Walled Struct.* **2018**, *123*, 70–81. [CrossRef]
34. Lewis, W.J.; Jones, M.S.; Rushton, K.R. Dynamic relaxation analysis of the non-linear static response of pretensioned cable roofs. *Comput. Struct.* **1984**, *18*, 989–997. [CrossRef]
35. Zheng, X.Y.; Zheng, H.D.; Lei, Y.; Chen, H. Nonlinear stochastic responses of a newly developed floating wind-solar-aquaculture system. *Ocean. Eng.* **2021**, *241*, 110055. [CrossRef]
36. Zheng, H.D.; Zheng, X.Y.; Lei, Y.; Li, D.A.; Ci, X. Experimental validation on the dynamic response of a novel floater uniting a vertical-axis wind turbine with a steel fishing cage. *Ocean. Eng.* **2022**, *243*, 110257.
37. Duan, Y.F.; Wang, S.M.; Yau, J.D. Vector form intrinsic finite element method for analysis of train–bridge interaction problems considering the coach-coupler effect. *Int. J. Struct. Stab. Dyn.* **2019**, *19*, 1950014. [CrossRef]
38. Higham, D.J.; Higham, N.J. *MATLAB Guide*; Society for Industrial and Applied Mathematics: Philadelphia, PA, USA, 2016.
39. Veletsos, A.S.; Darbre, G.R. Dynamic stiffness of parabolic cables. *Earthq. Eng. Struct. Dyn.* **1983**, *11*, 367–401. [CrossRef]
40. Loredo-Souza, A.M.; Davenport, A.G. The effects of high winds on transmission lines. *J. Wind. Eng. Ind. Aerodyn.* **1998**, *74*, 987–994. [CrossRef]
41. Takeuchi, M.; Maeda, J.; Ishida, N. Aerodynamic damping properties of two transmission towers estimated by combining several identification methods. *J. Wind. Eng. Ind. Aerodyn.* **2010**, *98*, 872–880. [CrossRef]

energies

MDPI

Article

Collapse Mechanism of Transmission Tower Subjected to Strong Wind Load and Dynamic Response of Tower-Line System

Junkuo Li [1], Fan Gao [1], Lihuan Wang [1], Yaning Ren [1], Chuncheng Liu [2,*], Aiquan Yang [2], Zhao Yan [2], Tao Jiang [2] and Chengbo Li [2]

1 State Grid Hebei Economic Research Institute, Shijiazhuang 050000, China; ljunkuo@163.com (J.L.); gfan@163.com (F.G.); wlihua@163.com (L.W.); ryaning@163.com (Y.R.)
2 School of Civil Engineering and Architecture, Northeast Electric Power University, Jilin 132012, China; y_aiquan@163.com (A.Y.); yzhao@163.com (Z.Y.); jtao@163.com (T.J.); lchengbo@163.com (C.L.)
* Correspondence: lccheng@neepu.edu.cn

Abstract: Transmission towers are prone to collapse under strong wind load, resulting in significant economic losses. In order to investigate the collapse mechanism and failure modes of the transmission tower under strong wind load and whether the wind vibration factor can greatly reflect the increasing effect of the fluctuating wind, the finite element method (FEM) is utilized to analyze the ultimate bearing capacity of a typical 220 kV transmission tower. The results show that the collapse of the tower under strong wind loads is usually due to the buckling of the leg members. When the reference wind speed is equal to 27 m/s, a small part of the main leg members reaches their yield strength, while the diagonal members are still in the elastic range, and the deformation of the transmission tower is unapparent at this wind speed. When reference wind speed is equal or greater than 30 m/s, the growing variety of main legs is totally into the plastic yield stage, and the overall deformation of this tower is visible. Therefore, the transmission tower is collapsed due to the large deformation caused by the elastic-plastic buckling of leg members. Based on the aforementioned study, a finite element model involving three transmission towers and four span transmission lines is established to analyze the dynamic response of the tower-line system below fluctuating wind. Results show that the wind-induced coefficients designed by current code not only notably satisfy the stress response of tower components subjected to fluctuating wind loads in the elastic phase but also accurately assess the collapse displacement of the transmission tower. The increasing effect of displacement on the top tower under fluctuating wind, unfortunately, could not considerably reply with the investigated factor, and the load-carrying capacity of the transmission tower in the plastic phase can be overestimated by static calculation results.

Keywords: transmission tower; tower-line system; collapse mechanism; wind load; dynamic response

check for updates

Citation: Li, J.; Gao, F.; Wang, L.; Ren, Y.; Liu, C.; Yang, A.; Yan, Z.; Jiang, T.; Li, C. Collapse Mechanism of Transmission Tower Subjected to Strong Wind Load and Dynamic Response of Tower-Line System. *Energies* **2022**, *15*, 3925. https://doi.org/10.3390/en15113925

Academic Editors: Li Tian, Jianyong Han and José A.F.O. Correia

Received: 11 March 2022
Accepted: 13 May 2022
Published: 26 May 2022

1. Introduction

The electric power system plays an important role in the development of national economy [1]. Remarkably, the stability of transmission tower as a vital component of the electric power system is related to the safety of the whole electric power system. In the event of a collapse of a transmission tower, the overall transmission line could suffer a serious blackout, bringing inconvenience to economic development and people's daily life. In addition, a number of secondary disasters caused by the collapse of transmission towers could result in huge economic and property losses to a country.

According to released statistics, strong wind is the major reason of transmission tower collapse in all types of natural disasters. Dempsey et al. have found that 80% to 90% of transmission tower system failures have been caused by high wind speed [2]. In 1960s, Davenport et al. initially proposed the concept of gust factor and equivalent static load method and carried out relevant calculations, which developed the empirical formula of downwind fluctuating wind speed spectrum [3]. Ozono et al. investigated the vibration characteristics of the tower-line system in the plane via a large number of experiments and utilized the FEM to study the coupling effect of transmission line span, conductor mass and boundary conditions on the wind-induced dynamic response of tower-wire coupled system [4]. Simultaneously, Yasui et al. employed a hybrid truss-beam model to analyze the wind-induced dynamic response of the transmission-line tower coupled system in the time domain and then proposed a method to determine the aerodynamic damping parameters of transmission towers and wires [5]. Mara et al. investigated the effect of wind direction on self-supported transmission tower in power transmission lines, and results revealed that the most unfavorable wind direction for the lines is lateral, which was perpendicular to the transmission line corridor direction [6]. Battista et al. established a finite element model of the tower-line system and analyzed the dynamic response and stability of the system under the simulated wind fields in the time and frequency domains, respectively [7]. Dasi et al. have studied the wind vibration response of the tower-line system via simulating conductors and ground lines with catenary cable elements [8]. Structural damage detection utilizing dynamic vibration data has also been the subject of much research attention by Salawu et al. in the past few decades [9,10]. Most of the vibration-based methods in the literature are verified by simple structural systems, such as two-dimensional trusses [11], shear building models [12,13], beams [14–18], and plates [19–22]. Alminhana et al. developed a special-purpose nonlinear dynamic analysis technique in order to investigate the multispan line sections under progressive failure scenarios and validated the method through full-scale tests [23]. Loredo-Souza et al. proposed a novel approach for wind tunnel modeling of transmission lines and carried out a relative test on conductors so as to investigate their dynamic behavior under strong wind [24].

Deng Hongzhou et al. regarded the Jiangyin long span tower-line system as a project background and conducted experimental studies on wind vibration at different wind speeds for single tower and tower-line system in the uniform and turbulent flow fields, separately [25]. The results showed that the conducting wire had little effect on the self-vibration frequency of the tower, but the damping of the tower-line system was remarkably improved compared with that of the single tower. With background of the Jiaojiang longs span linear tower, Lou Wenjuan et al. conducted wind tunnel tests on the transmission tower, employed the numerical calculation analysis based on the test, and then proposed a simplified calculation method for the transverse wind response of the transmission tower [26]. Yan Zhitao et al. carried out a wind tunnel test on the air-elastic model of the 1000 kV Sutong long span transmission tower-line coupled system [27]. By increasing the stiffness matrix of the transmission wire model, not changing the drape of the line and deriving the accurate aerodynamic similarity ratio of the wire, the air-elastic model of the tower-line system was designed to test the wind-induced response of the single tower and the tower-wire coupled system, which revealed the mechanism of tower-line coupling mechanism in terms of displacement, frequency, and damping. Guo Yong et al. established a spatial finite element model of the tower-line system with the Zhoushan long span transmission tower as the project background, conducted a time-domain analysis of the wind vibration response of the system, and also performed a wind tunnel experiment of the aeroelastic model of the tower-line coupled system [28]. Cheng Zhijun et al. carried out wind tunnel tests with aeroelastic model of an overhead transmission line, observed the wind vibration response of transmission towers at different wind speeds and wind angles, and finally conducted a qualitative analysis of the tower-line coupling effect [29]. Xie Qiang et al. performed a failure analysis of a transmission tower and carried out static nonlinear buckling analysis and dynamic analysis to assess the ultimate load-carrying capacity and the most vulnerable parts of the tower based on the field investigation of a damaged transmission line following Typhoon Mujigae in 2015 [30]. Niu Huawei et al. conducted wind performance assessment for a typical single-circuit tower-line system subjected to strong wind loads, proposed a probabilistic assessment method for the aforementioned system under strong wind, and then carried out wind tunnel tests to obtain the capacity spectrum curves of the tower-line system by incremental dynamic analysis (IDA) [31]. An Liqiang et al. utilized numerical simulation method to analyze the damage mechanism and failure modes of 13 transmission towers in 110 kV transmission lines following the super strong typhoon in Hainan province and to deeply study the influence of typhoon load characteristics on the structural force of the tower [32]. Based on the Tian-Ma-Qu stress-strain relationship, Bi Wenzhe performed a continuous collapse simulation of the tower-line system under the down-strike storm winds, which found that the collapse of the tower in this case was mainly caused by the significant local deformation of the tower in a short period of time [33]. Xu et al. and Qu et al. verified vibration control performance of friction dampers installed in a truss TV tower through numerical analysis, but their study is limited to the numerical analysis and lacks details for real implementation. Then, aeolian vibration of the transmission line, which is induced by the interaction between the transmission line motion and wind, is controlled using the stockbridge-type damper, a kind of tuned mass damper [34–39].

In order to ensure the operational safety of transmission lines under strong winds, it is critical to accurately predict the vulnerable parts in a transmission tower, which reveals collapse mechanism of this tower and provides scientific references for the wind-resistant design of transmission lines, tower retrofitting and maintenance. In this paper, a typical 220 kV transmission tower subjected to strong wind load was utilized as a project example to study the collapse mechanism and failure mode. Meanwhile, the analysis of wind vibration response of tower-line system under dynamic wind load was also carried out to investigate the dynamic response characteristics of the transmission tower under the impact of fluctuating wind so as to provide valuable reference for the wind-induced design of the high-voltage transmission tower.

2. Collapse Mechanism and Failure Mode of Transmission Tower

2.1. Investigation Description

A latticed 220 kV transmission tower designed by State Gird Corporation of China is studied in this paper. The tower is located in Hebei province where belongs to the IV meteorological zone. The parameters of this meteorological condition are as follows: design wind speed is 25 m/s, temperature is 25 degrees C above zero, and ice thickness is 0. The model number of the investigated tower is 2B2-ZM2. Its nominal height is 30 m. The wind span, weight span, and equivalent span are 410 m, 550 m, and 450 m, respectively. The properties of the conductor and ground line are shown in Table 1.

Table 1. Properties of conductor and ground line.

Type (Parameters)	2×LGJ-400/35 (Conductor)	JLB20A-150 (Ground Line)
Cross sectional area (mm^2)	425.24	148.07
Diameter (mm)	26.82	15.75
Density (per unit length) (kg/km)	1349	989.4
Maximum tension (N)	78,960	37,698
Breaking tension value (%)	25	100

2.2. FEM Models

2.2.1. FEM Elements and Material Properties

The transmission tower is a spatial structure system composed of main leg members, diagonal members, and redundant members. The main leg members and diagonal members are generally connected by bolts and gusset plates in order to make the nodes have a strong restraining effect. According to released research results, the difference between the calculation results of rigid connection and semirigid connection is not significant, and they can be utilized for FEM analysis. Therefore, the rigid node model is used in the numerical simulation to simplify the calculation [40–42]. Hence, the transmission tower is modeled in ABAQUS software, where tower members are modeled using B31 elements assuming rigid connections. In addition, tower members are structured with equal-legged angles, the yield strength of main leg members is 345 MPa, and the yield strength of diagonal members as well as redundant bracing members is 235 MPa. Last, the damping coefficients α_d and β_d in tower-line coupled system are totally based on Rayleigh formula, where the α_d and β_d of transmission towers are 0.2 and 0 compared to conductors with 0.02 and 0. The elastic modulus of both kinds of steel and density are taken as 206 GPa and 7850 kg/m^3, respectively. Specific information about the angles of model tower is shown in Table 2.

Table 2. Information of angles in model tower.

Type	Number	Members Specification (mm)	Cross Sectional Area (mm^2)	Angle Density (kg/m)	Radius of Gyration i_{y0} (mm)
Q345	1	L125 × 10	2437	19.130	2.48
Q345	2	L125 × 8	1975	15.500	2.50
Q345	3	L110 × 10	2126	16.69	2.17
Q345	4	L110 × 8	1724	13.53	2.19
Q345	5	L100 × 8	1564	12.28	1.98
Q345	6	L100 × 7	1380	10.83	1.99
Q345	7	L90 × 7	1230	9.66	1.78
Q345	8	L80 × 7	1086	8.53	1.58
Q345	9	L75 × 6	880	6.91	1.49
Q235	10	L75 × 5	741	5.82	1.50
Q235	11	L70 × 5	688	5.40	1.39
Q235	12	L63 × 5	614	4.82	1.25
Q235	13	L63 × 4	498	3.91	1.26
Q235	14	L56 × 5	542	4.25	1.10
Q235	15	L56 × 4	439	3.45	1.11
Q235	16	L50 × 5	480	3.77	0.98
Q235	17	L50 × 4	390	3.06	0.99
Q235	18	L45 × 4	349	2.74	0.89
Q235	19	L40 × 4	309	2.42	0.79
Q235	20	L40 × 3	236	1.85	0.79

2.2.2. Selection of Stress-Strain Relationship

In this paper, the stress–strain relationship of angles is the double broken line model of the yield strength of 345 MPa and 235 MPa with Mises yield and kinematic hardening criterion. When the steel reaches yield strength, there is an increasing number of slip surfaces and the material changes from yielding to strengthening, exhibiting elastic–plastic properties. The stress–strain relationship is shown as Figure 1.

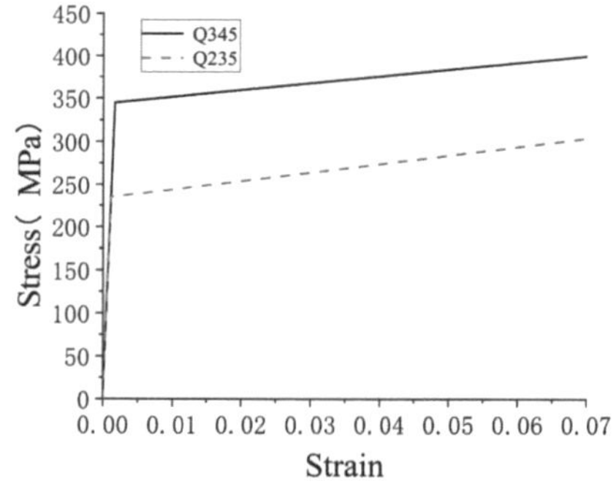

Figure 1. The stress-strain relationship for angles in the FEM models.

2.2.3. Modal Analysis of the Transmission Tower

In order to verify the reasonableness of the transmission tower, the self-vibration characteristics of the structure are usually calculated when the structural design is carried out.

The first three modes and the corresponding frequencies of the transmission tower obtained by the modal analysis in this paper are shown in Figure 2. As can be seen from Figure 2, the distribution of the first three modes of the transmission tower is reasonable. The first mode is out of plane bending deformation, the second mode is in-plane bending deformation, and the third mode is out of plane torsional deformation. There is no local vibration phenomenon, indicating that the finite element model of the tower is correct.

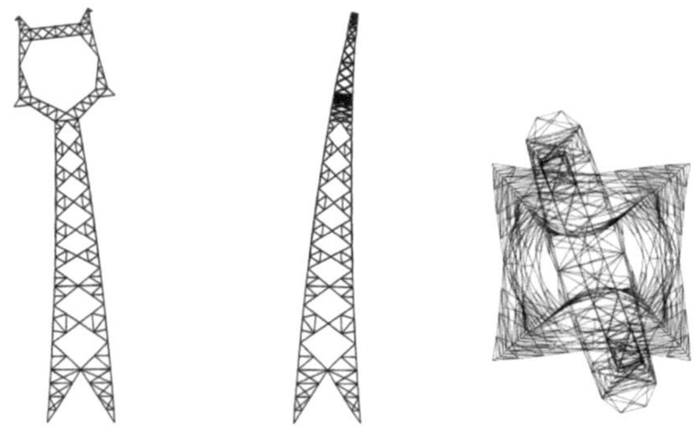

mode 1 (freq = 2.095 Hz) mode 2 (freq = 2.107 Hz) mode 3 (freq = 4.856 Hz)

Figure 2. The first three modes and the corresponding frequencies of the tower.

2.3. *Loads Calculation and Application*

Structure loads of the studied tower under the strong wind can be divided into horizontal and vertical loads according to the different direction. The direction of the wind load applied to the structure is perpendicular to the direction of the conductor and ground wire, and the vertical loads refer to the load acting along the direction of gravity. The wind loads are calculated separately according to the load code for the design of overhead transmission line [43].

2.3.1. Calculation of Vertical Loads

The vertical loads are composed by the gravity loads amongst transmission towers, conductors, ground lines, insulator strings, and metal fittings, respectively. The dead load of the transmission tower itself is considered by defining the material density and gravitational acceleration, which is automatically calculated by the ABAQUS software. The characteristic value of the self-weight load of conductors and ground lines should be calculated according to Equation (1).

$$G = n\gamma_1 A L_V \tag{1}$$

where, n is the number of subconductor per phase of the conductor. L_V is the vertical span from conductor and ground line, unit: m. γ_1 is ice-free vertical specific load between the conductor and ground line, unit: N/(m·mm^2). A is cross-sectional area for conductor and ground wire, unit: mm^2.

2.3.2. Calculation of Horizontal Loads

Based on the height of the transmission tower and cross-sectional location, the transmission tower is divided into 7 panels and set 7 simulation points. Referring to the load code for the design of overhead transmission line [43] to calculate the value of the wind load, the most unfavorable wind angle is 90°, which provided considerable impact on the transmission line. Therefore, only calculated the wind direction angle of 90°. The tower segmentations and simulating points are as shown in Figure 3.

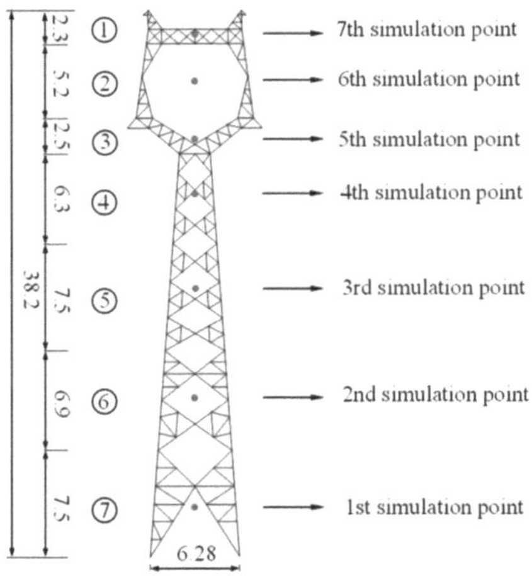

Figure 3. Divisions and simulating points of the studied tower (unit: m).

When the wind direction is perpendicular to the conductor and ground wire, the characteristic value of wind loads applied to the transmission lines can be expressed by Equations (2) and (3).

$$P_D = \alpha_L \cdot W_0 \cdot \mu_z \cdot \mu_{sc} \cdot \beta_c \cdot d \cdot L_p \cdot B \cdot sin^2\theta \quad (2)$$

$$W_0 = V^2/1600 \quad (3)$$

In Equation (2), α_L is the span reduction factor. μ_z is the wind pressure height variation coefficient. μ_{sc} is the body coefficient. When line diameter is less than 17 mm, it should be taken as 1.1. When the wire diameter is greater than or equal to 17 mm, it should be taken as 1.0. β_c is the wind gust factor of conductor and ground line. d is the diameter of the conductor and ground wire or the calculated outside one in case of ice cover. Split conductor is taken as the sum of the outside diameter of all subconductors. L_p is the horizontal span of the tower. B is the increase coefficient of ice load, when the thickness of ice coating is 5 mm and 10 mm, and B is 1.1 and 1.2, respectively. For the case of no ice load, B is 1.0, and for calculating tension of conductor, B is 1.0, where θ is the angle between the wind direction and direction of conductor and ground line. In Equation (3), W_0 is the characteristic value of the reference wind pressure, and V is the basic wind speed.

Wind load of transmission tower could be predicted by Equation (4).

$$P = W_0 \cdot \mu_Z \cdot \mu_{sc} \cdot B \cdot A \cdot \beta_Z \quad (4)$$

where, β_Z is the wind vibration factor of the tower at height Z (m) given by Equations (5)–(8).

$$\beta_{zi} = 1 + 2g \cdot \varepsilon_t \cdot I_{10} \cdot B_{Zi} \cdot \sqrt{1+R^2} \quad (5)$$

$$B_{Zi} = \frac{m_i\phi_{1i}}{\mu_{Si}\mu_{Zi}A_i} \cdot \frac{\sqrt{\sum_{j=1}^{n}\sum_{j'=1}^{n}(\mu_{Sj}\mu_{Zj}\phi_{1j}\overline{I}_{Zj}A_j)(\mu_{Sj'}\mu_{Zj'}\phi_{1j'}\overline{I}_{Zj'}A_{j'})coh_Z(z_j,z_{j'})}}{\sum_{j=1}^{n} m_j\phi_{1j}^2} \quad (6)$$

$$R^2 = \frac{\pi}{6\zeta_1}\frac{x_1^2}{(1+x_1^2)^{4/3}} \quad (7)$$

$$x_1 = \frac{30f_1}{\sqrt{k_w W_0}} \quad (8)$$

where, g is the peak factor, taking 2.5; ε_t is the pulsation reduction coefficient of tower under wind load; I_{10} is the nominal turbulence intensity at 10 m height, taking 0.14 for Class B ground roughness; B_{Zi} is the background factor; R^2 is the resonance factor; ϕ_1 is the first-order mode of the tower; $coh_{Z}(z_j, z_{j'})$ is the vertical coherence function; ζ_1 is the first-order damping ratio of the tower; and f_1 is the first-order vibration frequency of the tower.

To analyze the mechanical properties of the transmission tower under strong wind, the effect of wind load on the insulator string at the conductor line should also be taken into account. The characteristic value of the insulator string wind load can be obtained by Equation (9).

$$P_J = n \cdot \lambda_1 \cdot W_0 \cdot \mu_Z \cdot \mu_{S1} \cdot B_3 \cdot A_1 \quad (9)$$

In Equation (9), n is the number of insulator strings in the vertical wind direction; λ_1 is the wind load shielding reduction factor for downwind insulator strings; A_1 is the calculated value of wind pressure aera of sing insulator string; and B_3 is increased factor of wind load of insulator string over ice. The calculation results of wind load applied to the tower, conductors, and ground lines are illustrated in Tables 3 and 4.

Table 3. Wind loads of tower sections (the wind speed is 25 m/s and the wind direction angle is 90°).

Tower Section Number	Height above Ground (m)	Height of Tower Section (m)	Height Variation Factor of Wind Pressure μ_Z	Wind Vibration Factor of Tower β_Z	Shape Factor of Tower Section μ_S	Projection Area A_f (m²)	Characteristic Value of Basic Wind Pressure W_0 (Pa)	Characteristic Value of Wind Load (kN)	Design Value of Wind Load (kN)
1	37.0	2.3	1.481	5.860	2.217	0.528	0.391	3.432	4.805
2	33.1	5.9	1.430	1.537	2.217	2.154	0.391	4.100	5.740
3	29.4	1.8	1.380	2.642	2.217	1.596	0.391	3.441	4.817
4	24.9	6.3	1.308	1.316	2.356	3.635	0.391	5.758	8.061
5	18.0	7.5	1.190	1.116	2.356	6.018	0.391	7.352	10.293
6	10.8	6.9	1.021	1.059	2.356	7.120	0.391	7.083	9.916
7	3.6	7.5	1.000	1.012	2.356	9.467	0.391	8.818	12.345

Table 4. Wind loads on the conductor, ground line, and insulator.

Wind Speed (m/s)	Upper Conductor Wind Load P_{D1} (N)	Lower Conductor Wind Load P_{D2}, P_{D3} (N)	Ground Line Wind Load P_B (N)	Upper Insulator Wind Load P_{JD1} (N)	Lower Insulator Wind Load P_{JD1}, P_{JD2} (N)
15	4567	4344	1497	314	297
20	8118	7720	2662	558	528
25	12,679	12,056	4158	871	826
26	13,712	13,039	4497	942	893
27	14,786	14,059	4850	1016	963
28	15,900	15,118	5216	1093	1036
30	18,249	17,351	5987	1255	1189

2.3.3. Load Application on Transmission Tower

Before loading, the tower foot of the FEM model is set as a fixed constraint boundary, and the wind loads of the tower are applied to the corresponding nodes of the transmission tower body and cross arm. In addition, the wind loads and weight of conductor wires, ground wires and insulators, are applied as point forces to the corresponding positions of the transmission tower. The load application on the transmission tower is as shown in Figure 4.

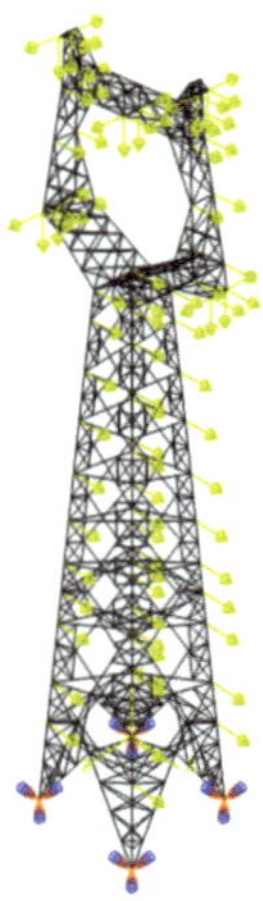

Figure 4. Load application on transmission tower.

2.4. Analysis of Failure Mode of Transmission Tower

When the transmission tower is subjected to external loads, main legs, and diagonal bracings are mainly carried the combined effect of axial forces and bending moment so as to result members in the buckling failure and potential deficiency of local stiffness and strength, which cause the collapse of the transmission tower. Hence, it is necessary to analyze the collapse of transmission towers.

To simulate the vulnerable member and mechanical characteristics of the transmission tower under strong winds, there are seven different wind speeds among 15 m/s, 20 m/s, 25 m/s, 26 m/s, 27 m/s, 28 m/s, and 30 m/s setting in this paper. The deformation of the transmission tower and the variation rule of the member internal force are analyzed by calculating the wind loads at different wind speeds, so as to obtain the maximum stress of the members and the top displacement of the transmission tower at different wind speeds. The results are shown in Table 5.

Table 5. Calculation results of displacement and internal force of transmission tower at different speed.

Wind Speed (m/s)	Tower Top Displacement (m)	Maximum Stress (MPa)	Maximum Stress Position
15	0.093	130	Main members in the 6th panel on the pressure side
20	0.165	207	Main members in the 6th panel on the pressure side
25	0.258	307	Main members in the 6th panel on the pressure side
26	0.280	330	Main members in the 6th panel on the pressure side
27	0.302	345	Main members in the 6th panel on the pressure side
28	0.329	345	Main members in the 5th panel on the pressure side
30	0.378	346	Main members in the 5th panel on the pressure side

As shown in Table 5, the transmission tower is still in elastic stage when the wind speed rises to 26 m/s. When the speed increases to 27 m/s, the member with maximum stress, the lower part of leg members in the 5 panel on the pressure side, proceeds to the plastic stage. Although the deformation begins to accelerate and this member even gradually withdraw from work, the maximum stress bracing is not buckled at this time, leading to stress redistribution in the transmission tower. After the structure enters the plastic stage, the stress of the main leg on the pressure side of the 5th panel increases significantly due to the leg member in the 6th panel on the same side reaching the yield strength. Figure 5 presents that leg members with the most significant change in stress are the No. 1 and No. 2 in the 5th panel. The stress variation curves of No.1 and No. 2 are shown in Figure 6. It can be seen from Figure 6 that the No. 2 rod will become the second yielding member on the compression side, and the stress of No. 1 member will increase sharply after the maximum stress member reaches yield strength. The stress–strain nephogram of vulnerable members is shown in Figure 7. It can be seen that due to the symmetry structure of the transmission tower, when the wind direction is perpendicular to the direction of the transmission line, and the wind blows to the tower at a 90 degree angle, the stresses of the two main legs is relatively close to each other, which indicates that the main members are more significantly affected by the axial forces and bending moments. Therefore, the possibility of lateral torsional behavior of the structure under wind loads is unlikely.

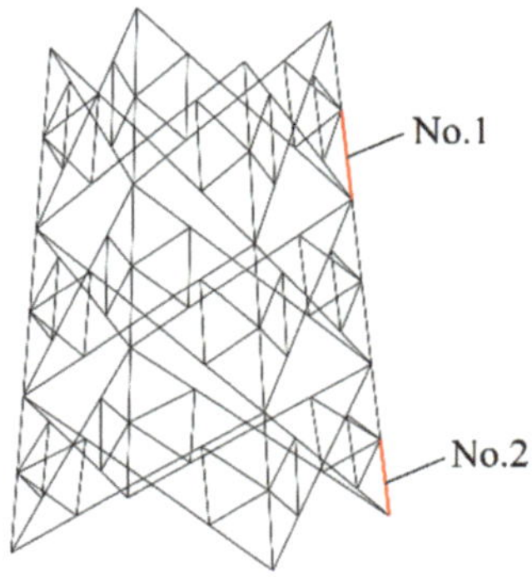

Figure 5. Positions of No. 1 and No. 2.

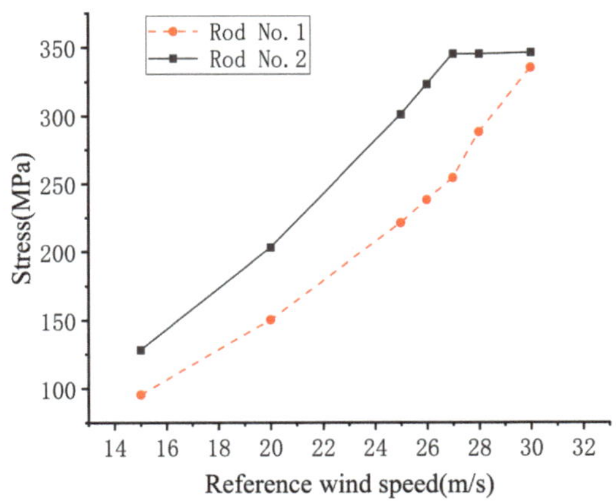

Figure 6. Stress variation curves of No. 1 and No. 2.

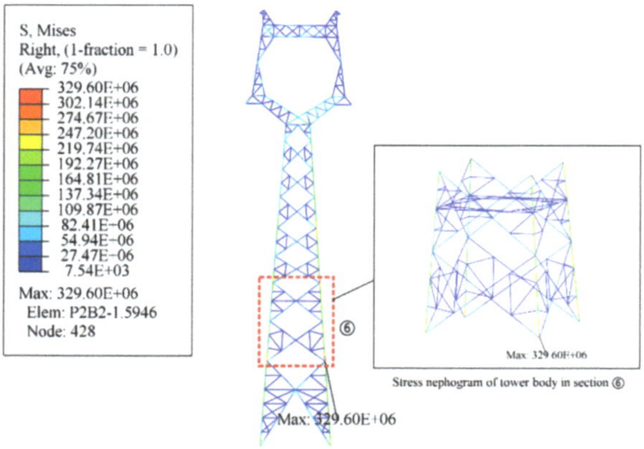

Figure 7. Stress nephogram of the transmission tower (V10 = 26 m/s).

The variation curve of transmission tower top displacement with wind speed is shown in Figure 8. When the wind speed exceeds to 30 m/s, the displacement of top tower changes abruptly. Here, the buckling analysis of transmission tower can be carried out. Firstly, the eigenvalue buckling analysis is performed on the transmission tower. Secondly, the obtained first-order buckling mode was applied to the tower as an initial defect. Finally, the nonlinear buckling analysis was carried out on the tower after updating the model. The results showed that vulnerable members in the studied tower was complied with buckling failure. Before the collapse of the transmission tower, the ultimate displacement of top tower is 0.378 m which can be taken as collapse displacement of the tower. Similarly, the wind speed of 30 m/s was determined as the critical collapsed wind speed in the same structure. The stress nephogram of the transmission tower and position of the failed member under this speed are shown in Figure 9. Since the main leg on the pressurized side of the 6th panel yielded in the first time, this member is determined to be the one that initially failed.

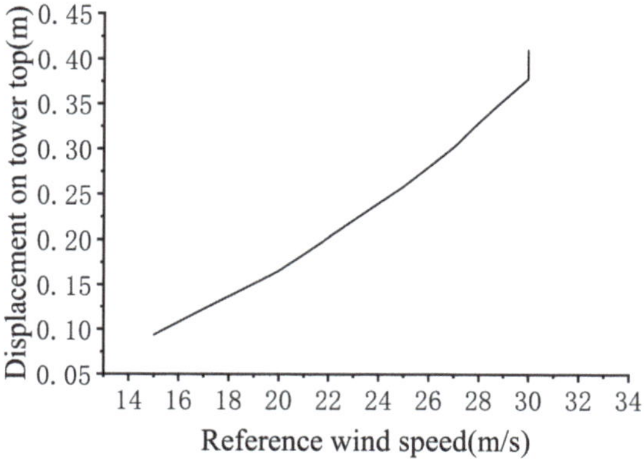

Figure 8. The variation curve of transmission tower top displacement with wind speed.

The stress–wind speed variation curve of a failure member is shown in Figure 10. After entering the plastic stage, the stress of failed member is not increased significantly but remains near the limit of yielding. As the leading load-bearing component, the buckling of the leg member could lead the whole tower to be in the failure mode, which contributes to the redistribution of structural internal forces. As the wind load continues to be applied, load paths of the leg member would be destroyed due to the buckling of its surrounding members one after another, resulting in an escalation in strain and a sharp decrease in load-carrying capacity and eventually collapse of the transmission tower. The failure mode of the transmission tower is shown in Figure 11.

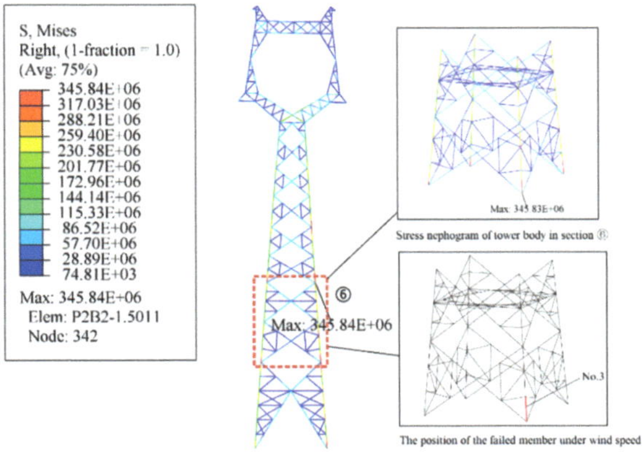

Figure 9. Stress nephogram of the failure member (V10 = 30 m/s).

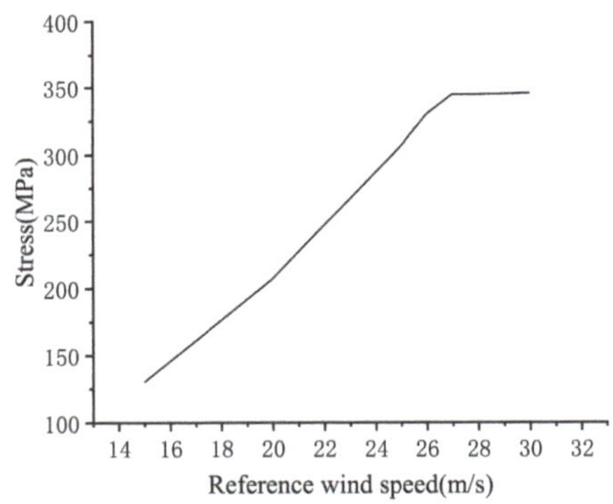

Figure 10. Stress–wind speed variation curve of buckling member.

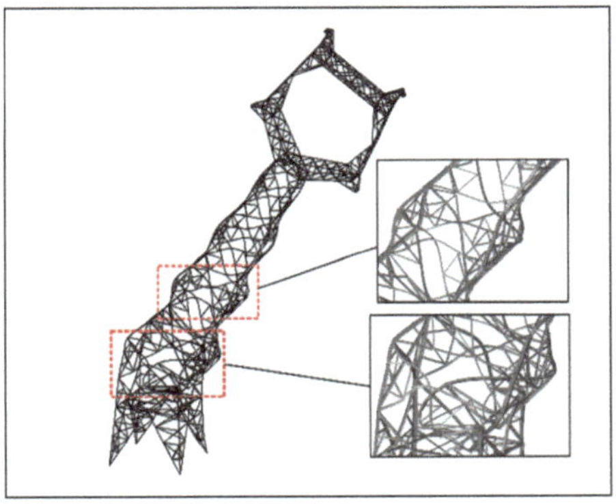

Figure 11. Failure mode of transmission tower under strong wind.

3. Analysis of Wind-Induced Response of Tower-Line System

Composed by transmission towers and conductors, ground wires, and metal fittings, high-voltage transmission lines are long-span and tail-flexible structural systems, which are susceptible to wind-induced dynamic excitation. Although the wind vibration coefficient and wind load adjustment factor are introduced in current codes to calculate the wind vibration response of transmission towers by equating the wind load into static load, it is still difficult to accurately reflect the time and amplification effect of the fluctuating wind. Therefore, it is especially important to analyze the wind vibration response of the tower-line system under the fluctuating wind. In order to precisely calculate the dynamic response of transmission towers under fluctuating wind loads, three towers and four transmission lines is modeled by the ABAQUS software for wind vibration response analysis based on the 1.1 project background as an example.

3.1. Finite Element Model of Tower-Line System

The transmission tower is in accordance with the same material properties and modeling approach as the single tower described above. To ensure the accuracy of the results, the initial configuration of conductors and ground wires is acquired by Equation (10), and the precise geometry for its normal operation is determined using an iterative form finding method. The form finding of conductors and ground lines is shown in Figure 12.

$$y = \frac{\sigma_0}{\gamma}(\mathrm{ch}\,\frac{\gamma}{\sigma_0}x - 1) \tag{10}$$

where σ_0 is the horizontal stress of the overhead line; γ is the specific loading, x is the horizontal coordinate of conductors and ground lines; y is the longitudinal coordinate of conductors and ground lines.

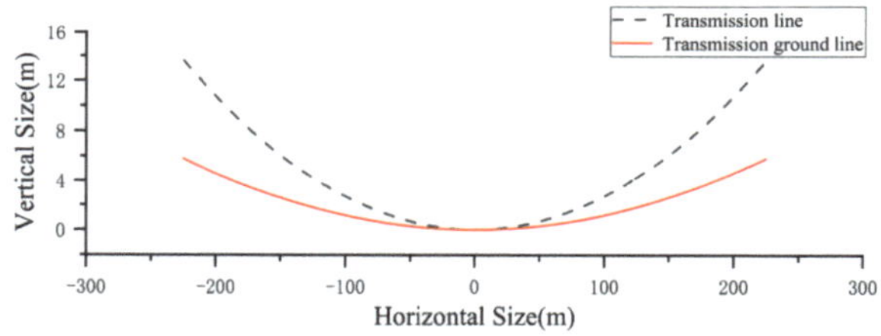

Figure 12. Form finding of conductor and ground line.

Due to only sustain axial tensile stress, the transmission line should be simulated using truss elements and ignoring the elements' compression. The conductors could be hinged to the insulators as same as the transmission tower and wire connection method. The FEM model of tower-line system is shown in Figure 13.

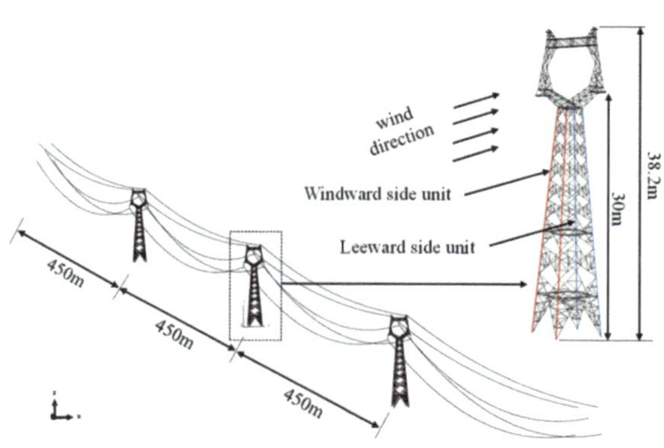

Figure 13. FEM model of tower-line system.

3.2. Simulation of Wind Load

The simulation of fluctuating wind is the basis for the analysis of the dynamic characteristics of the tower-line coupled system. Through the simulation of fluctuating wind speed, the wind load applied to the tower-line system can be obtained, which can lay the solid the foundation for dynamic response analysis of this system.

The atmospheric boundary layer wind consists of mean wind and fluctuating wind, and the wind speed at t and z which is the moment and height respectively can be described as Equation (11).

$$v(z,t) = \bar{v}(z) + v_f(z,t) \tag{11}$$

where, $\bar{v}(z)$ is the mean wind speed, and $v_f(z,t)$ is the fluctuating wind speed. Remarkably, the mean wind speed is calculated according to the equation of the variation of wind speed with height described by the exponential function proposed by Davenport, as shown in Equation (12).

In this case, the mean wind speed is calculated according to the equation of the variation of wind speed with height described by the exponential function proposed by Davenport, as shown in Equation (12).

$$\bar{v}_{z1} = \bar{v}_{z2}(z_1/z_2)^a \tag{12}$$

where z_1 and z_2 are heights; $\bar{v}_{z1}$ and $\bar{v}_{z2}$ are mean wind speeds; and a is the ground roughness index.

Fluctuating winds are then simulated using the Davenport wind speed spectrum, which is calculated as Equation (13).

$$\frac{nS(n)}{v_*^2} = \frac{4x^2}{(1+x^2)^{\frac{4}{3}}} \tag{13}$$

$$x = 1200\frac{n}{\bar{v}_{10}} \tag{14}$$

In Equation (13), n is the fluctuating wind speed frequency; $S(n)$ is the Davenport power spectral density function; and v_* is the flow shear velocity which are shown in $v_*^2 = K\bar{v}_{10}^2$, where $\bar{v}_{10}$ is the mean speed at 10 m height, and K is the constant coefficient associated with the geomorphology, taking 0.00215.

To consider the impact of spatial correlation, the fluctuating wind simulation is performed in MATLAB software by means of the harmonic synthesis method. The spatial correlation coefficient of fluctuating wind speed can be calculated by the following Equation (15).

$$\gamma_{ij}(n) = \exp\left(\frac{-nC_z^2(z_i - z_j)}{2\pi V(z_i)}\right) \tag{15}$$

In Equation (15), n is the frequency of fluctuating wind; C_z^2 is the attenuation coefficient in the vertical direction at any two points in space; $\bar{v}_{(z_i)}$ is the mean wind speed at i point; z_i and z_j symbolize the spatial coordinates of i point and j point in Z-axis direction.

In order to validate the accuracy of the time-history curves of the fluctuating wind, the highest and lowest points of the fluctuating wind speed power spectrum compared with the target spectrum are listed at 27 m/s the reference wind speed. Simultaneously, the time–history curves of wind speed are shown in Figure 14.

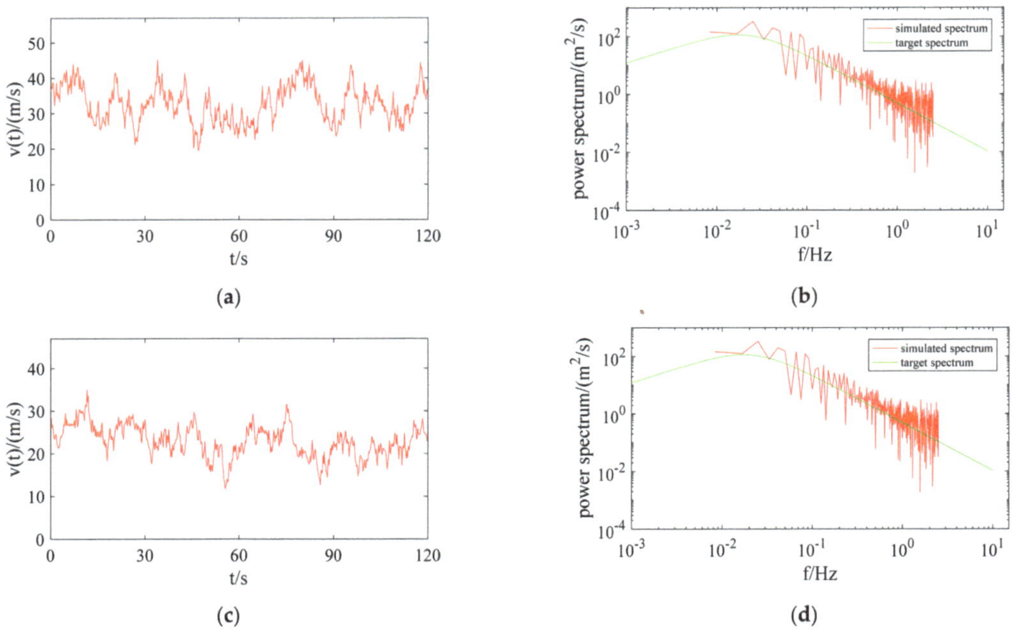

Figure 14. Time-history and power spectrum curves of fluctuating wind. (**a**) Time-history curve of wind speed at the highest point, (**b**) power spectrum curve at the highest point, (**c**) time-history curve of wind speed at the lowest point, (**d**) power spectrum curve at the lowest point.

Figure 14 shows that the power spectrum of fluctuating wind speed agrees well with the target spectrum and can be utilized for numerical calculations. The time–history curve of wind speed is converted into a time–history curve of wind load according to the load code for the design of overhead transmission line [43]. The FEM analysis of the tower-line system is carried out in Abaqus software. Specifically, the dynamic wind load of each calculation panel is equally distributed to each loading point, where the wind load lasted for 120 s with a step size of 0.2 s and wind angle of 90° at which the direction is perpendicular to the line direction. The dynamic response analysis of the tower-line system under the wind load with different wind speeds can be achieved from the response of the tower line system at different moments.

Based on the aforementioned static study, when wind speed reaches to 26 m/s, the entire transmission tower is in an elastic state. When wind speed rises to 27 m/s, the stress maximum member, in lower part of the main on the compression side of the 6th panel, is carried into the plastic phase, revealing that it gradually withdraws from work with the increasing deformation. Therefore, in the dynamic analysis, the structural response at a reference wind speed value of 25 m/s is calculated first, and this reference wind speed value is gradually increased in the subsequent calculations.

3.3. Numerical Results of Wind-Induced Vibration Response

By analyzing the dynamic response of the tower-line system under the fluctuating wind load, the effect of wind-induced vibration of transmission towers is obtained under the impact of the tower-line system. The results show that the collapse possibility of transmission towers could be increased due to the coupling effect of tower-line system and randomness of fluctuating wind loads. Figures 15 and 16 show the time–history curves of the maximum stress and top displacement of the transmission tower, respectively.

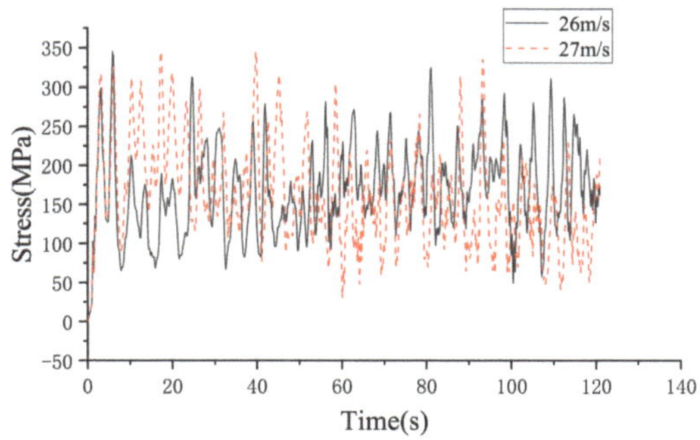

Figure 15. The maximum stress–time curve of transmission tower.

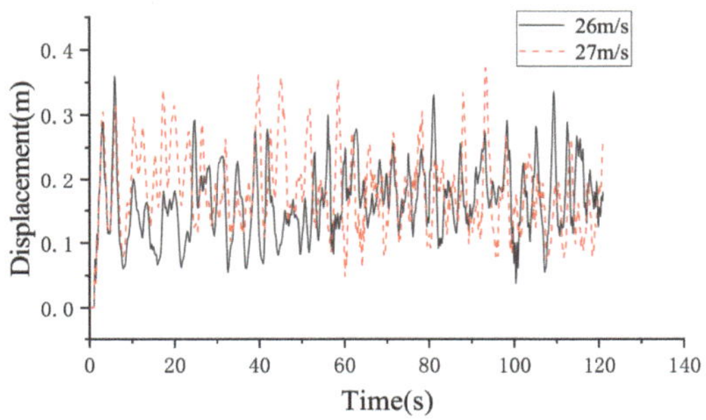

Figure 16. The top displacement–time curve of transmission tower.

Based on the aforementioned analysis, it is found that the tower-line system is completely in elastic stage when wind speed rise to 25 m/s, and the maximum stress position is alternated at the bottom of the 5th and 6th panel. As shown in Figure 15, when wind speed increases to 26 m/s, each transmission tower is in normal working condition, and the load-carrying capacity is still relatively stable. However, the lower leg member of 6th panel reaches the yield strength at the moment of 5.8 s, while the other moments were still in the elastic stage. When wind speed is 27 m/s, the stress of the main member on compressed side of the 6th panel attaches to yield. When wind speed rise to 28 m/s, the leg member at bottom of the 6th panel reached the ultimate stress immediately, while the finite element calculation is not to satisfy the convergence condition, which indicates that the tower could collapse at this time.

Figure 16 is the displacement–time curve on top of the tower. Combined with the stress–time curve, it can be found that, when the displacement on top of the tower is 0.36 m, the transmission tower has parts into yield. When the displacement is 0.4 m, the FEM model is not convergent, which indicates collapse of the transmission tower. Hence, it can be found that the tower-line system would be in a safe and stable cperation when wind speed is lower than 27 m/s. When the wind speed attaches to 27 m/s, the leg bracing on the pressure side of the 6th panel reaches yield strength. When the wind speed rises to 28 m/s, the leg bracing on the pressure side of the 6th panel attaches to the ultimate stress, which leads to the collapse of the tower. Meanwhile, it can be determined that the leg member on the pressure side of the 6th panel would be the vulnerable part of the tower to collapse and failure under the wind load with 90° wind angle.

Comparing the wind vibration response calculation results with those under static wind load, it can be found that there are few differences between the two in terms of failure mode, and yield wind speed would remain the same level. Compared with the static analysis, however, the collapse of transmission towers could be significantly earlier in the dynamic analysis.

Since the transmission tower is in the elastic stage under wind speed of 25 m/s, the dynamic initial wind load is calculated from the specific point when V_{10} = 25 m/s. The comparison of the response of the transmission tower under static and dynamic wind load is shown in Table 6.

Table 6. Comparison of wind-induced response of transmission tower under static and dynamic wind load.

Wind Speed (m/s)	Static Wind Load		Dynamic Wind Load		Ratio of Increased Displacement (%)	Ratio of Increased Stress (%)
	Maximum Displacement on Tower Top (m)	Maximum Stress of Tower Member (MPa)	Maximum Displacement on Tower Top (m)	Maximum Stress of Tower Member (MPa)		
25	0.258	307	0.297	297.6	15.1%	−3.1%
26	0.280	330	0.350	345	25.0%	4.5%
27	0.302	345	0.373	345	23.5%	0.0%
28	0.329	345	0.400	500	21.6%	44.9%

4. Conclusions

This paper presents the ultimate load-carrying capacity of a 220 kV transmission tower, collapse mechanism, and failure mode of this tower under strong wind. The static and dynamic FEM analyses of the tower-line system have been carried out in ABAQUS software. Specifically, dynamic analysis of the tower-line system is carried out to understand its dynamic response under fluctuating wind, which is once in contrast with the response beneath equal static wind load. Based on the numerical results obtained from the kind of transmission tower in this paper, the following conclusions can be summarized as follows:

1. From the analysis of the single tower under equivalent static wind load, it is found that, when the wind speed is 27 m/s, the maximum stress bracing is once the leg member on the compression side of the 6th panel, at which this member has entered into the plastic stage. With the enlargement of deformation, the leg member in the 5th panel reaches yield strength. When the wind speed is higher than 30 m/s, the main leg in the 6th panel buckles, which in the end renders the transmission tower to collapse. The ultimate displacement on the pinnacle of the tower earlier than failure is 0.378 m.
2. Combined with three towers and four lines, the tower-line system is modeled to analysis the dynamic time–history response under fluctuating wind. It is found that the maximum stress bracing of the center transmission tower in tower-line system be viewed as the leg member in the 6th panel on pressure side. When the wind speed rises to 27 m/s, the leg member of the 6th section on the pressure side reaches yield strength. When the wind speed attaches to 28 m/s, the aforementioned member has a large local deformation, which symbolizes the buckling failure of this member.
3. By comparing the failure mode of transmission tower under static wind load with that under dynamic wind load, it is found that the yield wind speed of both condition is at the same level, the failure mode and collapse displacement are close to each other, indicating that the wind vibration factor according to the load code for the design of overhead transmission line [43] not only greatly reflect the stress response of the tower bracing in elastic stage under fluctuating wind but also accurately assess the collapse displacement of transmission tower. Compared with the dynamic analysis, however, in the dynamic analysis the maximum displacement on the top tower is significantly larger than that, and the collapse of the transmission tower is earlier, which reveals that the wind vibration coefficient cannot greatly reflect the increasing effect of the fluctuating wind on the displacement of the top of the tower. Remarkably, the static calculation results overestimated the load-carrying capacity of the transmission tower in the plastic stage. Therefore, the wind resistance design and retrofitting bracing and even maintenance of the transmission tower should take into account the increasing effect of the displacement which are caused by the fluctuating wind. The research

results of this paper provide important reference value for the wind resistance design, reinforcement and maintenance of the same type of transmission tower.

Author Contributions: J.L.: conceptualization, project administration. F.G.: methodology, investigation. L.W.: investigation, resources. Y.R.: resources and data curation. C.L. (Chuncheng Liu): supervision, writing—review & editing. A.Y.: formal analysis. Z.Y.: software, visualization. T.J.: software, data curation. C.L. (Chengbo Li): writing—review & editing. All authors have read and agreed to the published version of the manuscript.

Funding: This research was funded by the National Natural Science Foundation of China (Grant No. 51278091) and the technology cost project of Hebei Huizhi Electric Power Engineering Design Co., Ltd. (Grant No. HZHTCG2021-09).

Data Availability Statement: Not applicable.

Acknowledgments: The authors gratefully acknowledge Hebei Huizhi Electric Power Engineering Design Co., Ltd. and School of Northeast Electric Power University that have contributed to the research results reported within this paper.

Conflicts of Interest: The authors declare that they have no known competing financial interests or personal relation-ships that could have appeared to influence the work reported in this paper.

References

1. Lei, X.; Fu, X.; Xiao, K.; Wang, J.; Nie, M.; Li, H.N.; Xie, W.P. Failure Analysis of a Transmission Tower Subjected to Wind Load Using Uncertainty Method. *Proc. CSEE* **2018**, 266–274. [CrossRef]
2. Dempsey, D.; White, H. Winds wreak havoc on lines. *Transm. Distrib. World* **1996**, *48*, 32–37.
3. Davenport, A.G. Gust loading factors. *J. Struct. Div.* **1967**, *93*, 11–34. [CrossRef]
4. Ozono, S.; Maeda, J. In-plane dynamic interaction between a tower and conductors at lower frequencies. *Eng. Struct.* **1992**, *14*, 210–216. [CrossRef]
5. Yasui, H.; Marukawa, H.; Momomura, Y.; Ohkuma, T. Analytical study on wind-induced vibration of power transmission towers. *J. Wind Eng. Ind. Aerodyn.* **1999**, *83*, 431–441. [CrossRef]
6. Mara, T.G.; Hong, H.P. Effect of wind direction on the response and capacity surface of a transmission tower. *Eng. Struct.* **2013**, *57*, 493–501. [CrossRef]
7. Battista, R.C.; Rodrigues, R.S.; Pfeil, M.S. Dynamic behavior and stability of transmission line towers under wind forces. *J. Wind Eng. Ind. Aerodyn.* **2003**, *91*, 1051–1067. [CrossRef]
8. Desai, Y.M.; Yu, P.; Popplewell, N.; Shah, A.H. Finite element modelling of transmission line galloping. *Comput. Struct.* **1995**, *57*, 407–420. [CrossRef]
9. Salawu, O. Detection of structural damage through changes in frequency: A review. *Eng. Struct.* **1997**, *19*, 718–723. [CrossRef]
10. Carden, E.P.; Fanning, F. Vibration based condition monitoring: A review. *Struct. Health Monit.* **2004**, *3*, 355–377. [CrossRef]
11. Lam, H.F.; Yuen, K.V.; Beck, J.L. Structural health monitoring via measured Ritz vectors utilizing artificial neural networks. *Comput.-Aided Civ. Infrastruct. Eng.* **2006**, *21*, 232–241. [CrossRef]
12. Yuen, K.-V.; Lam, H.-F. On the complexity of artificial neural networks for smart structures monitoring. *Eng. Struct.* **2006**, *28*, 977–984. [CrossRef]
13. Lam, H.F.; Ng, C.-T. The selection of pattern features for structural damage detection using an extended Bayesian ANN algorithm. *Eng. Struct.* **2008**, *30*, 2762–2770. [CrossRef]
14. Kim, J.-T.; Ryu, Y.-S.; Cho, H.-M.; Stubbs, N. Damage identification in beam-type structures: Frequency-based method vs. mode-shape-based method. *Eng. Struct.* **2003**, *25*, 57–67. [CrossRef]
15. Lam, H.; Lee, Y.; Sun, H.; Cheng, G.; Guo, X. Application of the spatial wavelet transform and Bayesian approach to the crack detection of a partially obstructed beam. *Thin-Walled Struct.* **2005**, *43*, 1–21. [CrossRef]
16. Lam, H.; Ng, C.; Veidt, M. Experimental characterization of multiple cracks in a cantilever beam utilizing transient vibration data following a probabilistic approach. *J. Sound Vib.* **2007**, *305*, 34–49. [CrossRef]
17. Lam, H.F.; Ng, C.-T. A probabilistic method for the detection of obstructed cracks of beam-type structures using spatial wavelet transform. *Probabilistic Eng. Mech.* **2008**, *23*, 237–245. [CrossRef]
18. Lam, H.F.; Ng, C.T.; Leung, A.Y. Multicrack Detection on Semirigidly Connected Beams Utilizing Dynamic Data. *J. Eng. Mech.* **2008**, *134*, 90–99. [CrossRef]
19. Yam, L.; Li, Y.; Wong, W. Sensitivity studies of parameters for damage detection of plate-like structures using static and dynamic approaches. *Eng. Struct.* **2002**, *24*, 1465–1475. [CrossRef]
20. Loutridis, S.; Douka, E.; Hadjileontiadis, L.; Trochidis, A. A two-dimensional wavelet transform for detection of cracks in plates. *Eng. Struct.* **2005**, *27*, 1327–1338. [CrossRef]

21. Yin, T.; Lam, H.-F.; Chow, H.-M. A Bayesian Probabilistic Approach for Crack Characterization in Plate Structures. *Comput. Civ. Infrastruct. Eng.* **2010**, *25*, 375–386. [CrossRef]
22. Lam, H.F.; Yin, T. Statistical detection of multiple cracks on thin plates utilizing measured dynamic response. *Eng. Struct.* **2010**, *32*, 3145–3152. [CrossRef]
23. Alminhana, F.; Mason, M.; Albermani, F. A compact nonlinear dynamic analysis technique for transmission line cascades. *Eng. Struct.* **2018**, *158*, 164–174. [CrossRef]
24. Loredo-Souza, A.; Davenport, A. A novel approach for wind tunnel modelling of transmission lines. *J. Wind Eng. Ind. Aerodyn.* **2001**, *89*, 1017–1029. [CrossRef]
25. Deng, H.Z.; Xu, H.J.; Duan, C.Y. Experimental and numerical study on the responses of a transmission tower to skew incident winds. *J. Wind Eng. Ind. Aerodyn.* **2016**, *157*, 171–188. [CrossRef]
26. Wenjuan, L.; Bingnan, S.; Jinchun, T. Wind test and numerical computation on wind-induced vibration for tall lattice tower. *J. Vib. Eng.* **1996**, *9*, 318–322.
27. Zhao, S.; Yan, Z.T.; Li, Z.L.; Dong, J.Y.; Zhong, Y.L. Investigation on wind tunnel tests of an aeroelastic model of 1000 kV Sutong long span transmission tower-line system. *Proc. CSEE* **2018**, *38*, 5257–5265.
28. Yong, G.; Bing, N.S.; Yin, Y. Time-domain analysis on wind-induced dynamic response of long span power transmission line system. *China Civ. Eng. J.* **2006**, *39*, 12–17.
29. Zhang, J.; Xie, Q. Failure analysis of transmission tower subjected to strong wind load. *J. Constr. Steel Res.* **2019**, *160*, 271–279. [CrossRef]
30. Cheng, Z.J.; Fu, G.; Lou, W. Research for the wind force on high-rise latticed tower. *J. Exp. Mech.* **2000**, *15*, 51–55.
31. Li, X.; Zhang, W.; Niu, H.; Wu, Z.Y. Probabilistic capacity assessment of single circuit transmission tower-line system subjected to strong winds. *Eng. Struct.* **2018**, *175*, 517–530. [CrossRef]
32. An, L.; Wu, J.; Zhang, Z.; Zhang, R. Failure analysis of a lattice transmission tower collapse due to the super typhoon Rammasun in July 2014 in Hainan Province, China. *J. Wind Eng. Ind. Aerodyn.* **2018**, *182*, 295–307. [CrossRef]
33. Wen-zhe, B.I.; Li, T. Study on the collapse failure of transmission tower-line system under downburst. *Eng. Mech.* **2022**, *39*, 1–6.
34. Xu, Y.-L.; Qu, W.L.; Chen, Z.H. Control of Wind-Excited Truss Tower Using Semiactive Friction Damper. *J. Struct. Eng.* **2001**, *127*, 861–868. [CrossRef]
35. Qu, W.; Chen, Z.; Xu, Y. Dynamic analysis of wind-excited truss tower with friction dampers. *Comput. Struct.* **2001**, *79*, 2817–2831. [CrossRef]
36. Markiewitz, M. Optimum dynamic characteristics of stock bridge dampers for dead-end spans. *J. Sound Vib.* **1995**, *188*, 243–256. [CrossRef]
37. Tesar, A.; Kuglerova, J. Tuned vibration control of overhead line conductors. *Int. J. Numer. Methods Eng.* **2000**, *48*, 1215–1239. [CrossRef]
38. Diana, G.; Cigada, A.; Belloli, M.; Vanali, M. Stockbridge-type damper effectiveness evaluation: Part i-comparison between tests on span and on the shaker. *IEEE Trans. Power Deliv.* **2003**, *18*, 1462–1469. [CrossRef]
39. Diana, G.; Manenti, A.; Pirotta, C.; Zuin, A. Stockbridge-type damper effectiveness evaluation: Part II-the influence of the impedance matrix terms on the energy dissipated. *IEEE Trans. Power Deliv.* **2003**, *18*, 1470–1477. [CrossRef]
40. Li, Z.; Han, Y.; Liu, C. A Study on the Bearing Capacity of the Transmission Tower-Leg with Semi-rigid Joints. *Prog. Steel Build. Struct.* **2021**, *23*, 25–31.
41. Qian, C.; Shen, G.; Guo, Y.; Xing, Y.L. Influence of semi-rigid connections on wind-induced responses of transmission towers. *J. Zhejiang Univ. (Eng. Sci.)* **2017**, *51*, 1082–1089.
42. Jiao, A.; Li, Z.; Liu, H. Study on ultimate strength of semi-rigid K-type joints in UHV transmission tower. *J. Build. Struct.* **2014**, *35*, 53–60.
43. *DL/T 5551-2018*; Power Industry Standards of the People's Republic of China. Load Code for the Design of Overhead Transmission Line. China Planning Press: Beijing, China, 2018.

energies

MDPI

Article

Properties of Lightweight Controlled Low-Strength Materials Using Construction Waste and EPS for Oil and Gas Pipelines

Hao Liu [1], Kaizhi Liu [2], Yiqi Xiao [1], Peng Zhang [1], Meixia Zhang [1,*], Youzeng Zhu [1], Kaixin Liu [1], Tianshuo Xu [1] and Rui Huang [1]

[1] Faculty of Engineering, China University of Geosciences, Wuhan 430074, China; cugwuhanly@163.com (H.L.); hhlxyq1172195103@163.com (Y.X.); cugpengzhang@163.com (P.Z.); youzengzhu@163.com (Y.Z.); 15207162405@163.com (K.L.); xtsxts11@cug.edu.cn (T.X.); 17755473906@163.com (R.H.)

[2] General Contracting Branch of China Communications Construction Co., Ltd., Beijing 100088, China; liukaizhi2022@126.com

* Correspondence: zhangmeixia@cug.edu.cn

Abstract: Due to its particularity and importance, long-distance oil and gas pipelines need to be well protected from damage by backfill materials. In this study, construction waste and expanded polystyrene (EPS) were used to replace conventional fine aggregate, and ethylene vinyl acetate-resin (EVA) was used to modify the surface of EPS to prepare lightweight controlled low strength materials (CLSM). Lightweight CLSM was tested in mechanics and physics and its microstructure was studied using microscopic analysis methods. The results revealed that the surface modification of EPS by EVA could greatly improve the compatibility of EPS with inorganic cementitious materials and prepare CLSM with a fluidity greater than 200 mm. EPS and cement content in cementitious materials play an important role in the development of material strength. When the volume ratio of EPS to construction waste was 2, and the content of cement in the cementitious materials was 35%, CLSM's unconfined compressive strength at 28 days was only 0.48 MPa. In order to obtain the lightweight CLSM that meets the mechanical properties, the EPS content should not be too large. It can be concluded from the microscopic analysis that the increase of EPS content will lead to poor pore uniformity of the specimen, forming a loose mesh structure of defects, which is not conducive to the development of strength. In this study, EPS and construction waste are used to provide a green idea for preparing lightweight controlled low strength materials, which provides a reference for the backfill protection of the material in oil and gas pipelines in the future.

Keywords: oil and gas pipelines; lightweight controlled low strength materials; expanded polystyrene; construction waste; ethylene vinyl acetate-resin surface modification

Citation: Liu, H.; Liu, K.; Xiao, Y.; Zhang, P.; Zhang, M.; Zhu, Y.; Liu, K.; Xu, T.; Huang, R. Properties of Lightweight Controlled Low-Strength Materials Using Construction Waste and EPS for Oil and Gas Pipelines. *Energies* **2022**, *15*, 4301. https://doi.org/10.3390/en15124301

Academic Editors: Li Tian and Jianyong Han

Received: 17 April 2022
Accepted: 6 June 2022
Published: 11 June 2022

1. Introduction

The backfill quality of pipelines has an important influence on pipeline operation. The loose backfilling around the pipeline can badly influence pipe performance and service life. First, there is a void around the pipeline; then the pipeline breaks and the surrounding soil is lost. Groundwater flows into the pipeline and oil leaks out of the pipeline. Finally, pavement cracking, subsidence, and collapse accidents are caused. A typical illustration of the incompact backfill is shown in Figure 1.

Due to the particularity and importance of oil and gas pipelines, long-distance oil and gas pipelines need to be greatly protected from damage [1]. There is an urgent need for a new backfill protection material with a certain compressive strength, impact resistance, and energy absorption. It can meet the safety and stability of the pipeline during the construction process and service period, thereby reducing the construction period, project cost, and construction difficulty [2].

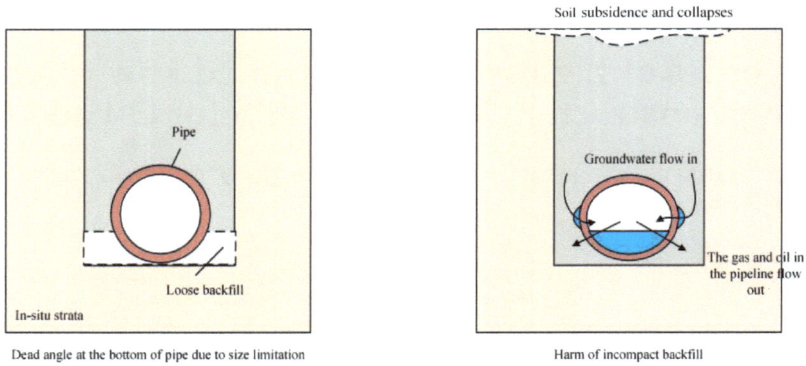

Figure 1. A typical illustration of the incompact backfill.

Controlled low strength materials (CLSM) is a type of self-compacting and high fluidity low strength material defined by ACI 229 [3]. Unlike the conventional backfill method, it can flow fill, greatly saving resources and costs [4–6]. CLSM is usually made of cement, fly ash, fine aggregate, water, and other chemical additives, of which the fine aggregate is mainly river sand. However, with the shortage of building materials, some experts have tried to replace sand with green, sustainable materials such as fine aggregate. Young-sang Kim et al. [7] prepared CLSM by using excavated soil—the results showed that the stability of CLSM could be improved by adding an appropriate amount of excavated soil, but it was unfavorable for the strength of CLSM if too much was added. Etxeberria Miren et al. [8] determined that CLSM made of recycled fine aggregate instead of natural aggregates had excellent properties and was characterized by self-compaction and easy excavation. Bhaskar Chittoori [9] et al. carried out a comprehensive study on the material properties using highly plastic clay as a kind of fine aggregate, and the final CLSM had a good economy. Waste glass powder and hydrated lime as fine aggregate were also used to prepare CLSM and achieved a compressive strength of up to 1.95 MPa [10].

At the same time, some researchers have also studied the use of lightweight materials to replace the fine aggregate to produce lightweight CLSM with excellent working performance. Her-Yung Wang et al. [11] used rubber particles to prepare CLSM, which greatly reduced the density of CLSM materials and improved the ability of materials to resist external shocks. Saofee Dueramae et al. [12] studied lightweight alkali-activated CLSM and revealed that the compressive strengths of this CLSM rapidly increased within the first 12 h, and the adherence at the interface resulted in reducing the UCS of the CLSM. The density of lightweight CLSM was less than 900 kg/m^3.

With the rapid development of society, a large amount of construction waste continues to be generated. Construction waste mainly refers to the solid waste generated during the renovation and expansion of a project and the demolition of old buildings, mainly including waste bricks, waste concrete, and other wastes [13]. The resource treatment of construction waste can reduce environmental pollution and is of great significance for sustainable development [14,15]. At the same time, expanded polystyrene (EPS) is widely used in the packaging of various commodities due to its low price—according to statistics, China discarded as much as 1.6 million tons of waste EPS plastic products in 2020. Discarded EPS is difficult to degrade and is a huge environmental hazard [16], particularly because EPS has the characteristics of being lightweight, having heat preservation and good sound insulation. Using EPS to prepare light CLSM has broad application prospects in oil and gas pipeline protection [17,18], but the compatibility between EPS and cementitious materials is poor, and the interfacial adhesion between them is small [19]. Many scholars have found that using ethylene vinyl acetate resin (EVA) to modify the surface of EPS can greatly improve the performance of the material [20,21]. In this study, construction waste and EPS were used as fine aggregate, and cement and fly ash were used as cementitious materials. EVA was used to modify the surface of EPS to prepare the lightweight CLSM.

The effects of EPS content and cementitious materials on the fluidity, strength, and stress–strain of CLSM were investigated by experiments. Finally, we used scanning electron microscope (SEM) analysis and energy dispersive spectroscopy (EDS) to study the microstructure of lightweight CLSM.

2. Experimental Investigation

2.1. Materials

The construction waste recycled aggregate (hereinafter referred to as construction waste) used in this experiment was a fine material processed and produced by a renewable resource company in Wuhan. The particle size grading of recycled aggregate is shown in Figure 2. The physical properties of the natural fine sand and recycled aggregate are shown in Table 1. Compared with conventional natural fine aggregate, the water absorption rate of this construction waste was 3.4 times that of natural river sand. The compressive strength of cement was 42.5MPa, and the fly ash is Class F third grade fly ash. The chemical compositions of cement and fly ash are shown in Table 2. In the test, EVA was used to modify the surface of EPS, and the appearance of it is a milky white viscous liquid; its physical and chemical properties are shown in Table 3. The particle diameter of EPS was 1 mm, and the bulk density was 30 kg/m^3.

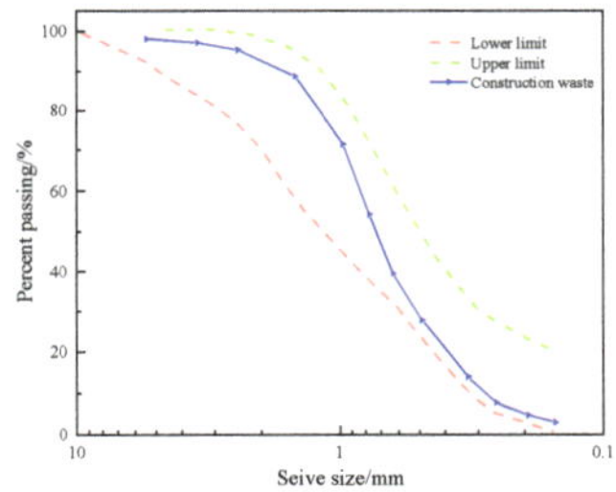

Figure 2. Particle size grading of recycled aggregate.

Table 1. Physical properties of the natural fine sand and recycled aggregate.

Property	Aggregate Type	
	Natural Fine Sand	Recycled Aggregate
Apparent density/kg/m^3	2.60	2.44
Moisture content/%	8.05	7.03
Water absorption/%	4.6	15.66
Fineness modulus	2.2	2.69

Table 2. Chemical composition of the cement and fly ash.

Material	Chemical Compositions/%									
	SiO_2	Al_2O_3	Fe_2O_3	CaO	K_2O	TiO_2	MgO	Na_2O	SO_3	Other
Cement (PA)	18.16	3.57	4.03	63.35	1.11	0.29	2.66	-	2.20	4.63
Fly ash (FA)	53.97	31.15	4.16	4.01	2.04	1.13	1.01	0.89	0.73	0.91

Table 3. Physical and chemical properties of the EVA.

	Solid Content	pH	Free Monomer (wt%)	Viscosity (mPa·s)	Appearance
Result	55.49%	4.9	0.0539	3250	Milky white

2.2. EPS Surface Modification Process

EPS is a kind of hydrophobic material. The interface bonding between it and the cementitious material is weak. The addition of EVA improving the material properties of cement slurries is worth studying. Seyed Ali Ghahari et al. [22] found that cellulose nanocrystals can improve the fracture behavior of cementitious materials. In order to improve the cohesion between the EPS material and the mixture, EVA emulsion is used to modify the surface of EPS. With special mixing technology [23] and processing [24], it can improve the dispersion of EPS in CLSM and enhance the mechanical properties of lightweight CLSM.

First, put all the EPS into the concrete mixer, add 50% EVA emulsion and 10% water, and after mixing evenly, add 20% cement and 20% fly ash to the mixer so that the surface of EPS is evenly coated with a layer of gel just like a shell formation; at the same time, after stirring the remaining water, cement, fly ash, and EVA emulsion evenly, gradually pour it into the EPS with good shell formation and then conduct CLSM-related performance tests. The schematic diagram of EPS' surface modified by EVA is shown in Figure 3.

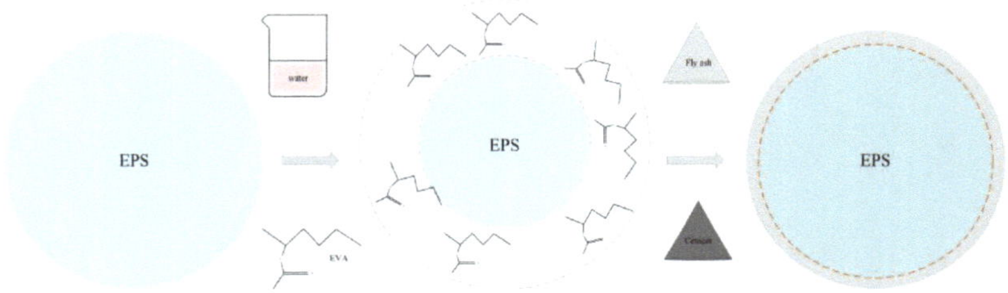

Figure 3. EVA and EPS shell-making diagram.

2.3. Test Method

In this study, the physical and mechanical properties of the CLSM included its fluidity, bleeding rate, wet density, and unconfined compressive strength.

The fluidity of fresh CLSM was measured by flow cone test with the height of 150 ± 3 mm and inside diameter of 76 ± 3 mm according to ASTM D6103 [25]. The wet density and unconfined compressive strength are tested with reference to Test Methods for Basic Properties of Building Mortar (JGJ/T 70-2009) [26]. The wet density test was used to determine the mass per unit volume of the mixture. The capacity tube was made of metal with inner diameter of 108 mm, net height of 109 mm, and volume of 1 L. First, the mass of the capacity tube was measured to be m1, and then the fresh CLSM was filled with the capacity tube and compacted. The total mass of the mixture and the tube was measured to be $m2$. The measurement results were accurate to g. The wet density of CLSM was calculated as follows:

$$\rho = \frac{m2 - m1}{V} \times 1000 \ (\mathrm{kg/m^3})$$

where $m2$ = the total mass of the mixture and the tube (g), $m1$ = the mass of the capacity tube (g), V = the mass of the capacity tube (cm^3).

The specimen with unconfined compressive strength was made into a size of 70.7 mm × 70.7 mm × 70.7 mm. The unconfined compressive strength of CLSM at different days was tested by a YAW-4605 pressure testing machine. Each sample contained three samples and the average values were the final result. The microscopic test included SEM scanning electron microscope inspection and EDS energy spectrum analysis. The residues after the unconfined compressive strength test were collected, dried, and observed by scanning electron microscope. Then the same sample surface was analyzed by EDS. As found in previous experiments, the drying temperature should not be too high to avoid damaging the cement hydration products.

2.4. Mix Proportion of CLSM

In this study, EVA was added using the optimal ratio of EPS modification [19] (the mass ratio of EVA to EPS was 2). According to the EPS weight in the test, the weight of EVA was adjusted in the same proportion, and the moisture in EVA was included in the total water consumption. In this study, the mixture of cement and fly ash was used as a binder at three different cement:fly ash ratios (7:13, 10:10, and 13:7 by weight). The EPS was then added into the binder with designated EPS:construction ratios (1.5:1, 1.75:1, and 2:1 by volume). In this study, three different binder ratios and three kinds of EPS volume ratio of construction waste were used. The total number of experiments was 9 Groups. The feasibility of preparing lightweight CLSM with EVA as light aggregate after surface modification of EPS was studied, and the effect of cement ratio in cementitious materials on the performance of CLSM was also critical. The details of the mixing proportions of lightweight CLSM are shown in Table 4.

Table 4. Mixing proportion of the EVA modified lightweight CLSM.

Experiment-Groups	Content (kg/m^3)					EPS: Construction Waste (Volume Ratio)	Fluidity of CLSM (mm)
	Cement	Fly Ash	Water	EPS	Construction Waste		
1(W7V1.5) *	140	260	317.13	8.75	400	1.50	340.0
2(W7V1.75)	140	260	316.00	10.00	400	1.75	320.0
3(W7V2)	140	260	314.88	11.25	400	2.00	285.0
4(W10V1.5)	200	200	317.13	8.75	400	1.50	330.0
5(W10V1.75)	200	200	316.00	10.00	400	1.75	307.5
6(W10V2)	200	200	314.88	11.25	400	2.00	280.0
7(W13V1.5)	260	140	317.13	8.75	400	1.50	320.0
8(W13V1.75)	260	140	316.00	10.00	400	1.75	305.0
9(W13V1.75)	260	140	314.88	11.25	400	2.00	252.5

Note: * W7V1.5 represents the mass ratio of cement to fly ash of 7:13 and the volume ratio of EPS to construction waste of 1.50.

3. Results

3.1. Fluidity

The fluidity of the lightweight CLSM measured in the test is shown in Figure 4. The fluidity values were between 245 mm and 350 mm, and all greater than 200 mm, which were classified as high fluidity CLSM. For the content of fly ash in the cementitious material at 35% and the volume ratio of EPS to construction waste at 2, the fluidity of CLSM was 245 mm. With the increase of cement content and EPS content in the cementitious material, the fluidity of CLSM gradually decreased. There are two main reasons: (1) fly ash is a spherical form and has a better morphological effect than cement, which plays a role in lubricating and reducing water in CLSM. When the cement content increases, the relative fly ash content decreases, which reduces the fluidity of CLSM; (2) after surface modification of EPS, with the increase of EPS content, which is equivalent to the increase of the total volume of aggregate and the relative decrease of moisture content, the fluidity of CLSM is reduced.

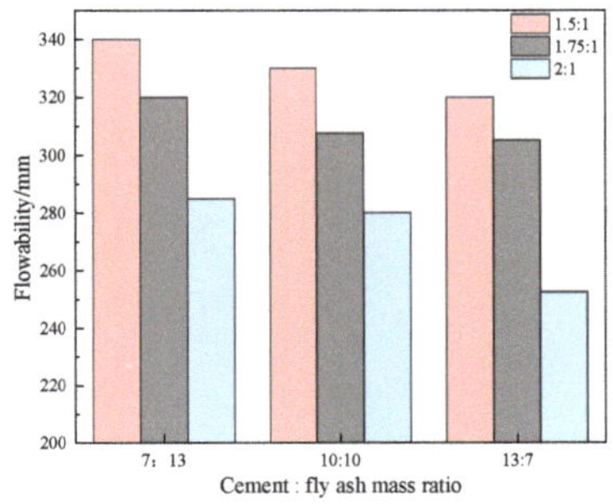

Figure 4. The fluidity of the lightweight CLSM.

3.2. Bleeding Rate

The bleeding rate of the lightweight CLSM was very small and almost zero; fresh CLSM was relatively uniform. It was clearly seen from Figure 5 that bleeding on the surface sample was small. There are two reasons for the condition: (1) The water absorption rate of the construction waste is relatively large; after adding EPS, there is no coarse aggregate in the CLSM mixture, so it is well-graded and integrated as a whole. (2) The EVA molecule is hydrophilic and can inhibit the bleeding and segregation of CLSM. The bleeding rates of 2, 3 are demonstrated in Figure 5: although the bleeding rates are large, there is little surface moisture, which shows the good uniformity and working performance of CLSM.

(**a**) (**b**)

Figure 5. The bleeding rates of the lightweight CLSM: (**a**) Group 2; (**b**) Group 3.

3.3. Wet Density

The wet density of the lightweight CLSM is shown in Figure 6, whose values were between 780 to 1032 kg/m^3. Comparing Group 1 with 3, the volume ratios of EPS to construction waste were from 1.5:1 to 2:1, and the wet densities were 995 and 780 kg/m^3, respectively, which was reduced by 21.6%. As can be seen from Figure 5, with the increase of fly ash contents in the cementitious material and EPS content, the wet density of CLSM was constantly decreasing. The content of fly ash and EPS had an effect on the wet density of CLSM, but the EPS content played a major role in the lightness of CLSM.

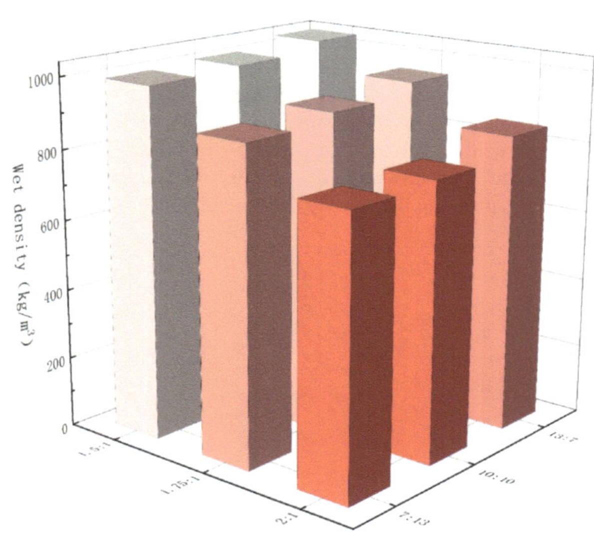

Figure 6. The wet density of the lightweight CLSM.

3.4. Unconfined Compressive Strength

The unconfined compressive strength test results of lightweight CLSM are shown in Figure 7. The content of fly ash in cementitious materials from 65 to 35% and the average unconfined compressive strength of CLSM at 3-d was 0.27 MPa and 0.76 Mpa, respectively, the unconfined compressive strength of lightweight CLSM increased by 180%. It can be concluded that cement in cementitious materials mainly controls the development of material strength.

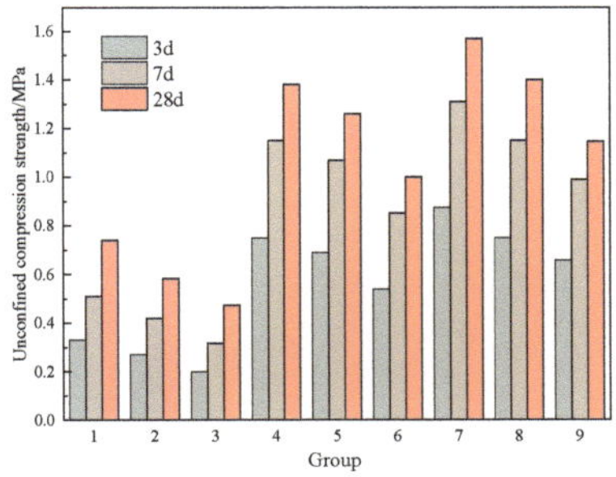

Figure 7. The unconfined compressive strength of lightweight CLSM.

Comparing Groups 4, 5, and 6, with the increase of EPS content in the material, the unconfined compressive strength of CLSM at 28 days was 1.38 MPa, 1.26 Mpa, and 1.01 MPa, respectively. The strength of CLSM with EPS content ratio of 1.5:1 and 1.75:1 was 8.7 and 26.8% lower than that with an EPS content ratio of 2.0:1. It means that the strength of CLSM decreases with the increase of EPS content. In order to ensure that the lightweight CLSM can meet the mechanical properties, the EPS content cannot be too large.

Based on the 3-days unconfined compressive strength, the UCS of Groups 1 and 7 at different curing periods was compared. The 28-days UCS of Group 1 was 123.9% higher than that of 3-days and the 28-days strength of Group 7 was 79% higher than that of 3-days. It can be concluded that with the increase of fly ash content in cementitious materials, the early strength development of CLSM decreases, but fly ash is beneficial to the later strength development of CLSM. The reason is that compared with cement, fly ash has lower volcanic ash characteristics and slower hydration reaction, which is unfavorable to the development of early strength of materials [27] and the corresponding increase in the setting time of CLSM.

It can be seen from Figure 8 that EPS particles are uniformly dispersed in CLSM. After EVA modification, the compatibility of EPS in CLSM is improved, and EPS has good cementation with aggregates and cement hydration products.

Figure 8. Failure samples after an unconfined compressive strength test.

3.5. Stress–Strain Curve

The stress–strain curves of Groups 4 and 6 at 28-days are shown in Figure 9a,b. Compared with the stress–strain curve of recycled aggregate concrete in the literature, it can be seen that the ultimate strain of recycled concrete is far less than that of lightweight CLSM. When the recycled concrete reaches the peak stress, the specimen is destroyed and the stress decreases rapidly, which shows brittle failure. However, the peak stress of light CLSM decreases slowly, showing plastic failure, and it has large plastic deformation. Comparing Group 4 with 6, it is concluded that with the increase of EPS content, the plastic strain of lightweight CLSM increases and the plasticity of CLSM increases.

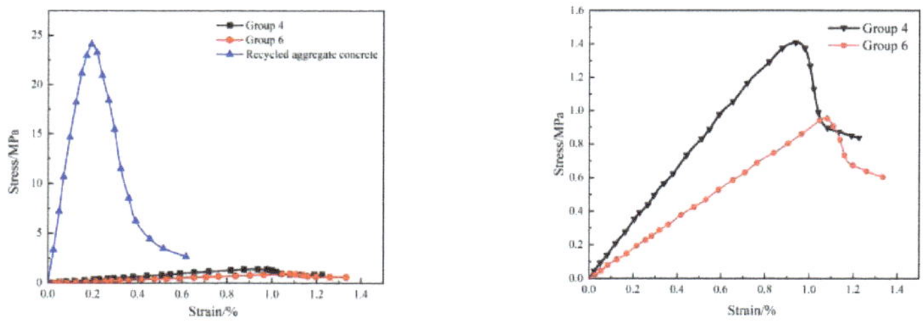

(**a**) (**b**)

Figure 9. The stress–strain curve of lightweight CLSM. (**a**) Comparison of lightweight CLSM and recycled aggregate concrete; (**b**) the stress–strain curves of Groups 4 and 6.

3.6. Microstructure

The microstructure of lightweight CLSM was observed by SEM, and the mechanism of EVA modified EPS was further studied. Microscopic photos of CLSM after EVA modified EPS are shown in Figure 9. In the preparation of the light CLSM using a special mixing process and EVA surface modified EPS, the EPS surface was wrapped with a shell structure and the shell surface was very rough. The cement hydration products can be seen in Figure 10. EPS and cement hydration products were effectively connected, and the compatibility between them was good. After modification, the stability of EPS in CLSM was improved, and the mechanical properties of lightweight CLSM were improved.

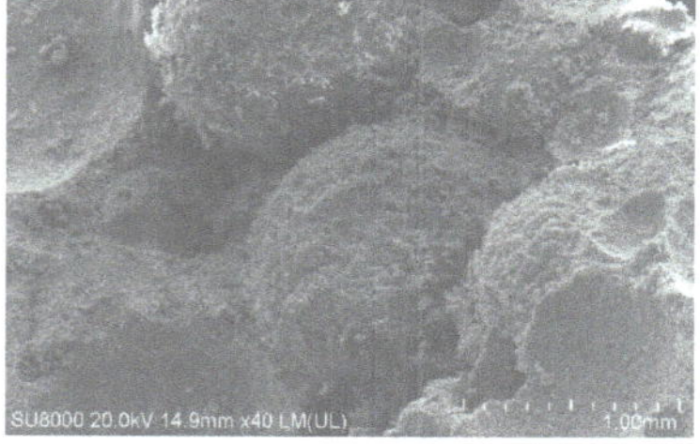

Figure 10. SEM Morphology of CLSM after EVA modified EPS.

The SEM morphology photograph of Group 4 is shown in Figure 11a. There were many tiny pores in light CLSM filled with cement hydration products. Flocculation C-S-H and needle-rod AFt are stacked together to form a connection structure around the aggregate. SEM Morphology photograph in Group 4 is shown in Figure 11b. It is obvious that with the increase of EPS content, the pores in CLSM gradually increased. Compared with Figure 11a, the pore uniformity became worse, and there was a loose network structure with defects. The hydration products between pores cannot be filled effectively, and these structures influence the strength development of lightweight CLSM. The strength of CLSM decreases with the increase of EPS content.

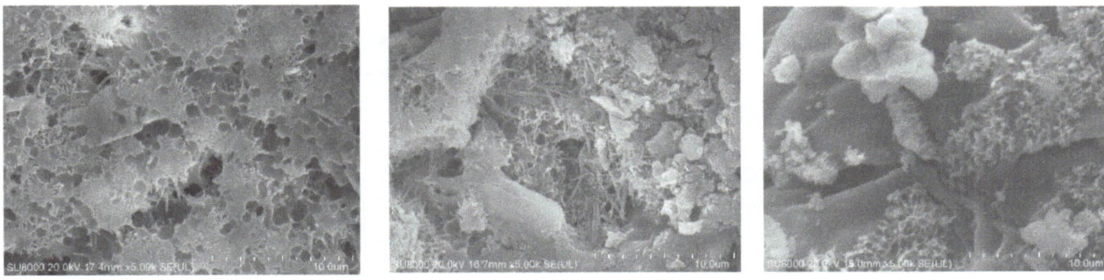

(**a**) (**b**) (**c**)

Figure 11. The SEM morphology photograph of lightweight CLSM. (**a**) Group 4; (**b**) Group 6; (**c**) Group 9.

The SEM morphology photograph in Group 9 is shown in Figure 11b. Compared with Figure 11a, with the increase in cement content, the number of pores in the CLSM structure decreased significantly. Flocculation C-S-H and plate C-H were well developed, and the pores were well filled. With the increase of cement content, the hydration process of cement was accelerated, forming a dense structure relative to Figure 11a,b, and improving the mechanical properties of lightweight CLSM.

EDS energy spectrum analysis was performed on lightweight CLSM samples of Groups 4 and 7. The test results are shown in Figures 12 and 13, and the elemental composition of lightweight CLSM is summarized and shown in Table 5. Table 5 shows that the lightweight CLSM mainly contains C, O, Mg, Al, Si, Ca, K, and P elements. Analysis and verification of SEM morphology photographs of cement hydration products are C-H, AFt, and C-S-H. With the increase of cement content, the cement hydration products in light CLSM increase and the pores in light CLSM structure decrease, so the strength increases. Microscopic analysis confirms the results of the unconfined compressive strength test.

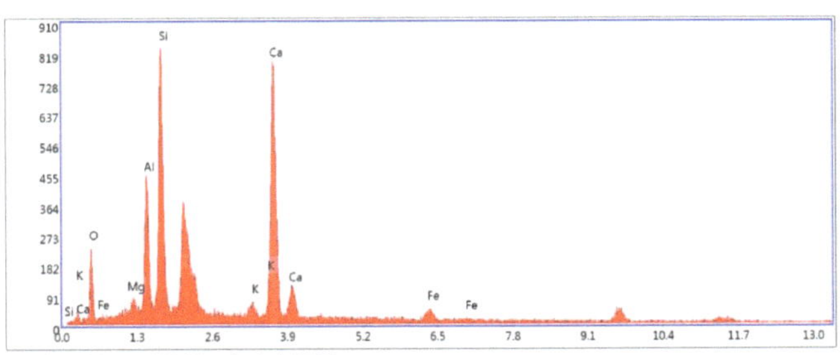

Figure 12. Group 4.

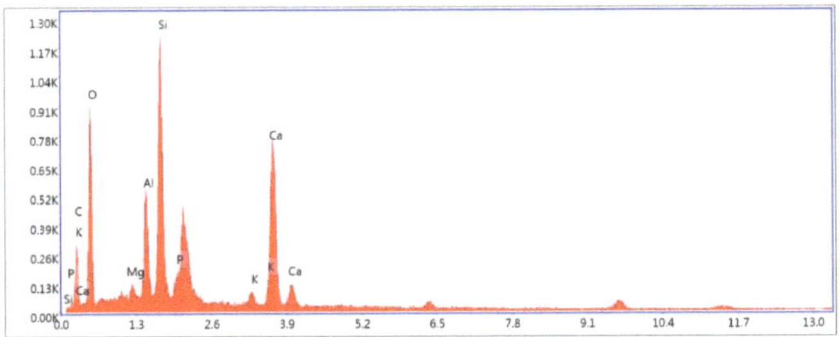

Figure 13. Group 7.

Table 5. The element composition of CLSM.

Group	Atomic Percent (%)							
	C	O	Mg	Al	Si	Ca	K	P
4	19.82	42.90	1.24	5.14	11.72	15.46	0.73	2.99
7	-	29.83	0.98	9.42	19.92	35.09	1.12	-

4. Conclusions

1. In this study, EVA was used to modify the surface of EPS. The shell structure wrapped by hydration products was formed on the surface of EPS. The compatibility between EPS and hydration products was improved and the mechanical properties of lightweight CLSM were enhanced.
2. Construction waste mainly refers to the solid waste generated during the renovation of the project and the demolition of old buildings. The lightweight CLSM prepared from EPS and construction waste has high fluidity, which was greater than 200 mm.
3. Adding EPS and fly ash content can reduce the wet density of CLSM, and EPS played a major role in reducing the density of CLSM.
4. EPS content and cement content in cementitious materials are important factors affecting the strength development of CLSM. When the volume ratio of EPS to construction waste was 200% and the cement content in the cementitious material was 35%, the 28-days unconfined compressive strength of CLSM was only 0.48 MPa. It can be concluded that the volume ratio of EPS to construction waste is 1.50, which is unfavorable for the compressive strength of CLSM.
5. From SEM morphology photographs, it can be concluded that the increase of EPS content will lead to the deterioration of pore uniformity in CLSM, forming a loose network structure with defects, which is unfavorable to the strength development. EDS spectrum analysis verified that the substances in SEM images were cement hydration products C-H, AFt, and C-S-H. And with the increase of EPS content, the content of hydration products in lightweight CLSM decrease, which leads to a decrease in strength.
6. The use of EPS and construction waste to prepare lightweight CLSM provides a green idea for the production of lightweight controlled low strength materials, which provides a reference for the backfill protection of the material in oil and gas pipelines in the future.

Author Contributions: Conceptualization, H.L., M.Z. and P.Z.; data curation, H.L. and Y.X.; formal analysis, H.L. and Y.X.; funding acquisition, M.Z., K.L. (Kaizhi Liu) and P.Z.; investigation, H.L. and M.Z.; methodology, H.L. and Y.X.; validation, M.Z., H.L., Y.Z. and T.X.; visualization, M.Z., K.L. (Kaizhi Liu) and P.Z.; writing—original draft, H.L., Y.X. and P.Z.; writing—review and editing, H.L., K.L. (Kaizhi Liu), R.H. and K.L. (Kaixin Liu) All authors have read and agreed to the published version of the manuscript.

Funding: This research was supported by Xinjiang Uygur Autonomous Region Science and Technology Major Project (No.: 2020A03003-1).

Data Availability Statement: Not applicable.

Conflicts of Interest: The authors declare no conflict of interest.

References

1. Lilly, M.T.; Ihekwoaba, S.C.; Ogaji, S.O.T.; Probert, S.D. Prolonging the lives of buried crude-oil and natural-gas pipelines by cathodic protection. *Appl. Energy* **2007**, *84*, 958–970. [CrossRef]
2. Ijaola, A.O.; Farayibi, P.K.; Asmatulu, E. Superhydrophobic coatings for steel pipeline protection in oil and gas industries: A comprehensive review. *J. Nat. Gas Sci. Eng.* **2020**, *83*, 103544. [CrossRef]
3. *ACI 229R–13*; ACI, Controlled Low Strength Materials. American Concrete Institute: Farmington Hill, MI, USA, 2013.

4. Riviera, P.P.; Bertagnoli, G.; Choorackal, E.; Santagata, E. Controlled low-strength materials for pavement foundations in road tunnels: Feasibility study and recommendations. *Mater. Struct.* **2019**, *52*, 72. [CrossRef]
5. Do, T.M.; Do, A.N.; Kang, G.O.; Kim, Y.S. Utilization of marine dredged soil in controlled low-strength material used as a thermal grout in geothermal systems. *Constr. Build. Mater.* **2019**, *215*, 613–622. [CrossRef]
6. Kaliyavaradhan, S.K.; Ling, T.C.; Guo, M.Z.; Mo, K.H. Waste resources recycling in controlled low-strength material (CLSM): A critical review on plastic properties. *J. Environ. Manag.* **2019**, *241*, 383–396. [CrossRef]
7. Kim, Y.S.; Do, T.M.; Kim, H.K.; Kang, G. Utilization of excavated soil in coal ash-based controlled low strength material (CLSM). *Constr. Build. Mater.* **2016**, *124*, 598–605. [CrossRef]
8. Etxeberria, M.; Ainchil, J.; Pérez, M.E.; González, A. Use of recycled fine aggregates for Control Low Strength Materials (CLSMs) production. *Constr. Build. Mater.* **2013**, *44*, 142–148. [CrossRef]
9. Chittoori, B.; Puppala, A.J.; Raavi, A. Strength and Stiffness Characterization of Controlled Low-Strength Material Using Native High-Plasticity Clay. *J. Mater. Civ. Eng.* **2014**, *26*, 04014007. [CrossRef]
10. Xiao, R.; Polaczyk, P.; Jiang, X.; Zhang, M.; Wang, Y.; Huang, B. Cementless controlled low-strength material (CLSM) based on waste glass powder and hydrated lime: Synthesis, characterization and thermodynamic simulation. *Constr. Build. Mater.* **2021**, *275*, 122157. [CrossRef]
11. Wang, H.; Chen, B.; Wu, Y. A study of the fresh properties of controlled low-strength rubber lightweight aggregate concrete (CLSRLC). *Constr. Build. Mater.* **2013**, *41*, 526–531. [CrossRef]
12. Dueramae, S.; Sanboonsiri, S.; Suntadyon, T.; Aoudta, B.; Tangchirapat, W.; Jongpradist, P.; Pulngern, T.; Jitsangiam, P.; Jaturapitakkul, C. Properties of lightweight alkali activated controlled Low-Strength material using calcium carbide residue—Fly ash mixture and containing EPS beads. *Constr. Build. Mater.* **2021**, *297*, 123769. [CrossRef]
13. Evangelista, L.; de Brito, J. Mechanical behaviour of concrete made with fine recycled concrete aggregates. *Cem. Concr. Compos.* **2007**, *29*, 397–401. [CrossRef]
14. Mália, M.; de Brito, J.; Pinheiro, M.D.; Bravo, M. Construction and demolition waste indicators. *Waste Manag. Res. J. A Sustain. Circ. Econ.* **2013**, *31*, 241–255. [CrossRef] [PubMed]
15. Tam, V.W.Y.; Tam, C.M. A review on the viable technology for construction waste recycling. *Resour. Conserv. Recycl.* **2006**, *47*, 209–221. [CrossRef]
16. Fernando, P.L.N.; Jayasinghe, M.T.R.; Jayasinghe, C. Structural feasibility of Expanded Polystyrene (EPS) based lightweight concrete sandwich wall panels. *Constr. Build. Mater.* **2017**, *139*, 45–51. [CrossRef]
17. Li, H.; Du, T.; Wu, X. Research on recycled aggregate concrete for construction waste recycling. *J. Huazhong Univ. Sci. Technol.* **2001**, *6*, 83–84.
18. Tittarelli, F.; Giosuè, C.; Mobili, A.; di Perna, C.; Monosi, S. Effect of Using Recycled Instead of Virgin EPS in Lightweight Mortars. *Procedia Eng.* **2016**, *161*, 660–665. [CrossRef]
19. Petrella, A.; Di Mundo, R.; Notarnicola, M. Recycled expanded polystyrene as lightweight aggregate for environmentally sustainable cement conglomerates. *Materials* **2020**, *13*, 988. [CrossRef]
20. Feng, Y.; Qin, D.; Li, L.; Li, Y.; Wang, C.; Wang, P. EVA enhances the interfacial strength of EPS concrete: A molecular dynamics study. *J. Exp. Nanosci.* **2021**, *16*, 382–396. [CrossRef]
21. Yong, F.; Caihua, Y.; Kui, H.; Yujing, C.; Yu, L.; Taoli, Z. A study of the microscopic interaction mechanism of styrene–butadiene-styrene modified asphalt based on density functional theory. *Mol. Simul.* **2021**, *35*, 1–12. [CrossRef]
22. Ghahari, S.; Assi, L.N.; Alsalman, A.; Alyamaç, K. Fracture Properties Evaluation of Cellulose Nanocrystals Cement Paste. *Materials* **2020**, *13*, 2507. [CrossRef] [PubMed]
23. Zhao, X.; Tian, W.; Jiang, X. Study on microstructure and properties of EVA modified EPS concrete. *J. Build. Mater.* **2010**, *13*, 243–246.
24. Jin, Z.; Ma, B.; Su, Y.; Qi, H.; Lu, W. Preparation of eco-friendly lightweight gypsum: Use of beta-hemihydrate phosphogypsum and expanded polystyrene particles. *Constr. Build. Mater.* **2021**, *297*, 123837. [CrossRef]
25. *ASTM D6103-04*; American Society for Testing and Materials. Standard Test Method for Flow Consistency of Controlled Low Strength Material (CLSM). American Society for Testing and Materials Press: Conshohocken, AL, USA, 2004.
26. *JGJ/T 70—2009*; Ministry of Housing and Urban-Rural Development of the People's Republic of China. Standard for Testing Basic Performance of Building Mortar. China Architecture & Building Press: Beijing, China, 2009.
27. Fernández-Jiménez, A.; Palomo, A. Composition and microstructure of alkali activated fly ash binder: Effect of the activator. *Cem. Concr. Res.* **2005**, *35*, 1984–1992. [CrossRef]

energies

Article

Key Techniques for Rapid Jacking and Laying of Pipelines: A Case Study on 'Jingshihan' Gas Pipelines in China

Tianshuo Xu [1], Le Wang [2], Peng Zhang [1,*], Yuheng Zhou [1], Kaixin Liu [1], Xin Feng [1], Yongchun Qi [2] and Cong Zeng [1]

1 Faculty of Engineering, China University of Geosciences, Wuhan 430074, China; xtsxts11@cug.edu.cn (T.X.); zhouyh908@163.com (Y.Z.); 15207162405@163.com (K.L.); cugfengxin@163.com (X.F.); zengcong@126.com (C.Z.)
2 No. 4 Branch Company of China Petroleum Pipeline Engineering Co., Ltd., Langfang 065000, China; wangle@cnpc.com.cn (L.W.); qiyongchun-369@163.com (Y.Q.)
* Correspondence: cugpengzhang@163.com

Abstract: With the rapid growth of the oil and gas storage, transportation, and pipeline industries, it is necessary to improve the construction process of oil and gas pipelines. By combining the technical advantages of horizontal directional drilling and pipe jacking construction, the direct pipe laying method is suitable for pipeline crossing in different strata in the oil and gas, water conservancy and hydropower, and municipal industries due to its advantages of less construction land, high speed, and reversibility. Using the rapid jacking and laying of pipelines crossing Nanjuma River in the 'Jingshihan' gas pipeline double line project as a case study, this paper investigates the application status of the direct pipe laying method, summarizes the project, and introduces the construction of the working well, equipment selection, guiding control technology, supporting equipment installation, and drag reduction measures, as well as analyzes the influencing factors of thrust force and trajectory deviation combined with formation information.

Keywords: oil and gas pipelines; rapid jacking and pipeline laying; equipment selection; thrust force

check for updates

Citation: Xu, T.; Wang, L.; Zhang, P.; Zhou, Y.; Liu, K.; Feng, X.; Qi, Y.; Zeng, C. Key Techniques for Rapid Jacking and Laying of Pipelines: A Case Study on 'Jingshihan' Gas Pipelines in China. *Energies* **2022**, *15*, 2918. https://doi.org/10.3390/en15082918

Academic Editors: Li Tian and Jianyong Han

Received: 12 March 2022
Accepted: 13 April 2022
Published: 15 April 2022

1. Introduction

Oil and gas pipelines represent one of the main ways to transport oil and gas resources around the world. In order to further improve the safety of oil and gas transportation, it is imperative to optimize and upgrade the construction of oil and gas transportation pipelines. The stable supply of oil and gas is one of the key factors to ensure the steady development of the social economy, for which the construction of oil and gas transportation pipelines is extremely important [1]. The construction of pipelines is a highly systematic and professionally complex project. With the growth of the oil and gas pipeline industry, a series of challenges in the construction of actual storage and transportation pipelines have attracted the attention of the academic community and have gradually become an important research direction in this field [2].

During the buried installation of oil and gas pipelines, they cross roads, railways, waters, mountains, and other aboveground and underground structures [3]. To reduce the impact on existing buildings and environment, trenchless construction crossing has widely been used in long-distance pipeline crossing projects around the world. At present, the trenchless construction of oil and gas pipelines mainly includes horizontal directional drilling, pipe jacking, shield tunnel, and drill-blast tunnel methods [4].

Horizontal directional drilling construction has a great influence on the crossing distance, with long period of repeated hole expansion and serious mud pollution [5]. It has low adaptability to complex hard rock strata, and the use of counterweights and floating leads to a short life cycle of the pipeline and difficult maintenance in the later period. The pipe jacking method requires an excavation of the working well, the preliminary work is more complex, and the construction efficiency is significantly reduced when the measured construction distance is more than 800 m [6]. There are still problems such as increasing friction and a high risk of the pipeline becoming stuck when long distance passes are crossed through complex strata. Although shield construction has strong adaptability, its equipment structure is complex, the manufacturing, installation, and commissioning cycle is the longest, and the cost is high [7].

Considering the advantages and disadvantages of the several types of trenchless construction techniques described above for oil and gas pipelines, the rapid jacking and pipeline laying method came into being. This technology has fewer requirements for the construction site, can achieve drilling and pipe laying in one step, and has the characteristics of a short construction period, moderate cost, and strong applicability [8].

The rapid jacking and pipeline laying method combines pipe jacking and directional drilling technology. The principle is to use a boring machine for tunnel excavation at the front of the pipelines, connect the pipeline to be laid with the direct pipe laying machine, clamp the pipeline with the clamping device, and use the friction to thrust the pipelines and equipment forward under the action of the thrusting machine. The method combines the advantages of pipe jacking and horizontal directional drilling to realize the simultaneous progress of prefabricated pipelines and tunnel excavation, so that the prefabricated pipelines can be laid in one push. With its advantages of a small construction area, high speed, and retractability, it is suitable for pipeline crossing in sand, silt, clay, pebble, gravel, rock, and other strata in various programs such as oil and gas, water conservancy, and hydropower engineering, as well as municipalities, with a diameter range of 800–1500 mm [9,10]. The process of the direct laying method is outlined in Figure 1.

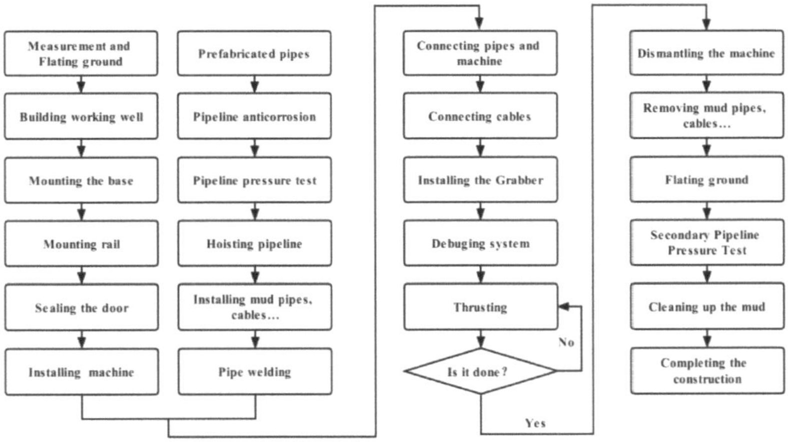

Figure 1. Flowchart of the rapid jacking and pipeline laying method.

The rapid jacking and pipeline laying method and its equipment have been used in Germany, the Netherlands, Italy, the UK, France, Canada, and other countries since it originated, and China introduced support for the rapid jacking and pipeline laying method in 2013. Table 1 lists some of the projects that used the rapid jacking and pipeline laying method. According to records, the longest pipeline length constructed using the DP method was 1400 m (4593 ft). This was a gas pipeline project in the Netherlands called the Pipeline Ommen–Esveld, crossing in Lochem (constructed in 2011), with a pipe diameter of 1219.2 mm. In September 2016, this method was applied for the first time in China and tested and verified in the project crossing the Chuanshan River in the reroute project of the Zhenjiang University Park of the west–east gas pipeline, with a crossing horizontal distance of 346 m [11]. In January 2017, this construction method was promoted and applied in Wuding River in the Shaanxi–Beijing Fourth Line Pipeline Project [12].

Table 1. Construction cases of rapid jacking and pipelines laying method.

Project Name	Country	Diameter (mm)	Crossing Length (m)	Soil Type	Year
Brine Outlet	Germany	1219	464	Sand, clay, stones, wood	2009
Pipeline Angerlo–Beuningen, crossing Valburg M&R	The Netherlands	1219	435	Sand, gravel, wood	2010
Collier Parkway crossing	United States	914.4	226	Sand, clay	2011
Pipeline Poggio Renatico–Cremona; Scolo Riolo crossing	Italy	1219	394	Clay, little sand	2011
——	United Kingdom	1219	440	Gravel, clay, mudstone, sandstone bedrock layer	2017
Wudinghe Crossing Project	China	1219	423	Fine sand	2017
The Thickwood Perimeter Sewer Project	Alberta	1200	470	Overburden clays, dense sands and gravels, oil sands, and limestone	2020

2. Project Overview

The double-track project of the 'Jingshihan' gas pipeline is divided into two sections. This paper deals with the first section (Zhuozhou–Dingxing section). The whole length of the line is about 47.2 km, the diameter is D1219 mm, and the design pressure is 10 MPa. According to the overall trend of the line, Nanhou Village, Dingxing County, Baoding City, Hebei Province passes through the 'Nanjuma' River, using the rapid jacking and pipeline laying method. The pressure at the crossing point is 10 MPa, the transportation temperature is 0–20 °C, and the regional grade is two. The steel pipe in the crossing section is a D1219 × 27.5 mm L555M longitudinal submerged arc-welded steel pipe, and the 3LPE strengthening outer coating is used for anticorrosion at room temperature. The grade of the crossing project is large. Considering the factors such as the buried depth of the curve, the geological conditions at the crossing point, the scour depth, and the position of the exploration line, the entry angle was set to 5°, and the excavation angle was set to 6°. The horizontal length of the crossing is 560 m. Considering the stress characteristics of pipelines and the capacity of the rapid laying and jacking machine, the curvature radius of the pipe crossing section is 1500 * D (D is the outer diameter of the pipe crossing section: 1219 mm), which is 1828.5 m. The construction site plan is shown in Figure 2.

Figure 2. Figure of overall layout of construction site.

The soils were classified according to Chinese standard of GB 50568-2010 [13]. The strata were mainly silt, fine sand, pebble layer, and silty clay. The subsurface conditions along the pipeline are shown in Figure 3, and the soil distribution and characteristics of each formation are summarized in Tables 2 and 3.

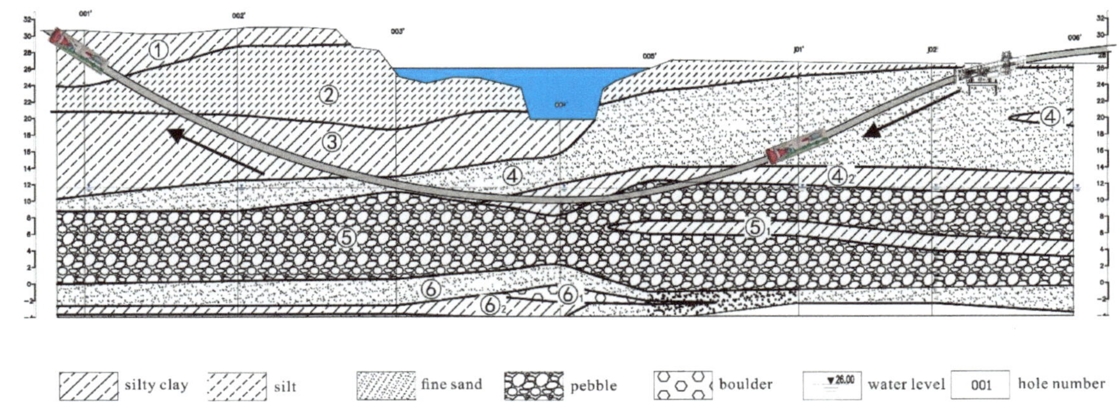

Figure 3. Geological profile of this project.

Table 2. Summary of soil properties.

Layer	Soil	Thickness (m)	Natural Moisture (%)	Density (g/cm^3)		Void Ratio	Saturation (%)	Atterberg Limits (%)	
				ρ_d	ρ_o	e_o	Sr	W_L	W_P
①	Silty clay	1.70–6.70	24.4	1.5	1.88	0.792	83.3	27.5	16.9
②	Silt	0.40–9.80	19.4	1.6	1.93	0.676	78.1	25.7	15.6
③	Silty clay	2.00–13.90	25.8	1.6	1.95	0.751	93.1	32.1	18.1
④$_2$	Silty clay	1.70–3.80	20.4	1.7	2.05	0.592	93.4	28.9	17.5
⑤$_1$	Silty clay	1.20–1.90	19.1	1.7	2.06	0.567	91.3	28.9	17.5

Table 3. Summary of average grain composition of pebble layer and sand layer.

Layer	Lithologic Characters	Particle Size (mm)					
		Pebble	Gravel	Grit			Powder
		>20	2–20	0.5–2	0.25–0.5	0.075–0.25	0.005–0.075
		Percentage (%)					
④	Fine sand	-	-	4.7	42.9	43.0	9.4
⑤	Pebble	50.0	18.8	21.4	9.8	-	-
⑥	Fine sand	-	-	10.0	23.5	53.5	13.0
⑦	Pebble	50.0	18.0	21.0	9.8	1.2	-
⑧	Fine sand	-	-	-	34.2	57.9	8.4

3. The Key Points and Challenges of This Project

According to the design drawings, geological data, construction specifications, and owners' regulations and requirements, through careful reading and analysis of construction data, and considering the characteristics, functions, and natural conditions of the project, the key points and challenges of this project are as follows:

(1) The friction resistance of long-distance pipelines is large, there is a risk of the pipe becoming stuck [6], and the requirement for lubrication grouting is high.

The crossing length of the pipeline in this project is long. Since the pipe thrust machine provides 500 t of thrust, it was calculated according to the grouting friction reduction coefficient that the requirements can be met only when lubrication slurry is injected abundantly and the lubrication effect is great. Because the oil and gas pipeline cannot be drilled on the pipe wall for annular grouting, it can only be grouted slowly through the front end of the boring machine. Therefore, this project has strict requirements for the formulation of lubricating mud and mud pressure.

(2) Through the cobble layer, the excavation face is easy to collapse, large pebbles are difficult to break, and the mud carrying debris poses a challenge.

The project has a pebble layer section of about 70 m, with a pebble particle size of 3 to 6 cm and a maximum particle size of 8 cm; the boring machine through this section of the excavation face is unstable and can easily collapse, and the mud lubrication effect is poor, which leads to difficulties, such as mud pipe accidents.

(3) It is difficult to control the trajectory when crossing complex strata.

The trajectory of this project is a vertical curve in the process of crossing, and it is complex to cross the stratum. Deviation in the vertical direction deviation of the boring machine can easily occur at the stratum boundary. In particular, when crossing the pebble layer, the attitude control of the boring machine is difficult, and the trajectory can easily deviate, making it difficult to correct the deviation.

4. Key Technology of the Construction

4.1. Driving Shaft Construction

Since the rapid jacking and pipeline laying method has low requirements in terms of the driving shaft structure, considering the factors such as construction period and cost, the steel sheet pile supporting structure featuring fast construction, low cost, and recyclability was selected. The foundation pit of the shaft was supported by an SP-U400 × 170 × 15.5 hot-rolled U-shaped steel sheet pile with a length of 9 m. The purlin and support were hot-rolled Q235B grade HW400 × 400 × 13 × 21 mm steel. The pipeline crossing space was reserved at the end. The back end of the foundation pit was sloped from 5° at the bottom of the pit, leaving enough suitable slope for the subsequent cut back. The length of the foundation pit was 22 m, the width was 6.8 m, the depth of the soil entry point of the pipeline axis was 3.0 m, and the installation space of the reserved hole seal and ground anchor was 1.5 m. The excavation depth of the initial foundation pit was 4.8 m, the pouring of concrete occurred at 30 cm, and the bottom plate was 4.5 m deep with respect to the original ground. The first purlin and support of the foundation pit went down 1.0 m from the top of the steel sheet pile, and three supports were arranged at the front end of 12 m with a spacing of 3.5 m. There was no support at the back end of the first layer. Affected by the limit of the pipe clamp stroke, the second purlin of the foundation pit was 3.3 m from the top of the steel sheet pile, and one pipe was set at the end of the foundation pit, with a spacing of 5 m.

After the overall structure was completed, H-shaped steel was installed at the bottom through the anchor, which was convenient for the subsequent installation of the pedestal of the thrust machine, as shown in Figure 4.

4.2. Equipment Selection

4.2.1. Selection of Boring Machine

The boring machine is used for tunnel construction, and its principle is similar to the slurry balance pipe jacking equipment. On the basis of pipe jacking construction, the dynamic configuration, curve guidance, lubrication system, and slurry transportation were optimized. According to the pipe jacking construction parameters, combined with the characteristics of pipeline construction, the overcut diameter [14] should be larger than the pipeline diameter. According to the actual situation, as gas pipeline with a diameter of 1219 mm, mainly passing through the fine sand layer, was selected in this project. The length of the main engine was about 14 m, the weight of the whole machine was about 200 t, and the excavation diameter was 1330 mm. Considering the risk of the pipe becoming stuck due to extreme geological conditions, the maximum thrust force of the equipment was 5000 kN. The rapid laying and jacking machine was composed of a power system, thrust section, and tail shield system (mud pump), driven by 90 kW hydraulic pressure. The head of the boring machine was equipped with a compound cutter head, and the head torque reached 120 kN·m. The specific parameters of the equipment are shown in Table 4.

Figure 4. Driving shaft structure and installation diagram.

Table 4. Specifications of the rapid laying and jacking machine.

Items	Parameter
Length of the total machine	14 m
Weight of the total machine	200 t
Maximum thrust force	5000 kN
Excavation diameter	1330 mm
Maximum construction speed	200 mm/min
Maximum thrust distance	2000 m
Cutter speed	0–9.6 r/min
Cutter opening ratio	47%

The tunnel through the complex strata, including silt, fine sand, pebble layer, and silty clay, for this project had a pebble layer section of about 70 m, with a pebble particle size of 3–6 cm and a maximum particle size of 8 cm; according to the characteristics of permeability, particle size distribution, groundwater pressure, technical parameters of the cutter head, and diameter of the pipelines, specific equipment selection and transformation were carried out. In addition, some unpredictable obstacles were expected in the process of pipeline crossing, such as large boulders, reinforced concrete pile foundation [15].

Therefore, the composite cutter head was specially designed for the particularity of the strata. To cut soft soils and break pebble and other unpredictable obstacles, a mixed cutter head with a diameter of 1330 mm (Figure 5) was used, which was installed with disc cutters for rock formations and scraper cutters for soft soil formations. Moreover, a cone-shaped crusher in the excavation chamber, with a function of secondary cutting, reduced the diameter of stones and other obstructions to a conveyable grain size during thrusting of the pipeline.

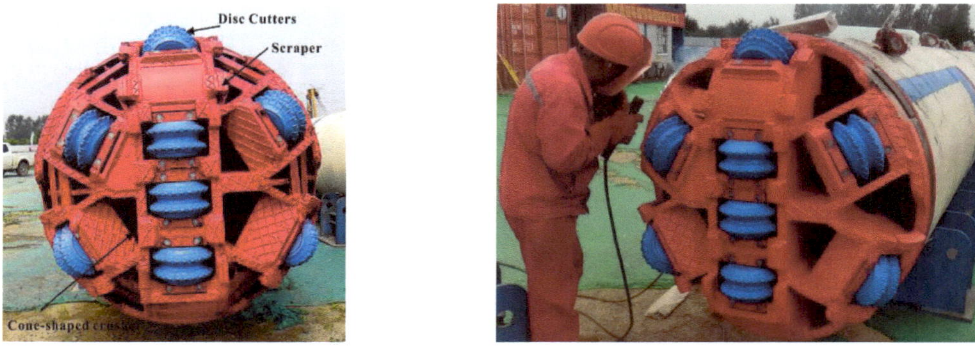

Figure 5. Layout form and modification of cutter head of the boring machine.

4.2.2. Selection of Thrust Machine

The thrust machine is a power device that clamps the pipe through a pipe-clamping device and then advances through a cylinder connected to the clamping device under thrust. The inner side of the clamping device is made from special vulcanized rubber, which will not cause damage to the pipeline coating under clamping. During construction, the pipe thrust machine mainly overcomes the resistance of the direct pipe laying machine to the heading face, the friction between the pipe wall and the rock and soil, the viscous force between the pipe and the slurry, and the friction caused by the pipe buckling. At present, there are three specifications of the thrust equipment, with corresponding thrusts of 3000 kN, 5000 kN, and 7500 kN. According to the calculation results of friction, combined with the existing experience with the rapid jacking and pipeline laying method, and consulting the construction data of similar projects abroad, the theoretical maximum thrust of this project was about 3000 kN. Considering that the large slope construction equipment needs a large pullback force when failure occurs, and the thrust machine should have 1.5–2 times the safety factor, this project selected the thrust machine shown in Figure 6, with a maximum thrust of 5000 kN, mainly composed of a pedestal, clamping device, and hydraulic cylinder.

4.3. Trajectory Control of Boring Machine

The rapid laying and jacking machine features a gyro-guided automatic measurement system. Before starting, the parameters of the designed excavation route are input into the computer in advance, and the measurement is tracked during the propulsion process. The measurement parameters are transmitted to the computer automatically in real time. The actual propulsion route and the design route are compared, and the processing results are fed back to the operating room in an intuitive way. The steering cylinder is controlled by the displayed data, so that the boring machine moves along the axis direction.

(1) The automatic measurement numerical display includes the longitudinal slope, lateral angle, plane deviation, elevation deviation, and incision mileage, to promote the monitoring of numerical changes in construction, with the timely adjustment of various parameters;
(2) When the boring machine starts to advance, the guide cylinder stroke is adjusted to avoid the sinking phenomenon after the header starts to enter the soil;
(3) The angle of the boring machine is checked after every 2 m of propulsion, and the attitude of the machine is adjusted when deviation occurs;
(4) After advancing 20 m, measurements are carried out. The allowable deviation of the pipeline position is ±30 mm, and the direction of the center line should be offset after advancing 50 m. When deviation occurs, it should be corrected in time. When the equipment is advancing, rectification should be carried out slowly, and a large

angle (30°) of rectification cannot be carried out at once, so as to prevent subsequent pipeline jamming and deformation.

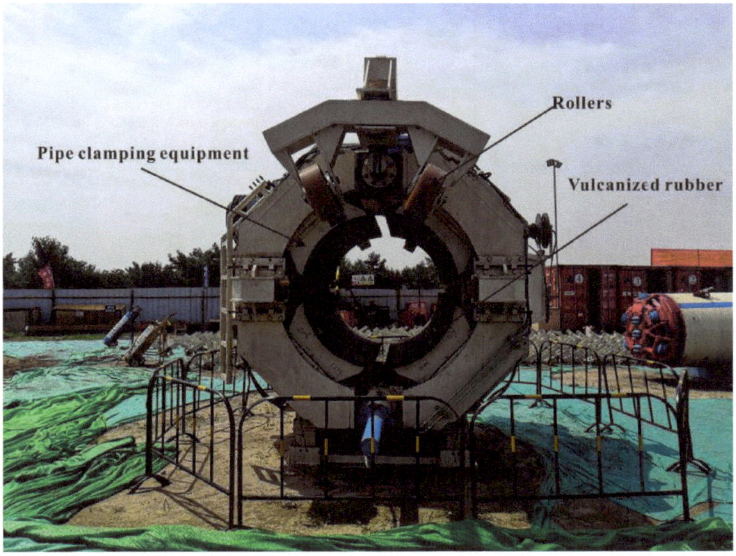

Figure 6. The thrust machine of the rapid laying and jacking machine.

The horizontal and vertical deviation of the pipeline is shown in Figure 7. Due to the vertical curve of the pipeline crossing in space and various strata, it was difficult to correct the deviation in the vertical direction, which was prone to error. Upon advancing to the position of 160.5 m, the front end of the boring machine entered the pebble layer, and the bearing capacity of the stratum with large porosity was low. The trajectory of the boring machine began to sink in the vertical direction, and it was difficult to correct the deviation. Due to the trajectory deviation, when jacking to the position of 178.5 m, the friction around the pipe reached 5365 kN, and the pipe thrust machine became unable to thrust the whole pipe. By using the cylinder retraction of the pipe thrust machine to provide tension, the pipe was pulled back to the position of 157 m of the boring machine, and then thrust again (Figure 8).

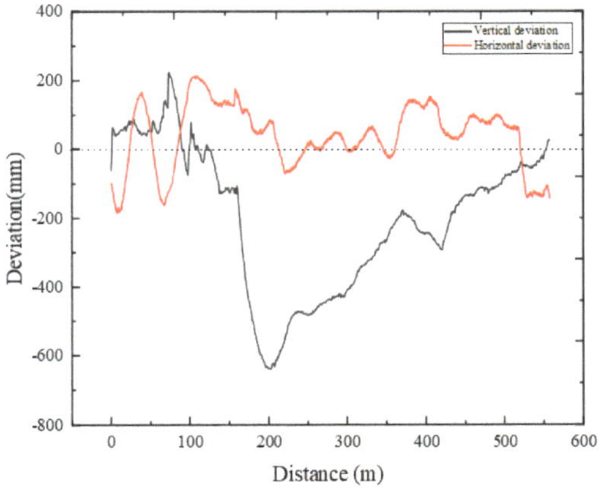

Figure 7. The deviation of the pipeline.

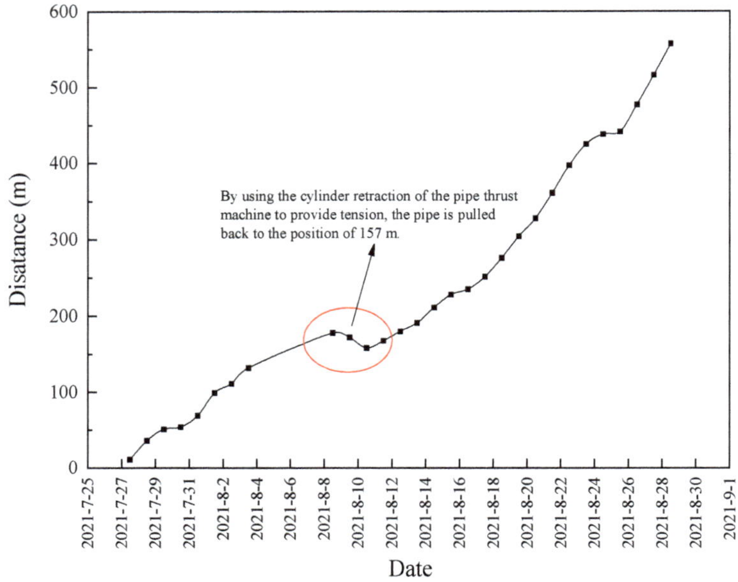

Figure 8. The thrust distance over time.

When the boring machine reached 373 m, the water and soil pressure in front of the head was large, and the slurry discharge was large, while the displacement could not be accurately determined. Therefore, formation loss was caused, and the attitude of the boring machine pointed downward again. Upon reaching 420 m, with the decrease in water and soil pressure, the bottom layer became gradually stable, and the propulsion trajectory was gradually corrected. When the boring machine reached the unearthed point, the vertical direction error was 24.0 mm, and the horizontal direction error was 144.0 mm (Figure 7). The feasibility of the guidance system was verified by measuring the trajectory.

4.4. Equipment Installation

4.4.1. Hole Sealing

The hole seal (Figure 9) should be firmly installed at the entry point of the originating well to prevent the leakage of water, sand, and mud in the annular space gap during excavation, so as to ensure that there is enough mud in the annular space gap to support the stability of the hole wall and lubricate the pipeline.

Before installation, the hole seal was first installed on the prefabricated frame of the hole according to the soil entry angle to ensure that the hole was perpendicular to the equipment and to prevent groundwater and sand from pouring into the starting well. A 2.2 m $\times$ 2.2 m square frame was welded to the steel plate side according to the soil entry angle. Then, the center coordinates of the soil entry point and the edge position were determined in the steel sheet pile. The prefabricated steel portal was welded to the 20 mm steel plate, and the existing portal sealing ring was installed.

The whole pipeline was thrust by the thrust machine, and the boring machine was slowly thrust into the hole. Thrusting was halted when the cutter head was about to reach the sealing position. The position of the fan seal pressure plate was adjusted so that the diameter of the ring formed by the fan pressure plate was larger than the diameter of the cutter head, thereby ensuring that the cutter head could pass. At the same time, the seal ring was smeared with grease to lubricate the sealing part to prevent the wear and damage of the seal ring. After adjustment, the hydraulic cylinder began thrusting the boring machine forward. The cutter head was slowly thrust into the sealing ring of the hole, and then the cutter head was thrust into the outer plate of the hole. Finally, the outer plate of the hole was removed with a crane, and it was verified whether the rubber seal of the hole was intact.

Figure 9. The hole seal of the rapid jacking and pipeline laying method.

When all inspections were correct, the operators in the starting foundation pit communicated with the driving driver in the well control room to control the propulsion speed and ensure that the attitude of the boring machine did not change. After everything was confirmed to be normal, the boring machine began to move forward formally.

4.4.2. Pedestal Installation

The ground anchor adopted the combination scheme of a steel sheet pile and H-shaped steel, the frame composed of the thrust machine pedestal (Figure 10) and H-shaped steel was fixed, and the steel sheet pile provided the horizontal and vertical components generated by the thrust machine. The pedestal was divided into two main parts, which were installed and fixed on the ground anchor at the bottom of the driving shaft. Each part of the pedestal could be mounted separately into the driving shaft using a crane.

4.4.3. Circulating Mud

The prefabricated length of the 'Nanjuma' River pipeline was 590 m, of which the actual length of the rapid jacking and pipeline laying crossing section was 571.64 m. The slope of the entry point of the crossing section was 5°; hence, it was necessary to set up one slurry supply pump and two slurry discharge pumps. The grouting pump was installed on one side of the slurry separation device, and one of the discharge pumps was installed on the tail shield (inside the pipe adapter). Another slurry pump was installed in the middle of the pipeline for mud transportation. The project mud-water separation equipment and pump (Figure 11) were installed at a distance of about 300 m from the starting point of the foundation pit to separate the residue in the circulating mud. The mud circulation equipment layout of the whole pipeline is shown in Figure 12.

Figure 10. The pedestal of the thrust machine.

(**a**) (**b**)

Figure 11. The project mud–water separation equipment and installed pump: (**a**) mud–water separation equipment; (**b**) installed pump.

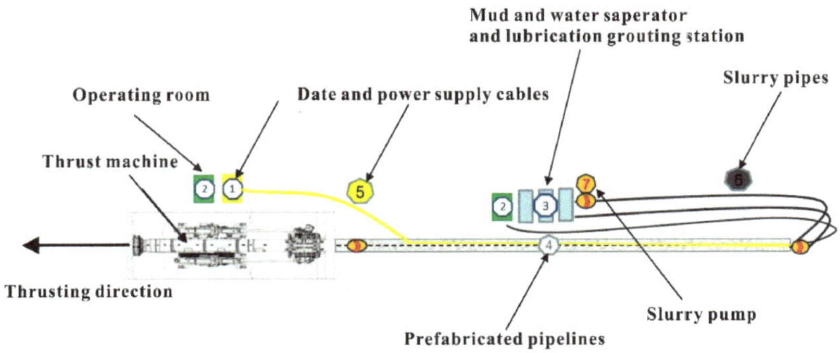

Figure 12. The mud circulation equipment layout of the whole pipeline.

In order to maintain the stability of the excavation face and carry pebble residue, the viscosity of the circulating mud funnel was controlled at 80 s and the specific gravity was controlled at 1.05–1.2 g/cm^3.

The control of the mud circulation pressure was determined by the actual situation of the excavated strata, mainly as a function of groundwater pressure, permeability coefficient, loose situation, and seepage quantity.

Under normal working conditions, the slurry pressure changes the input flow and output flow by setting the slurry pump on the ground and the rotational speed of the slurry discharge pump in the pipeline, so as to control the slurry pressure in the cutter head cabin and keep the slurry pressure in balance with the water and soil pressure on the excavation surface. In general, the mud pressure control value is

$$P = P_0 + 0.2bar, \quad (1)$$

where P is the boring machine mud control pressure, and P_0 is the combined value of water and soil pressure at the site of the boring machine.

At the same time, in the tunneling process, the mud pressure control value should be adjusted according to the specific geological conditions, burial depth, and surface subsidence monitoring results.

4.5. Measures to Reduce Friction Resistance

4.5.1. Grouting Lubrication

Lubrication grouting is mainly divided into boring machine lubrication grouting and hole supplementary grouting lubrication mud; the former is mainly set on the equipment of multichannel grouting hole grouting, while the latter is mainly set on the shaft above the storage tank connected to the hole sealing. The primary propulsion distance of this project was 570 m. In order to avoid mud pressure loss along the way when the propulsion distance was too long, the mud pump was installed in advance in the middle of the pipeline. Considering the cost, the formula of lubricating mud was determined by laboratory tests. The experimental equipment is shown in Figure 13.

4.5.2. Pipeline Sending Ditch

The sending mode of the steel pipe was mediated by the combination of a pipeline sending ditch and roller. The size of the sending ditch (Figure 14) was 2 m in the upper port, 1.5 m in the lower port, and 1.2 m in depth. Before digging the sending ditch, the slope of the sending ditch needed to be calculated, especially the section where the pipeline entered the hole, to ensure that the pipeline could enter the hole and pass through it smoothly. The pipeline was placed in the sending ditch in a floating state to avoid direct contact between the pipeline and the ground, effectively reducing its friction resistance, and ensuring that the pipeline was not damaged during the dragging process.

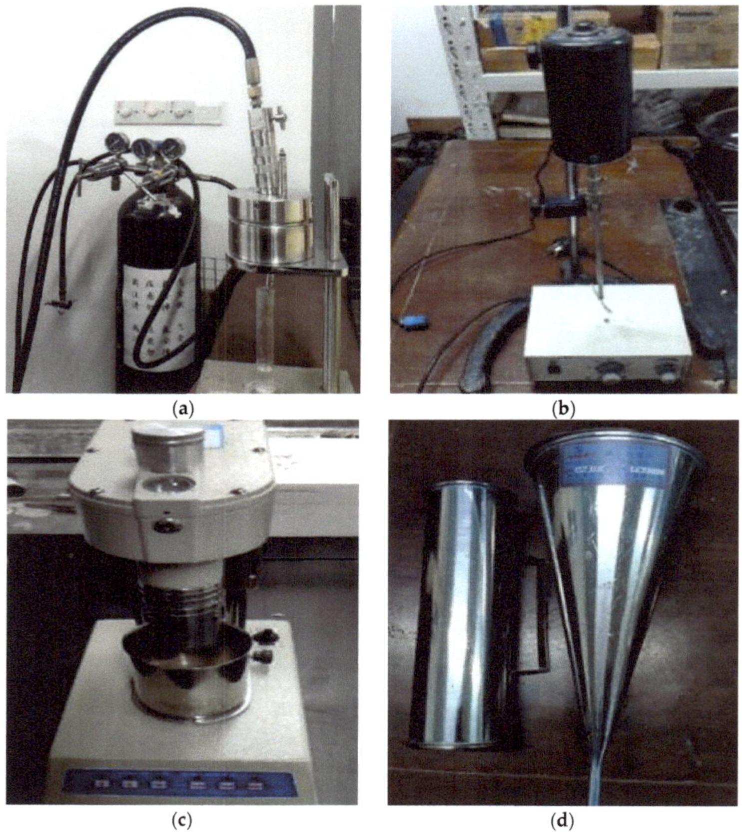

Figure 13. The experimental equipment of lubricating mud: (**a**) adhesion coefficient instrument, (**b**) electric agitator; (**c**) rotational viscometer; (**d**) funnel viscometer.

Figure 14. Pipeline sending ditch.

5. Analysis of Thrust Force

In the analysis of the thrust force of rapid jacking and pipeline laying (Figure 15), the pipe jacking force model can be derived from [16], as a function of the face resistance and the frictional resistance along the pipeline. Unlike pipe jacking, this method welds the pipeline together before starting to advance, and even a pipeline without the advancing tunnel would produce frictional resistance with the ground and the sending ditch.

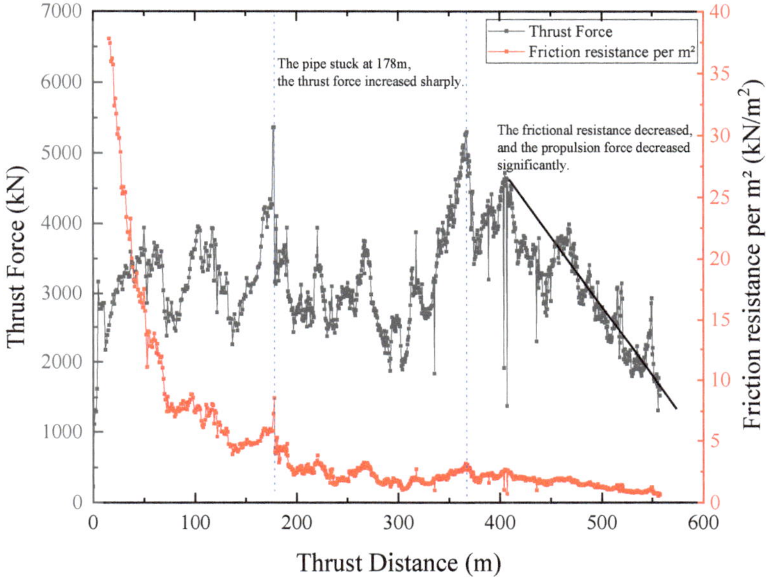

Figure 15. The thrust force of the construction.

The thrust force first increased at 178 m, when the boring machine was located at the junction of silty clay and the pebble layer; the front end of the boring machine entered the pebble layer, and the attitude changed. The vertical direction of the pipeline deviated from the axis greatly, resulting in a sharp increase in the friction around the pipe, which prevented the whole pipeline from re-thrusting.

When thrusting to 290 m, the boring machine was at the lowest point of the propulsion trajectory, and the propulsion force was at a low value. As the attitude of the boring machine was adjusted upward, the curvature became smaller, and the additional coefficient of the propulsion force became larger, resulting in a gradual increase in the propulsion force. At about 362 m, the water and soil pressure in front of the machine head reached a maximum of 0.7 MPa at this time, thereby quickly reattaining the thrust force. At this time, the boring machine was located at the junction of the silty clay and the fine sand layer, and the fine sand entered the annular gap at the front end of the pipeline, leading to a large friction resistance. Upon adding grouting to the annular gap, the frictional resistance began to decrease.

As the propulsion mileage increased and the burial depth gradually decreased, the thrust force decreased, and the frictional resistance per unit area gradually decreased to less than 2 kN/m^2. This was due to mud lubrication, stabilizing the soil and floating pipes. In the process of forming a stable mud jacket, the frictional resistance per unit area of the pipe wall gradually decreased to less than 2 kN/m^2. The presence of mud not only reduced the friction factor between the outer wall of the pipe and the soil, but also reduced the earth pressure acting on the pipe by infiltrating the surrounding soil to form a mixed mud jacket wrapping the pipe. Due to the length of the rapid laying and jacking machine, when the jacking distance was less than 100 m, it was observed that the friction resistance of the pipe in the state where the mud jacket was not fully formed was large, and it could even reach 2–4 times the friction resistance value in the stable state of the mud jacket.

6. Conclusions

6.1. Summary

The rapid jacking and pipeline laying method combines the advantages of horizontal directional drilling and pipe jacking, leading to the development of a new trenchless construction method, which can enable the pipeline to cross obstacles such as rivers, lakes, mountains, and roads. The successful application of this method in this project solved the problems of large pipeline axis deviation and complex construction geology. At the same time, the pipe pulling technology after pipe clamping was enabled, and the project was successful. During the construction, the following findings could be noted:

(1) Due to the process particularity of rapid jacking and pipeline laying, a support form with a short construction period, low cost, and easy removal could be selected in the construction of the launching shaft, effectively shortening the construction period and reducing the cost. After construction was completed, this was convenient to complete site leveling and equipment reuse.

(2) When crossing geologically diverse and complex conditions, especially through gravel layers, a slurry balance type boring machine should be selected, the opening rate of the boring machine should be appropriately increased, the hob and scraper should be used together, and the secondary crushing bin should be equipped to crush the larger solitary stone and unpredictable obstacles, carrying the face through the circulating mud.

(3) In the pipeline propulsion process, a gyroscope guidance system and propulsion fixed distance were used to control the propulsion trajectory in the curve propulsion process through a combination of manual measurement and review. Due to the complex geological conditions and the difficulty of correction, there was a certain error in the vertical direction in the propulsion process of this project, but there was almost no error in the horizontal direction, and the vertical and horizontal errors were controlled in 20 cm increments during completion. Therefore, this guidance system is suitable for any direct pipe laying method.

(4) When the thrust equipment provides thrust force, it can also provide the same magnitude of tension. When the pipe is stuck in the process of pipe propulsion, the pipe can be pulled back through the contraction of the cylinder of the thrust equipment at any time.

(5) In the process of laying oil and gas pipelines using the rapid jacking and pipeline laying method, there is no reserved grouting hole around the pipeline. The annular lubrication grouting is reserved through the grouting hole in front of the boring machine and the grouting hole in the hole. Therefore, the annular lubrication mud should be stopped regularly to supplement the slurry. Because of the overall advancement of the pipeline, rent reduction measures should be designed for the rear pipeline. An overhead and pipeline sending ditch floating pipeline can reduce the friction resistance and ensure that the outer coating of the pipeline is not destroyed.

(6) The construction unit price of the rapid jacking and pipeline laying method in this project was about 2000 USD/m, with about 40% allocated to pipe jacking and 20% allocated to shield construction, which is slightly higher than the cost of horizontal directional drilling. However, the project passed through a pebble layer, which is very difficult for horizontal directional drilling technology. Considering the technology using horizontal directional drilling with casing pipes, the cost would be greater than the rapid jacking and pipeline laying method, in addition to associated construction risks. Therefore, the rapid jacking and pipeline laying method was the most suitable construction technology for this project with its advantages of high construction efficiency, low cost, and small footprint.

6.2. Conclusions

Through the successful implementation of this project, it was verified that the direct pipe laying method has the advantages of low cost and fast speed in oil and gas pipeline crossing projects, which will become more and more favored by the trenchless industry, and will also become a powerful supplement to the construction methods of shield, pipe jacking, and directional drilling in the trenchless field, providing new ideas for the construction of oil and gas pipelines in China and abroad.

With the promotion of this technology, research on the force characteristics and thrust force calculation model of the rapid jacking and pipeline laying method will continue.

Author Contributions: Conceptualization, T.X. and C.Z.; Data curation, T.X., L.W. and Y.Z.; Formal analysis, T.X. and K.L.; Funding acquisition, L.W. and P.Z.; Investigation, T.X.; Methodology, T.X. and C.Z.; Project administration, L.W., P.Z. and Y.Q.; Validation, T.X. and X.F.; Visualization, X.F.; Writing—original draft, T.X., Y.Z. and K.L.; Writing—review & editing, Y.Z. and K.L. All authors have read and agreed to the published version of the manuscript.

Funding: This research received no external funding.

Institutional Review Board Statement: Not applicable.

Informed Consent Statement: Not applicable.

Data Availability Statement: Not applicable.

Conflicts of Interest: The authors declare no conflict of interest.

References

1. Wang, H. Oil and gas pipeline construction status analysis and optimization method exploration. *Chin. J. China Pet. Chem. Stand. Qual.* **2021**, *41*, 2.
2. Lu, H.; Matthews, J.; Iseley, T. How does trenchless technology make pipeline construction greener? A comprehensive carbon footprint and energy consumption analysis. *J. Clean. Prod.* **2020**, *261*, 121215. [CrossRef]
3. Najafi, M. *Trenchless Technology: Planning, Equipment, and Methods*; McGraw-Hill: New York, NY, USA, 2013.
4. Lu, H.; Behbahani, S.; Azimi, M.; Matthews, J.; Han, S.; Iseley, T. Trenchless Construction Technologies for Oil and Gas Pipelines: State-of-the-Art Review. *J. Constr. Eng. Manag.* **2020**, *146*, 03120001. [CrossRef]
5. Yan, X.; Ariaratnam, S.T.; Dong, S.; Zeng, C. Horizontal directional drilling: State-of-the-art review of theory and applications. *Tunn. Undergr. Space Technol.* **2018**, *72*, 162–173. [CrossRef]
6. Li, C.; Zhong, Z.; Liu, X.; Tu, Y.; He, G. Numerical simulation for an estimation of the jacking force of ul-tra-long-distance pipe jacking with frictional property testing at the rock mass–pipe interface. *Tunn. Undergr. Space Technol.* **2019**, *89*, 205–221. [CrossRef]
7. He, X.-C.; Xu, Y.-S.; Shen, S.-L.; Zhou, A.-N. Geological environment problems during metro shield tunnelling in Shenzhen, China. *Arab. J. Geosci.* **2020**, *13*, 87. [CrossRef]
8. Kantemirov, I.F.; Kabirov, R.R. Modern trenchless technologies in the construction of gas and oil pipeline crossings. direct pipe technology. *Synerg. Sci.* **2021**, *61*, 94–100.
9. Lamont, C.; Ross, M.; Lueke, J. Thickwood Trunk Sewer—Direct Pipe Installation. In *Pipelines*; Asce Library: Reston, VA, USA, 2021; pp. 312–320. [CrossRef]
10. Robison, J.L.; Arens, N.A. Direct Pipe Construction: An Engineering Perspective on Quality Assurance. In Proceedings of the Pipelines 2017 Conference, Phoenix, AZ, USA, 6–9 August 2017; pp. 227–238.
11. Wang, L. Construction Technology and Application of the Direct Pipe Method. *Chin. J. Tunn. Constr.* **2018**, *38*, 1566–1572.
12. Liu, X.M.; Tan, M.X.; Li, Q.Y.; Xiang, Y.Q. Application of direct pipe laying in sandy stratum large diameter pipe-lines. *Nat. Gas Oil* **2018**, *36*, 5–8.
13. *GB 50568-2010[S]*; Standard for Oil and Gas Field and Pipeline Investigation of Geotechnical Engineering. China Planning Press: Beijing, China, 2010; pp. 83–85.
14. Staheli, K. *Jacking Force Prediction: An Interface Friction Approach Based on Pipe Surface Roughness*; Georgia Institute of Technology: Atlanta, GA, USA, 2006.
15. Zhang, P.; Ma, B.; Zeng, C.; Xie, H.; Li, X.; Wang, D. Key techniques for the largest curved pipe jacking roof to date: A case study of Gongbei tunnel. *Tunn. Undergr. Space Technol.* **2016**, *59*, 134–145. [CrossRef]
16. Pruiksma, J.P.; Pfeff, D.; Kruse, H.M.G. The calculation of the thrust force for pipeline installation using the Direct Pipe method. In Proceedings of the 7th Pipeline Technology Conference, Hannover, Germany, 28–30 March 2012.

energies

Article

Seepage Performance of Fibre Bundle Drainage Pipes: Particle Flow Simulation and Laboratory Testing

Sifeng Zhang [1], Guozhang Ren [2], Guojian Zhang [3,*], Ziyin Ren [4], Chong Xia [5] and Yuan Gao [5]

1 School of Transportation Engineering, Shandong Jianzhu University, Jinan 250101, China
2 Shandong Hi-Speed Maintenance Group Co., Ltd., Jinan 250032, China
3 School of Surveying and Geo-Informatics, Shandong Jianzhu University, Jinan 250101, China
4 Shandong Transportation Research Institute, Jinan 250101, China
5 Shandong Hi-Speed Group Co., Ltd., Jinan 250014, China
* Correspondence: g_j_zhang@cumt.edu.cn; Tel.: +86-187-6401-8326

Abstract: Mining coal, oil and other energy will form much slope engineering, such as open-pit mine slope and oil depot slope. The groundwater seepage seriously affects the stability of these slope engineering projects. Drainage pipes are commonly used in slope engineering projects to reduce the risk of moisture decreasing soil stability. Such pipes are prone to blockage by soil accumulation after a period of operation, resulting in decreased drainage or complete failure. By installing fibre bundles in drainage pipes, drainage can be maintained under soil ingress. This paper conducted particle flow simulations of the influences of soil particles on the clogging of geotextile filters and drainage pipes under various influences and estimated their seepage rates. Higher water pressure, smaller flower hole intervals in the pipe, greater soil friction angles and smaller pipe inclination angles are less conducive to drainage. Under silting conditions, the seepage and drainage performance of a drainage pipe can be improved by installing a fibre bundle. Five types of fibre bundles were tested with plastic rope providing the best drainage effect. With plastic rope and cotton rope, the best drainage is achieved using uneven arrangements of fibre bundles. In contrast, nylon rope, hemp rope and polyester rope perform best when uniformly arranged. The greater the number of fibre bundles per unit cross-sectional area of pipe, the better the seepage conductivity. Seepage is also greater when the soil in the pipe has a higher sand content. These results provide a reference for the design and construction of more reliable drainage systems for slope engineering in wet areas.

Keywords: fibre bundle drainage pipe; numerical simulation; laboratory test; seepage performance; slope engineering

Citation: Zhang, S.; Ren, G.; Zhang, G.; Ren, Z.; Xia, C.; Gao, Y. Seepage Performance of Fibre Bundle Drainage Pipes: Particle Flow Simulation and Laboratory Testing. *Energies* **2022**, *15*, 7270. https://doi.org/10.3390/en15197270

Academic Editor: José António Correia

Received: 21 July 2022
Accepted: 21 September 2022
Published: 3 October 2022

1. Introduction

Groundwater in rock and soil increases the hydrodynamic and hydrostatic pressure of the rock and soil and also decreases its strength. About 90% of landslides are rain-induced, which demonstrates the important influence of water on slope stability [1]. Water is also an important influence on slope deformation and instability in slope engineering; therefore, a continuous and effective drainage system may be useful in guaranteeing slope safety [2]. The ordinary drainage pipes often used in slope engineering are prone to blockage by mud and sand in their middle and late stages of operation, resulting in poor drainage or complete failure [3]. During the construction design, it is assumed that the drainage system is operating normally. However, there may be risks to slope stability that are buried and remain unseen. Figure 1 is a block diagram of ordinary drainage pipes, while Figure 2 shows the arrangement of drainage pipes in a retaining wall.

Figure 1. A block diagram of the ordinary drainage pipe.

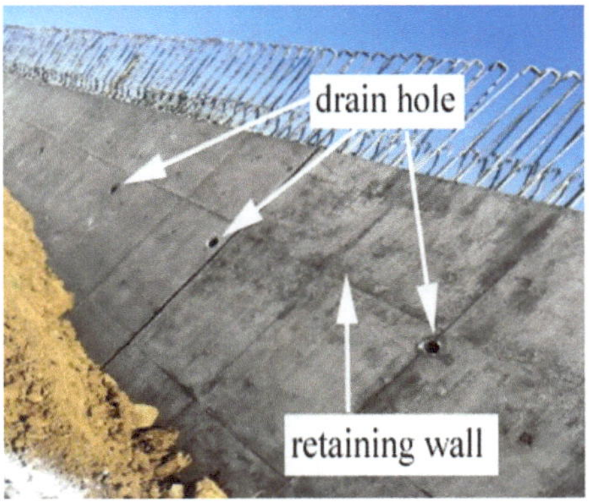

Figure 2. An arrangement diagram of the drainage pipe in the retaining wall.

There have been many studies on slope drainage based on laboratory tests, numerical modelling, field tests and theoretical analysis. For example, Jiang and Guo used comprehensive drainage measures, such as pavement and subgrade drainage, to achieve good results with certain practical value [4]. Jiang integrated subgrade and pavement drainage measures to ensure the integrity and systematicity of drainage facilities [5]. Akay used EPS geofoam as a soil substitute filler in slope rehabilitation works and proposed that the use of internal drainage systems increases the safety factor of slope systems under water seepage conditions [6]. Dong et al. proposed the use of no-fines concrete or inverted filter geotextiles to improve the inverted filter structure of drain holes in retaining walls, which proved very effective in engineering applications [7]. Tian and Ye et al. used physical simulation methods to study the anti-blocking characteristics of new drainage holes in highway slopes, and crystallization and blockage in under-construction tunnel drainage systems. This provided a reference for future field applications of new drainage holes [8,9]. Li Chunxiao et al. introduced a new type of composite road drainage network and showed that it significantly improved the road's performance and service life [10]. Zhang Sulei et al. used drainage tests to show that capillary permeable drainage pipes have good anti-blocking ability [11]. Singh et al. proposed a steel pipe system with drainage and reinforcement functions, which can quickly drain seepage water [12]. Guo et al. proposed a new drainage device that provides another way to solve the poor drainage of existing cutting slopes and retaining walls [13]. Moreover, Zhang Luqing et al. proposed fibre bundle seepage drainage holes, which provide long-term effective drainage of blocked drainage pipes [14]. Miura et al. studied drainage conditions under the influence of blocked drainage pipes.

There were two main causes of significant reductions in slope drainage volume. One was creep of the filter fibre and the other was blockage of the drainage pipes due to the deposition of fine particulate matter [15]. Under silting conditions, the initial discharge of the drainage pipe decreases greatly at first and then decreases slowly with silting at later stages [16,17]. Through numerical simulation, Crenshaw concluded that different types of drainage pipes have different drainage mechanisms and layouts, which lead to changes in drainage speed [18]. D'Acunto et al. used a 2D free boundary numerical model to simulate the drainage effect of drainage holes in a slope [19]. Cook et al. argued that the spacing of drainage holes and the permeability of slope soil are closely related to the variation in groundwater level [20]. Tao et al. found that the particles in low-porosity soil do not readily fall off [21].

To summarize, although some scholars have made preliminary explorations of fibre bundle seepage drainage in the field of slope drainage, their research methods are relatively simple. No studies have explored the seepage effect and the influences of fibre bundle drainage pipes via a combination of numerical simulations and laboratory tests. Hence, it is difficult to accurately determine the seepage mechanism of fibre bundle drainage pipes.

To make up for the deficiencies of existing research, this paper established a particle flow numerical model by using PFC to simulate the degree of blockage due to soil particles on a geotextile filter layer and in a drainage pipe. The influences of water pressure, hole spacing, soil internal friction angle and drainage pipe inclination angle were explored. Combined with laboratory tests, this paper probed the influences of fibre bundle type, arrangement, quantity, soil properties and drainpipe diameter on the seepage effect of fibre bundle drainage pipes under conditions where soil can accumulate in the pipe. The results reveal the seepage mechanism of fibre bundle drainage pipes. The purpose of this paper was to provide an optimization scheme for solving the problem of non-water diversion after drainage pipes become blocked.

2. Numerical Simulation of Particle Flow

2.1. Soil Particle Drainage Shedding Model

A drainage shedding model of soil particles was established using PFC2D software (5.00.25). The modelling steps were as follows: (1) Establish a closed wall "generate particles in the wall" to give the properties of particle homogeneous sand "to apply self-weight stress to complete the initial calculation. (2) Delete the surrounding closed wall to establish a vertical wall on the side, and establish two walls with a certain interval (interval distance = 1 cm) on the bottom surface (to simulate the flower hole in the drainage pipe). "Set up water pressure below the model height of 5 m" to apply self-weight stress to simulate the state of soil particles gushing out of the flower hole of the drainage pipe (that is, the wall gap). The size of the model was 4 m × 7 m. The physical and mechanical parameters of the soil particles are shown in Table 1. The soil particle size is shown in Table 2. The model is shown in Figures 3 and 4.

Table 1. The physical and mechanical parameters of soil particles.

Elastic Modulus/MPa	Poisson's Ratio υ	Cohesive Force/kPa	Internal Friction Angle/°	Natural Density/(kg/m^3)	Tensile Strength/kPa
35	0.3	20	20	1800	1.31

Table 2. The content of each particle size of soil.

>Particle size (mm)	10	2	0.5	0.25	0.075	0.005
Relative content (%)	0	4.5	16.9	52.4	85.9	99.9

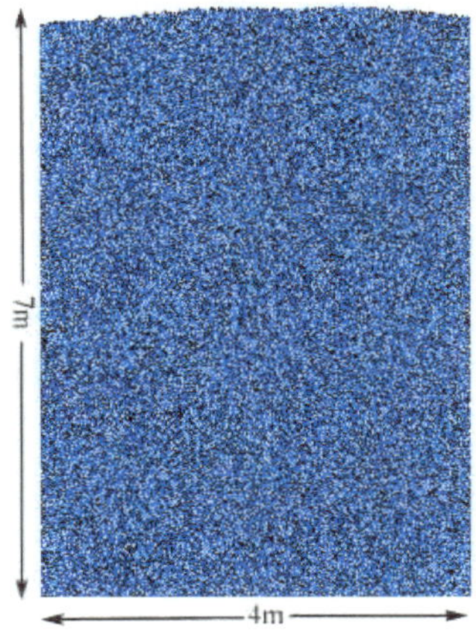

Figure 3. Soil particles.

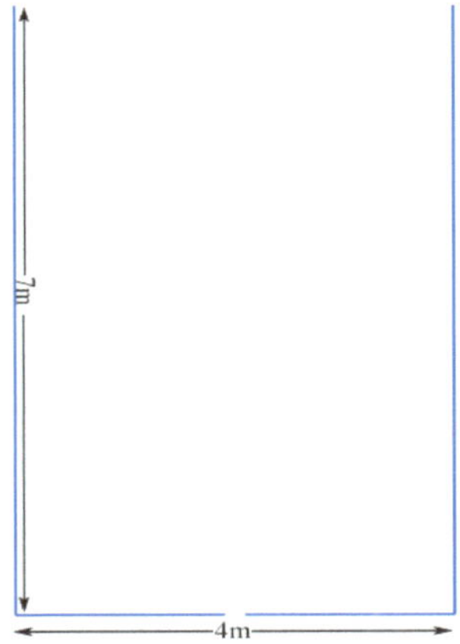

Figure 4. Wall.

2.2. Drainage Pipe Blockage Model

The drainage pipe model shown in Figure 5 was established, in which the length of the drainage pipe is 3 m and the diameter of the drainage hole is 0.1 m. The soil particles fall freely from the top of the drainage pipe at a distance of 0.6 m from its closed end. A wall model with a height of 0.01 m was established at 1.5 m in the centre of the drainage pipe to simulate a soil ridge blocking the outflow of particles.

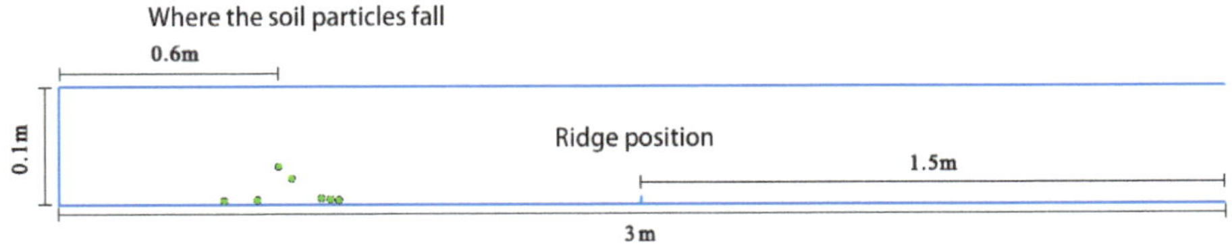

Figure 5. A drainage pipe model (unit: m).

3. Analysis of Simulation Results

3.1. Effect of Water Pressure on Drainage Shedding of Soil Particles

It can be seen from Figure 6 that the number of shedding soil particles increases gradually with increases in pore water pressure at the opening. The analysis shows that the greater the water pressure, the greater the hydraulic gradient between the soil along the height direction, the pore water pressure in the soil and the water pressure in the drainage pipe. The hydraulic gradient has a certain impact on blockage of the drainage filter layer. When the hydraulic gradient is high, finer particles are carried by seepage and a large number will be deposited in the interior and surface of the geotextile filter layer, resulting in blockage of the filter layer. So, the greater the hydraulic gradient, the more easily blockage of the drainage filter layer occurs and the worse the drainage effect [22].

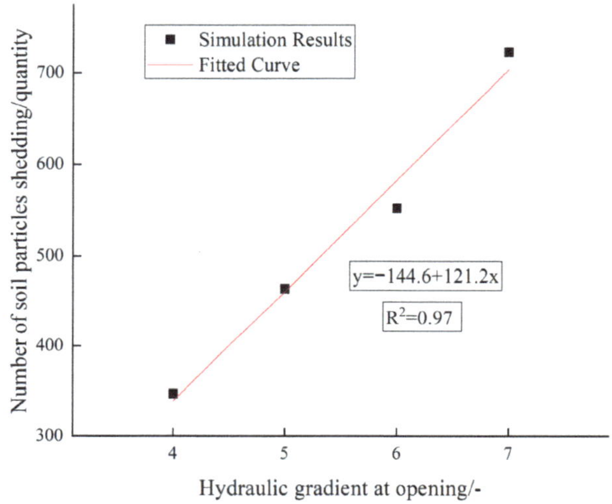

Figure 6. The effect of water pressure on drainage and shedding of soil particles.

3.2. Effect of Opening Spacing on Drainage Shedding of Soil Particles

As can be seen from Figures 7 and 8 and Table 3 that when the hole spacing is 0.5 cm, 1 cm, 2 cm, 3 cm, 4 cm and 5 cm, the number of soil particles shed is 1590, 980, 890, 865, 850, and 850, respectively. It reflects that the number of shed soil particles decreases with the increases in hole spacing. In the process of soil particle shedding at the front of the flower hole in the drainage pipe, a "micro soil arch" can easily form, which restrains soil particle shedding. If the hole spacing is too close, there is a trend of penetration between two similar "micro soil arches". As the opening distance continues to decrease, the two "micro soil arches" become connected, resulting in a sharp increase in the number of shed soil particles. This readily causes silting in the geotextile filter layer, which reduces drainage performance. It can also be seen from Figure 8 that there is an inverse logarithm function relationship between the shedding number of soil particles and the opening distance when the opening space is larger than 0.5 cm, and the optimal value (the curve inflection point) is about 3 cm. This analysis shows that the hole spacing in the drainage pipe should be about 3 cm. In addition, soil with different properties may have a different penetration range, causing the optimal position to also be different.

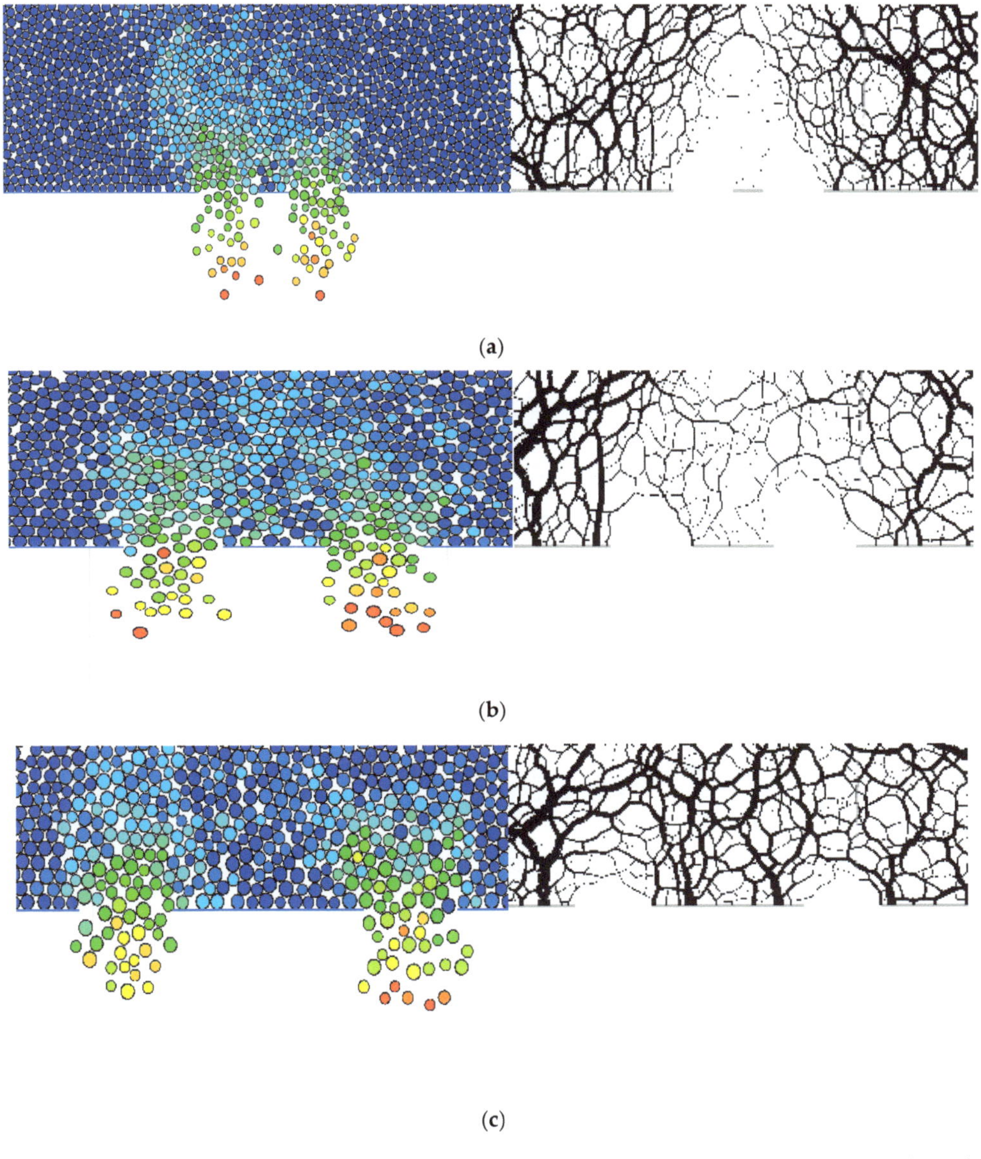

Figure 7. A displacement vector and contact force chain diagram of soil particles with different spacing. (a) opening spacing 0.5 cm. (b) opening spacing 1 cm. (c) opening spacing 2 cm.

Table 3. The number of soil particles shedding in different opening space.

Opening space	0.5 cm	1 cm	2 cm	3 cm	4 cm	5 cm
Number of soil particles shedding	1590	980	890	865	850	850

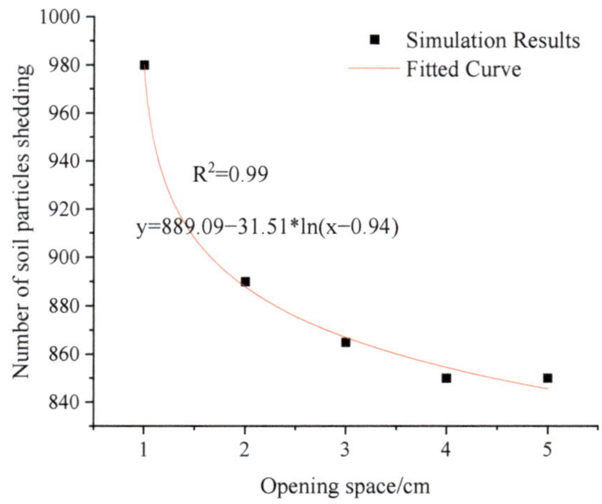

Figure 8. The effect of hole spacing on drainage shedding of soil particles.

3.3. Effect of Soil Friction Angle on Drainage Pipe Blockage

In the simulation, the soil friction angles were 20°, 30° and 40°, respectively, while the other soil parameters remained unchanged. It can be seen from Figure 9 that the greater the soil friction angle, the higher the soil particle accumulation height near the soil ridge. According to the analysis, the internal friction angle can be summarized as the sliding friction that needs to be overcome when the soil particles slide with each other, which is mainly affected by the roughness of the particle surfaces, the embedding of particles, and interlocking and detaching from the occlusal state. The greater the internal friction angle is, the stronger the occlusal action between particles and the less vulnerable they are to the impact of water flow. This makes them accumulate readily in the drainage pipe, resulting in blockage. In real applications, a suitable soil buried drainage pipe can be selected for better drainage.

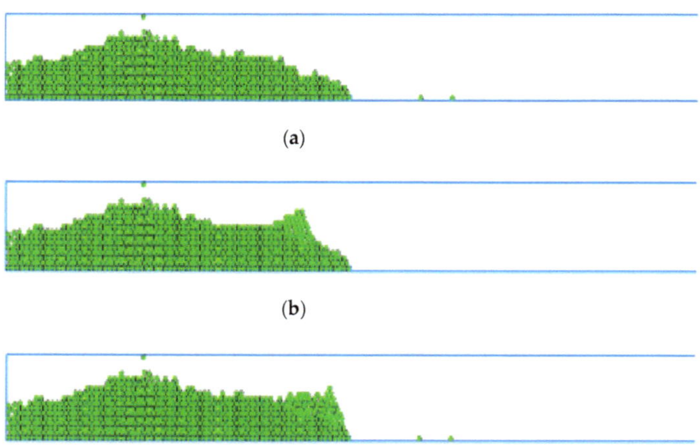

(**a**)

(**b**)

(**c**)

Figure 9. The effect of internal friction angle on blockage of drainage pipe. (**a**) Internal friction angle 20°. (**b**) Internal friction angle 30°. (**c**) Internal friction angle 40°.

3.4. Influence of Drainage Pipe Inclination Angle on Blockage

When the tilt Angle of the drainpipe is 0°, 5°, and 10° (Figure 10), the number of soil particles blocked is 248, 130, and 112, respectively. As can be seen from Figure 11, when the tilt Angle of the drainpipe increases from 0° to 5°, the number of soil particles blocked decreases from 248 to 130, which decreases by 118. When the tilt Angle of the drainage pipe increases from 5° to 10°, the number of soil particle clogging decreases from 130 to 112 by 18°. In other words, with the increase of the inclination Angle of the drain pipe, the number of clogged soil particles gradually decreases, and according to the slopes of the two line segments, the reduction amount of clogged soil particles decreases. This shows that blockage of a horizontal drainpipe is more serious than that of an inclined one. The analysis shows that after a drainage pipe is inclined, the sliding friction that needs to be overcome when the soil particles slide with each other is offset; hence, the particles do not readily accumulate under water flow. Therefore, in practical applications, drainage pipes should be inclined and buried properly.

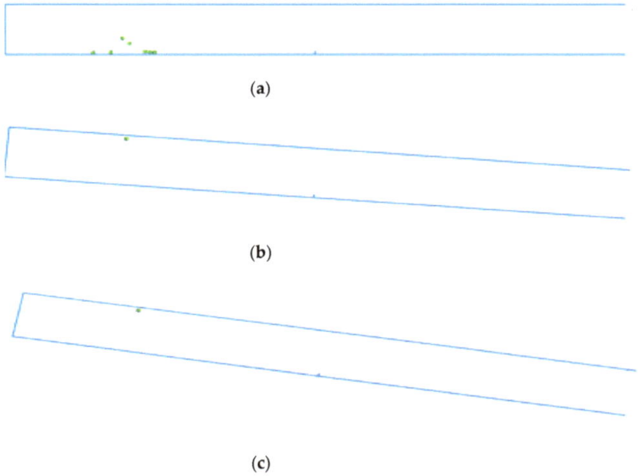

Figure 10. The different inclination angles of drainage pipes. (**a**) tilt angle 0°. (**b**) tilt angle 5°. (**c**) tilt angle 10°.

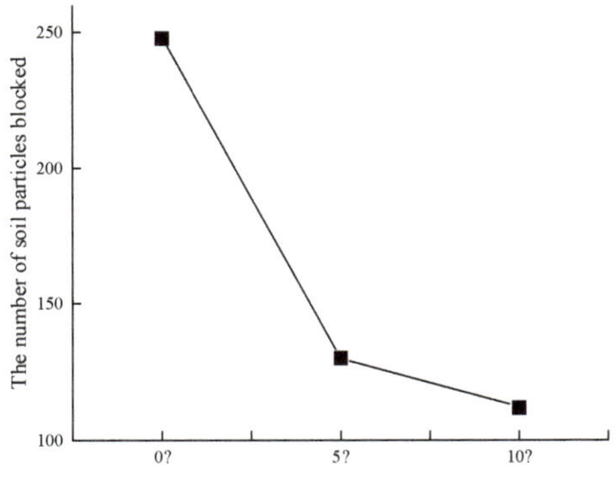

Figure 11. A statistical diagram of the number of clogged soil particles in the drainage pipe at different inclination angles.

4. Test Methods

With reference to the literature and actual projects, this study selected five types of fibre bundles with the same diameter: plastic rope, hemp rope, polyester rope, nylon rope and cotton rope. They were each arranged and fixed in a drainage pipe according to the design method. Then, the pipes were filled with soil and vibrated to simulate the blocked state. The test was started after the device was installed and debugged, and the actual seepage flows of the fibre-bundle drainage pipes under various influences were recorded. Finally, the influences of various factors on the drainage effect in these pipes were analysed according to the measured flow, as shown in the following section.

4.1. Test Program

The following three test programs were designed to consider the various factors that may affect the drainage performance of fibre bundle drainage pipes:

(1) Program for testing fibre bundle drainage pipe performance considering the influences of fibre bundle type and quantity.

In the experiment, the diameter of each fiber bundle was the same, and 6, 8, 10 and 12 fiber bundles were placed in each drain pipe to conduct a comparative test to find out the influence of different types and quantities of fiber bundles on the drainage effect, and to find out the fiber bundles with more economical types and quantities and better drainage effect. At the same time, a set of drainage pipes without fibre bundles was set up for comparison. Table 4 shows the material specifications and uses, and Figure 12 is a test diagram of a fibre bundle drainage pipe.

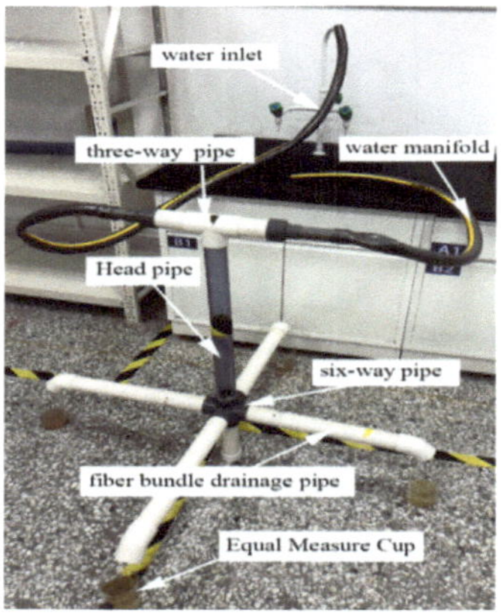

Figure 12. A test diagram of the fiber bundle drainage pipe.

Table 4. The material specifications and uses.

Devices and Materials	Uses	Material Quality	Diameter/mm	Length/mm
Fiber bundle	for seepage in pipes	polyester rope, hemp rope, cotton rope, nylon rope, plastic rope	2–3	50
Head pipe	provide 0.6 m high head pressure	transparent hard plastic tube	46	600
drainage pipe	simulate horizontal drilling	polyvinyl chloride (PVC)pipe	50	400
drainage pipe (Figure 13)	balance the atmospheric pressure	polyvinyl chloride (PVC)pipe	30	100
filter cloth (Figure 14)	simulated geotextile to prevent soil particle loss	plastic cloth	/	70

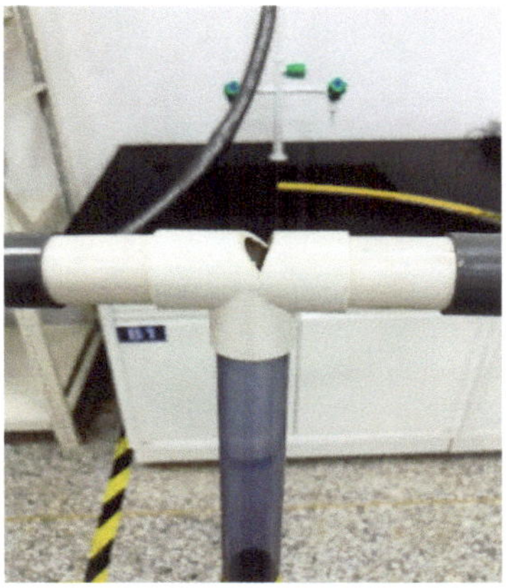

Figure 13. A diagram of a three-way drainage pipe balancing atmospheric pressure.

Figure 14. An arrangement of simulated geotextile for the fiber bundle drainage pipe.

(2) Program for testing fibre bundle drainage pipe performance considering fibre bundle arrangement.

A group of drainage pipes are not provided with fiber bundles. The other group drainpipes were set with plastic rope, twine rope, polyester rope, nylon rope and cotton rope respectively, and the layout of each fiber bundle can be divided into two types: uniform layout and uneven layout. The control test was carried out under the condition that the pipe diameter was the same, the tube was not splined and the tube was filled with silty clay. Through the above experiments, it is expected to find a better way for this kind of fiber bundle to conduct water.

(3) Program for testing fibre bundle drainage pipe performance considering the influences of drainage pipe diameter, soil properties and strand-shaped fibre bundles.

A test of drainage pipe performance was conducted with different pipe diameters, soil properties and strand-shaped fibre bundles (3 fibre bundles twisted into 1 strand) in the soil-filled state, so as to determine their influences on the seepage effect (measured as seepage flow rate). The specific test program is shown in Table 5.

Table 5. A design table of the drainage performance test program of the fiber bundle drainage pipe considering the influence of the drainage pipe diameters, soil properties and strand-shaped fiber bundles.

No.	Types of Ropes	Arrangement	Drainpipe Aperture/mm	Strands (Yes/No)	Soil Properties
1	cotton rope	Even/uneven	50	no	silty clay
2	cotton rope	Even/uneven	60	no	silty clay
3	plastic rope	Even/uneven	50	no	silty clay
4	plastic rope	Even/uneven	50	no	silty clay with 25% sand
5	hemp rope	Even/uneven	50	yes	silty clay
6	hemp rope	Even/uneven	50	no	silty clay

4.2. Test Method and Process

To measure the specific seepage flow rates of the fibre bundle drainage pipes under different influences in the soil-filled state, the specific test method was as follows. (1) Arrange and fix the fibre bundle in the drainage pipe, (2) install the test device, (3) conduct the test and record the seepage flow and (4) analyse the data.

Arrangement and fixation of fibre bundles. The type and number of fibre bundles were determined to ensure that the cross-sectional area of the fibre bundles was almost the same. The relative spatial positions of the fibre bundles were arranged according to the design plan. As shown in Figure 15, the left end of a fibre bundle was passed through the No. 1 sieve hole of the left sieve and then tied to ensure that it could not pass through the sieve again on the right side. The right end of the fibre bundle was then passed from the left side of the drainage pipe through the No. 1 sieve hole of the right sieve on the right side, and then a knot was tied at the sieve. These steps were repeated to install the designed number of fibre bundles in the corresponding position so that each fibre bundle was arranged straight in the pipe. Figure 16 shows the uniform and uneven distribution of fiber bundles in the pipeline.

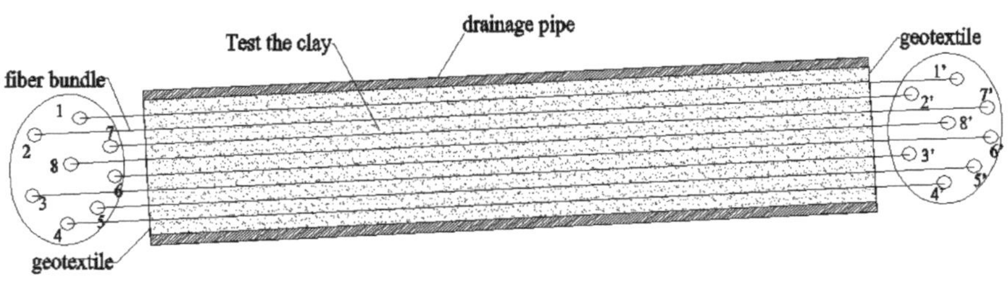

Figure 15. A schematic diagram of the fiber bundle drainage pipe arrangement.

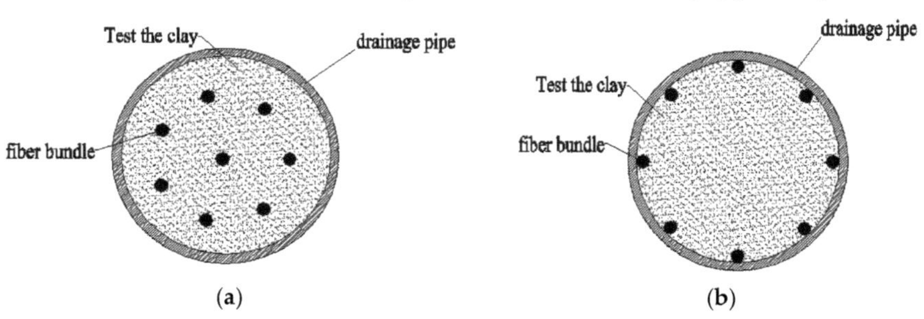

Figure 16. A schematic diagram of the fiber bundle in the drainage pipe. (**a**) even arrangement. (**b**) uneven arrangement.

Filling of the fibre bundle drainage pipe. A sieve was placed on the ground with the lower end of the drainage pipe placed against it. The pipe was filled with test soil from the upper outlet and vibrated with a stick while filling to ensure an even soil density and to achieve blockage of the pipe. The soil was filled up to the position of the upper outlet. Then, the pipe was turned over and the lower end was filled with test soil according to the same method. It was ensured that the fibre bundle was in a straightened state during soil-filling. The excess fibre bundle outside the pipe outlet was cut after filling, then simulated geotextile was wrapped around the pipe outlet and adhered to it with glue. The quality of the drainage pipe was then measured with a balance. The drainage pipe was installed on the side of a six-way pipe outlet, the four sides of which were filled with soil.

(1) Installation of the drainage pipe test device. The six-way outlet was installed with a head pipe with a design height on the top and a section of PVC pipe at the bottom. The lower part of the PVC pipe was plugged with a pipe cap to prevent water leakage. A three-way pipe was installed at the top of the head pipe with a water inlet pipe at one end of the three-way pipe and a water outlet pipe at the other end. A small opening was made on the three-way pipe to balance the atmospheric pressure and maintain the height of the water pressure. A measuring cup was placed under the outlet of each drainage pipe. Then, the faucet was turned on to observe whether the device leaked. If water leakage occurred, the faucet was turned off and the device was fastened again. The test timing was started when no water leakage occurred, the height of the water pressure was unchanged and the drainage pipe was stable.

(2) Recording of seepage flow rate and data analysis. The faucet was turned off and the measuring cups were removed when the designated test time was reached. The water volumes in the measuring cups under each fibre bundle drainage pipe were measured. The seepage flow rates of the drainage pipes were determined and analysed in relation to the different test factors.

5. Results

The test data were analysed to determine the parameters providing optimal drainage performance. Please note that the experimental data are the average of 8 group experiments.

5.1. Influences of the Type and Quantity of Fibre Bundles

Tests were conducted with different types and quantities of fibre bundles to observe the effects on seepage flow. This helped determine the type of fibre bundle that provides the most significant drainage effect.

It can be seen in Table 6 that the seepage flow was much greater in drainage pipes containing a fibre bundle than in an ordinary drainage pipe in the soil-filled state. This shows that a fibre bundle pipe provides better drainage than an ordinary blocked drainpipe. As the space in a drainage pipe is limited, the number of fibre bundles with drainage effect in the drainage pipe increased with an increasing number of fibre bundles, thereby helping to increase the seepage flow of the blocked drainage pipe. So, this type of fibre bundle has an ideal drainage effect. Figure 17 shows that pipes with unevenly arranged fibre bundles had a generally better seepage effect than pipes with evenly arranged fibre bundles (for plastic rope or cotton rope fibre bundles with the same number of fibres). This is because an uneven arrangement of this type of fibre provides better drainage and can increase the seepage flow of a blocked drainage pipe.

Table 6. The seepage flow of drainage pipes of different fiber bundles.

Material	Plastic Rope (1)		Plastic Rope (2)		Cotton Rope			Hemp Rope	Polyester Rope		Nylon Rope	
Arrangement	**Even**	**Uneven**	**Even**	**Uneven**	**Even**	**Uneven (1)**	**Uneven (2)**	**Even**	**Even**	**Uneven**	**Even**	**Uneven**
6	690	870	1060	1120	83	174	176	360	152	102	109	63
8	1269	1373	1278	1348	135	300	210	595	207	104	145	88
10	1456	1686	1682	1835	170	444	370	1000	306	137	167	136
12	1620	1824	1931	2120	245	576	509	1430	354	183	241	188

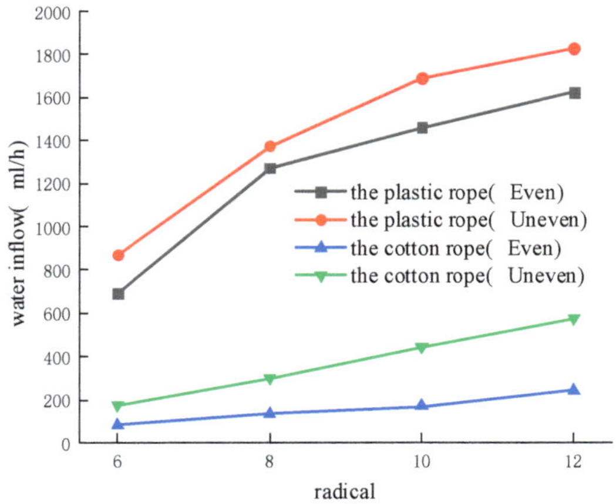

Figure 17. A seepage effect diagram of the plastic/cotton rope fiber bundle.

Table 6 indicates that water will drain out of a pipe when a hemp rope fibre bundle is soaked with water. When the number of fibre bundles increases from 6 to 12, the seepage flow increases from 360 mL/h to 1430 mL/h. This is an increase of nearly 1070 mL/h, showing greatly improved drainage efficiency. The seepage flows of the drainage pipes with polyester rope fibre bundles were low with 6, 8, 10 and 12 strands, indicating that this type of fibre bundle provides a bad drainage effect and is unable to significantly improve drainage efficiency compared with the soil-filled state. The nylon rope fibre bundles, arranged evenly or unevenly under a soil-filled state provided no significant increases in seepage flow compared with the control group (without fibre bundles), indicating that nylon rope fibre bundles provide a relatively weak drainage capability.

Figure 18 indicates that the seepage flow rate of the blocked drainage pipe increased with increasing numbers of fibre bundles. The quantity of hemp rope fibre had a great influence on the seepage flow rate. Figure 18 shows that the seepage flow rates of evenly arranged polyester rope fibre bundles were much greater than those of unevenly arranged ones, indicating that the former is more effective. Figure 19 shows that the seepage flow rates of nylon rope fibre bundles were greater when evenly arranged than when unevenly arranged, indicating that the former is more effective.

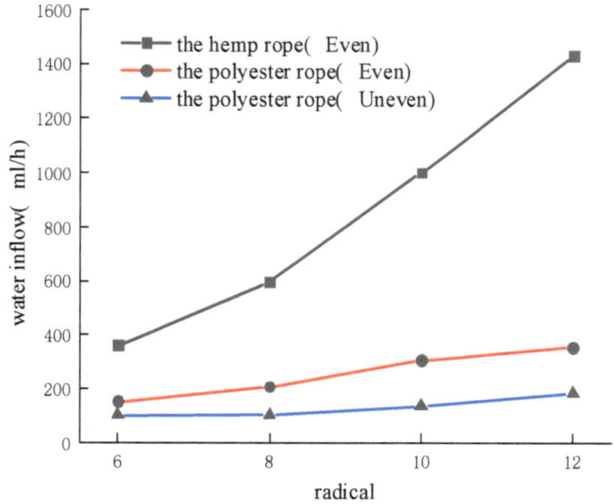

Figure 18. A seepage effect diagram of the Hemp/polyester rope fiber bundle.

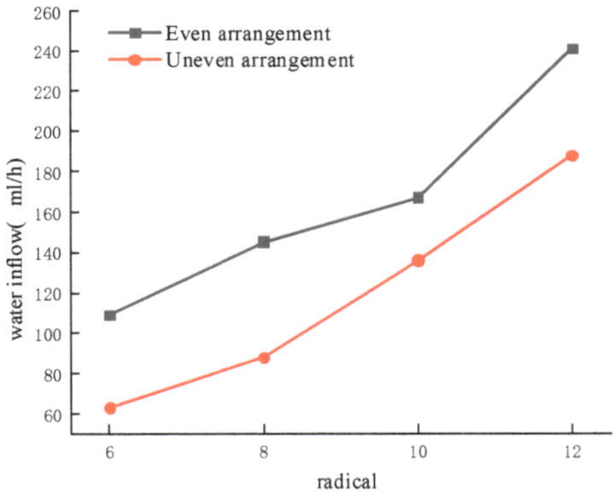

Figure 19. A seepage effect diagram of the nylon rope fiber bundle.

According to Figure 20, when various types of fibre bundles are evenly arranged in soil-filled drainage pipes, the seepage flow rate gradually increases with the number of fibre bundles. Plastic rope and then hemp rope fibre bundles had the best drainage effect. The various fibre bundles can be ranked from strongest to weakest effect as: plastic rope > hemp rope > polyester rope > cotton rope > nylon rope. The drainage effect of cotton rope will be greater than that of polyester rope when the fibre bundles are arranged to maximize their effect.

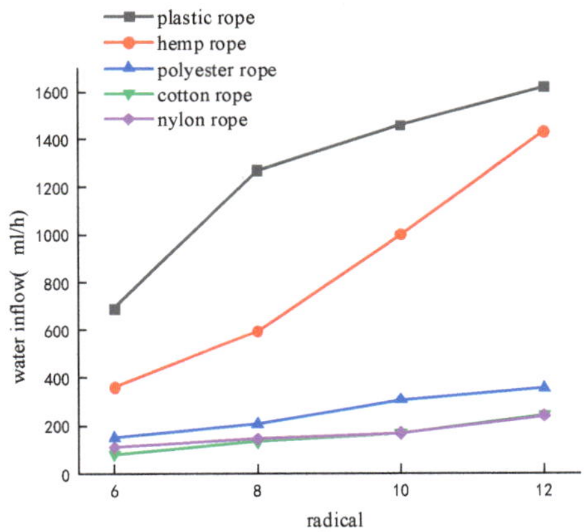

Figure 20. The comparison of seepage flow of different types of fiber bundles.

5.2. Influences of Pipe Diameter, Strand-Shaped Fibre Bundle and Soil Properties

For a given number of cotton rope fibre bundles arranged unevenly in soil-filled drainage pipes of different diameters, Table 6 and Figure 21 show that the average seepage flow rate is about 57 mL/h greater in the smaller diameter pipe than in the larger diameter one. Because the fibre bundles conduct permeation, using more fibre bundles per unit cross-sectional area of pipe provides better drainage. Table 6 and Figure 22 show that the seepage flow rate of a pipe with a strand-shaped hemp rope (with three hemp rope fibre bundles twisted into one strand) was less than that of a pipe with an unstranded hemp rope and provided poor seepage efficiency. Hemp rope fibre bundles that are not twisted into a strand have a larger contact area with the soil in the pipe than strand-shaped bundles for a given number of bundles; therefore, the former has the better seepage effect. At the same time, evenly arranged hemp rope fibre bundles provide a better seepage effect than unevenly arranged ones because they can conduct drainage better under the soil-filled state. According to Table 6, the plastic rope fibre bundle provides a better seepage effect when unevenly arranged than when evenly arranged. Therefore, the uneven arrangement method was chosen to compare the drainage function of fibre bundles under different soil properties. Figure 23 shows that in the soil-filled state, the seepage effect of the plastic rope drainage pipe with silty clay plus 25% sand is better than that of silty clay without sand. The higher the sand content of the soil in the pipe, the greater the seepage coefficient of the soil, the better the drainage effect of the fibre bundles and the higher the seepage flow.

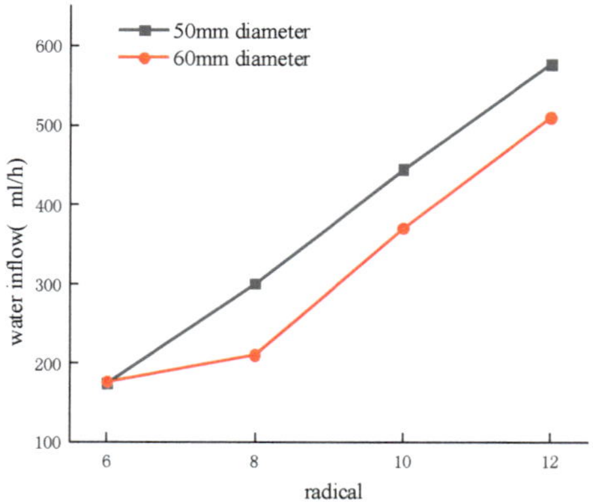

Figure 21. The comparison of seepage flow of the fiber bundles with different pipe diameters.

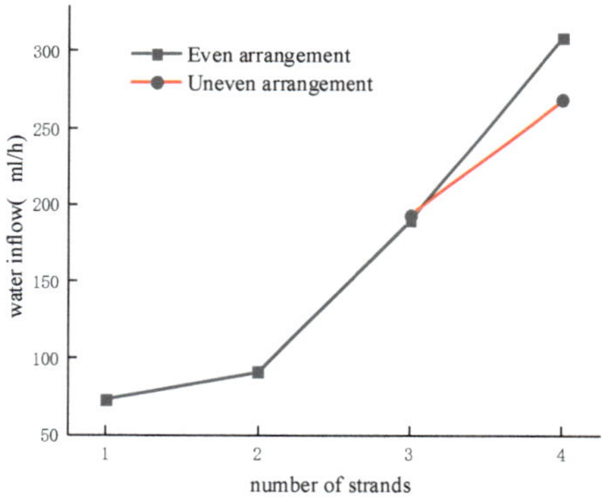

Figure 22. A seepage effect diagram of strand-shaped hemp rope fiber bundle.

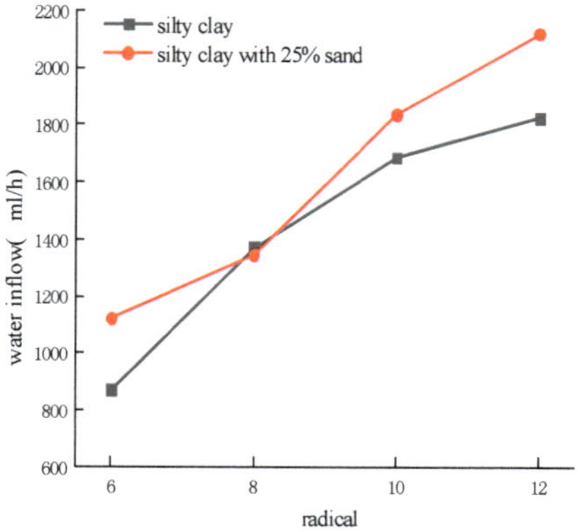

Figure 23. Seepage flow effect diagram of fiber bundles in different soil properties.

6. Discussion

This paper carried out engineering tests at a high-filled subgrade project at the intersection of Lushan Avenue and Zhangxin Road in Zibo City, Shandong Province. The project site receives heavy rainfall in summer. Therefore, drainage pipes were installed in the retaining wall for water drainage. To ensure the long-term performance of the drainage pipes and the safety of the subgrade, fibre bundle drainage pipes were used in the drainage structure of the retaining wall (Figure 24).

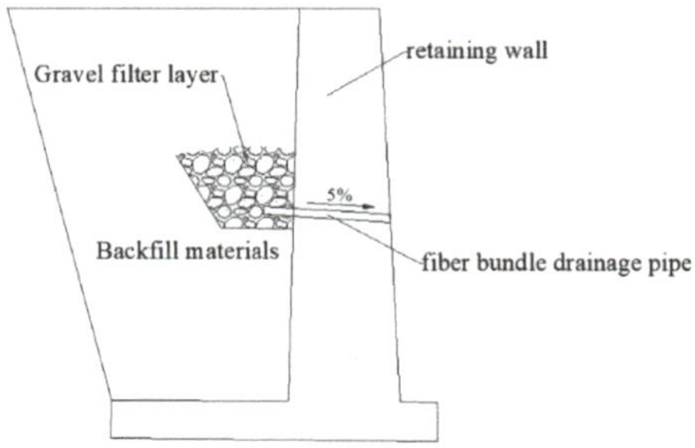

Figure 24. A schematic diagram of the waterproof and drainage structure of the retaining wall.

PVC pipes (diameter = 100 mm) with fibre bundle drainage were used. There were 20 fibre bundles arranged evenly in each pipe, which was outward-inclined by 5%. Figure 25 represents the site arrangement of the drainage pipes, while Figure 26 shows the fibre bundle arrangement.

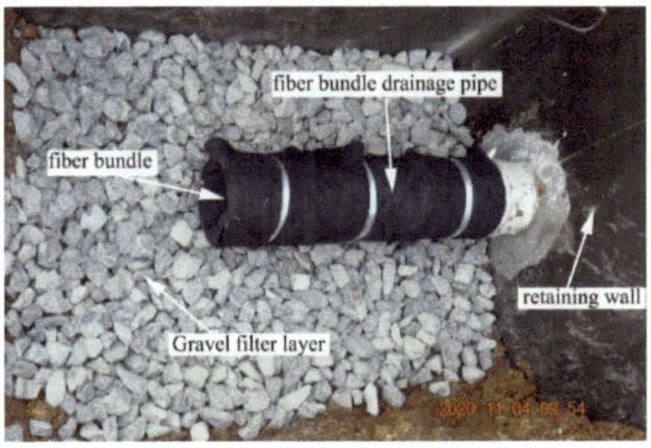

Figure 25. The site arrangement of the fiber bundle drainage pipe.

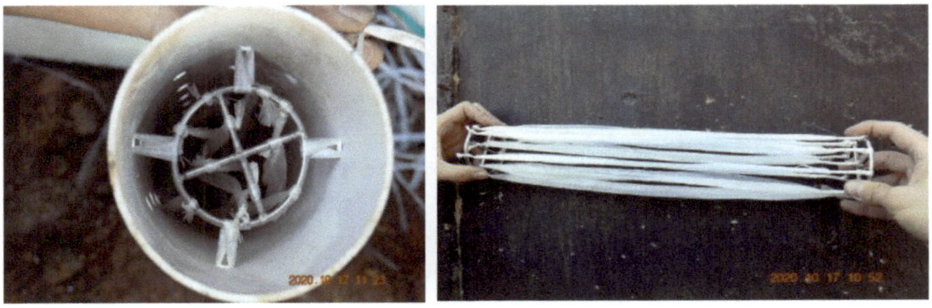

Figure 26. The arrangement of the fiber bundle in the drainage pipe.

These pipes showed a great drainage effect after being blocked by mud and sand during a later rainfall period. They greatly reduced the potential safety hazards of the project. It provides a reference for the design and construction of similar projects.

7. Conclusions

To solve engineering problems such as geotextile blockages and soil accumulation in ordinary drainage pipes, and to ensure smooth flow in slope drainage systems, this paper proposes the idea of installing fibre bundles in drainage pipes. Particle flow simulations of the influence of soil particles on the geotextile filter layer and of drainage pipe flow rates under different influences were carried out by PFC. The concrete seepage effect and influences on fibre bundle drainage pipes were also studied by laboratory tests. The main conclusions are as follows:

(1) The greater the water pressure, the greater the pore water pressure and internal hydraulic gradient in the soil, and the easier it is to block the drainage filter layer, resulting in poor drainage. If the distance between the flower holes in the drainage pipe is too close and the "micro-soil arches" are connected, the amount of soil particle shedding will increase sharply, which can easily cause silting in the geotextile filter layer and is not conducive to drainage.

(2) The greater the soil friction angle, the more easily a drainage pipe is blocked. Pipes can be buried in soil with a smaller internal friction angle to achieve better drainage. The greater the inclination angle of the drainage pipe, the more difficult it is for soil particles to accumulate within it; hence, drainage pipes should be appropriately tilted when buried.

(3) The laboratory tests show that fibre bundle drainage pipes provided better drainage than ordinary drainage pipes in the soil-filled state. In a fibre bundle drainage pipe with a diameter of 50 mm, the drainage improvement of fibre bundles can be ranked from strong to weak as: plastic rope > hemp rope > polyester rope > cotton rope > nylon rope.

(4) The plastic rope and cotton rope fibre bundles provided better drainage effects in soil-filled pipes when they were unevenly arranged, while the nylon rope, hemp rope and polyester rope fibre bundles provided better drainage when evenly arranged.

(5) The greater the number of fibre bundles arranged in drainage pipes of a given diameter, the better the drainage performance. For a given number of fibre bundles, the drainage effect is better in a pipe with a smaller diameter than a larger one; that is, a greater number of fibre bundles per unit cross-sectional area is more conducive to drainage.

(6) The higher the ratio of sand to soil in the drainage pipe, the better the seepage effect. Therefore, the service life of drainage pipes can be prolonged by filling the gravel filter layer behind the drainage pipe with medium-coarse sand.

Author Contributions: Conceptualization, S.Z. and G.Z.; methodology, G.Z. and G.R.; formal analysis, G.R.; investigation, Z.R. and C.X.; writing—original draft preparation, Y.G.; writing—review and editing, S.Z. and G.Z.; project administration, S.Z. and G.R.; funding acquisition, S.Z. All authors have read and agreed to the published version of the manuscript.

Funding: This study was supported by the National Natural Science Foundation (grant no. 51108252) and the Shandong Transportation Science and Technology Project (grant no. 2017B59).

Data Availability Statement: The data supporting reported results can be found by contacting the corresponding author (g_j_zhang@cumt.edu.cn).

Acknowledgments: The authors gratefully acknowledge the experimental site support from Zibo project department of Shandong Hi-Speed Group Co., Ltd. We also thank for the staff who work in the lab. They help us transport experimental materials and set up experimental equipment. At last, we are grateful to the editors and reviewers who have made a great contribution to the improvement of the article quality.

Conflicts of Interest: The authors declare no conflict of interest.

References

1. Ghiassian, H.; Ghareh, S. Stability of sandy slopes under seepage conditions. *Landslides* **2008**, *5*, 397–406. [CrossRef]
2. Wen, Q.Z. The value of rock and soil parameters of landslide. *J. China Foreign Highw.* **2009**, *3*, 45–47.
3. Shen, B.; Ai, C.L. Mountain Area Highway Drainage and Water Damage Survey and Analysis. *J. Highw. Transp. Res. Dev.* **2006**, *10*, 5–9.
4. Jiang, Y.H.; Guo, X.C. Drainage design for urban roads. *J. Hefei Univ. Technol. Nat. Sci.* **2007**, *10*, 1341–1344, 1348.
5. Jiang, T. Highway Drainage Techniques in Collapsible LOESS Areas. Master's Thesis, Chang'an University, Xi'an, China, 2006.

6. Akay, O. Slope stabilisation using EPS block geofoam with internal drainage system. *Geosynth. Int.* **2016**, *23*, 9–22. [CrossRef]
7. Dong, H.W.; Fu, G.H. Research on retaining wall drainage technology. *Highway* **2007**, *1*, 5–8.
8. Tian, Q.Y.; Qian, N.G.; Hao, Z.Y. Laboratory test study on mechanical silting of a new type of inclined drainage hole in highway slope. *Highway* **2016**, *7*, 39–44.
9. Ye, F.; Tian, C.M.; He, B.; Zhao, M.; Wang, J.; Han, X.B.; Song, G.F. Experimental Study on Scaling and Cloggingin Drainage System of Tunnels under Construction. *China J. Highw. Transp.* **2021**, *3*, 159–170.
10. Li, C.X.; Guo, S.C.; Gao, K. Application of geotechnical drainage materials in pavement and subgrade drainage. *J. China Foreign Highw.* **2005**, *6*, 133–136.
11. Zhang, S.L.; Feng, Q.Z.; Ying, G.G.; Zhang, D.L. Study of Causes of Highway Tunnel Leakage and Field Test of New Drainage Material. *J. Highw. Transp. Res. Dev.* **2013**, *10*, 86–91, 140.
12. Singh, J.; Horikoshi, K.; Mochida, Y.; Takahashi, A. Centrifugal tests on minimization of flood-induced deformation of levees by steel drainage pipes. *Soils Found.* **2019**, *59*, 367–379. [CrossRef]
13. Guo, Y.; Leng, W.; Nie, R.; Zhao, C.; Zhang, X. Laboratory evaluation of a new device for water drainage in roadside slope along railway systems. *Geotext. Geomembranes* **2018**, *46*, 897–903. [CrossRef]
14. Zhang, L.Q.; Yang, Z.F.; Liu, D.A.; Shang, Y.J. A new drainage technique of long-term efficiency for slope and underground engineering. *J. Eng. Geol.* **2007**, *3*, 395–398.
15. Miura, N.; Chai, J. Discharge Capacity of Prefabricated Vertical Drains Confined in Clay. *Geosynth. Int.* **2000**, *7*, 119–135. [CrossRef]
16. Santi, P.M.; Hewitt, K.; VanDine, D.F.; Barillas Cruz, E. Debris-flow impact, vulnerability, and response. *Nat. Hazards* **2010**, *56*, 371–402. [CrossRef]
17. Mininger, K.T.; Sand, P.M.; Short, R.D. 2010 AEG Outstanding Student Professional Paper Life Span of Horizontal Wick Drains Used for Landslide Drainage. *Environ. Eng. Geosci.* **2011**, *17*, 103–121. [CrossRef]
18. Crenshaw, B.A. Water Table Profiles in the Vicinity of Horizontal Drains. *Environ. Eng. Geosci.* **2004**, *10*, 191–201. [CrossRef]
19. D'Acunto, B.; Parente, F.; Urciuoli, G. Numerical models for 2D free boundary analysis of groundwater in slopes stabilized by drain trenches. *Comput. Math. Appl.* **2006**, *53*, 1615–1626. [CrossRef]
20. Cook, D.I.; Santi, P.M.; Higgins, J.D. 2007 AEG Student Professional Paper: Graduate Division: Horizontal Landslide Drain Design: State of the Art and Suggested Improvements. *Environ. Eng. Geosci.* **2008**, *14*, 241–250. [CrossRef]
21. Tao, J.L.; Tao, H. Factors Affecting Piping Erosion Resistance: Revisited with a Numerical Modeling Approach. *Int. J. Geomech.* **2017**, *17*, 04017097. [CrossRef]
22. Liu, Z.Y. Effect of Hydraulic Gradient on Clogging of Filter Layer of Landslide Drilling Drainage System. Master's Thesis, Zhejiang University, Zhejiang, China, 2020.

energies

Article

Numerical Simulation of Tail Over-Fire Air Supply of a Grate Biomass Boiler

Shidan Chi [1,*], Yan Liang [1], Weixi Chen [2], Zhen Hou [1] and Tao Luan [2,*]

1 Shandong Electric Power Engineering Consulting Institute Co., Ltd., Jinan 250061, China
2 School of Energy and Power Engineering, Shandong University, Jinan 250061, China
* Correspondence: chishidan.edu@hotmail.com (S.C.); prof.luantao@gmail.com (T.L.)

Abstract: By taking a 130 t/h water-cooled grate biomass boiler as the research object, the ANSYS software is applied to simulate the effects of the tail burnout air's incidence angle, wind speed, and pipe diameter on flow field distribution in the furnace, the movement of unburned carbon particles at the tail of the grate under cold operation. The influence rules of the incidence angle, pipe diameter, and wind speed of the tail burnout air on the combustion in the furnace and the movement of the tail particles were obtained. The results show that the setting of burnout air at the tail of the grate can better organize the air flow field at the grate tail, change the particle distribution, prevent small density particles from directly falling into the ash hopper, and prolong the residence time of particles on the grate. Under the study conditions, with an increase in the outlet velocity and the pipe diameter of the tail burnout air, as the movement degree and the burnout degree of grate particles increase, the boiler efficiency increases. With an increase in the incidence angle of the tail burnout air, the flow field in the furnace and the reduction ratio of the tail particles increases at first and then decreases, and the optimal incidence angle is 11–13°.

Keywords: grate boiler; grate tail burnout air; particle distribution; incidence angle; air distribution

check for updates

Citation: Chi, S.; Liang, Y.; Chen, W.; Hou, Z.; Luan, T. Numerical Simulation of Tail Over-Fire Air Supply of a Grate Biomass Boiler. *Energies* **2022**, *15*, 7664. https://doi.org/10.3390/en15207664

Academic Editor: Jaroslaw Krzywanski

Received: 26 July 2022
Accepted: 11 October 2022
Published: 17 October 2022

1. Introduction

Under the press of greenhouse gas emission effects and the escalating global energy demand, the study of biomass energy is currently receiving increased interest. Biomass is a kind of renewable energy with low pollution, a wide distribution, and a rich amount. It is regarded as one of the most potential and reliable renewable energy sources for future energy development [1]. Due to the adjustment of the global power structure, the proportion of renewable energy in new energy will increase from 25% to 40% by 2040 [2]. Biomass fuel is diversified, and can be converted into solid, liquid, and gas fuels. So far, the proportion of biomass energy has reached to 10–14% of the total global energy supply [3]. Meanwhile, due to the diversification of the biomass's composition, the combustion stability and combustion efficiency of biomass still need to be improved [4].

Currently, the most widely used biomass direct combustion technology in China is mainly fluidized bed combustion and grate furnace combustion [5]. The grate-fired boiler has the best adaptability, the least preparation requirements, and the highest efficiency; therefore, it has become the most popular technology among industrial boilers, and an effective process for the large-scale and efficient utilization of biomass fuel [6].

In recent years, numerical simulation has played an important role in the study of combustion characteristics and optimization design for industrial boilers. Zhang Jinhui [7] obtained flow field distribution on the grate through numerical simulation, and found that the moisture content, temperature, and density of particles on the grate had obvious faults at the junction of the drying zone and the combustion zone. Sun [8,9] and others have explored the change of bed combustion when the fuel particle diameter changes, by comparing the changes in temperature and gas emissions. Gómez [10] explored a transient model combining the CFD method and a set of Eulerian variables to simulate the biomass combustion through particle drop. The result found that the model can predict numerous processes that occur during the combustion of the solid phase, and the accuracy of the model is verified by experimental data. Kær [11] employed CFD (computational fluid dynamics) analysis in a straw burning grate furnace and found that grate combustion plays a key role in the whole process. The uneven mixing of gas components in the furnace will lead to a high concentration of CO in biomass combustion, and the unburned carbon content in fly ash is relatively high. Zhou Anqi [12] built a three-dimensional CFD furnace model according to industrial scale grate boilers. Different ratios of air supply were simulated and analyzed in this model. The results found that boiler thermal efficiency is proportional to the proportion of primary air when the primary air's ratio of the total air supply is between 43% and 50%. Meanwhile, the conversion rate of fuel has a relationship with the changing primary air volume. Shang Yuwei [13] et al. studied the influence of the primary air rate of a biomass grate furnace on bed combustion, and found that increasing the air supply in the fixed carbon combustion section is beneficial for shortening the combustion time and reducing the carbon content of ash. Liu Ruimei [14] et al. optimized and analyzed the factors of the grate furnace biomass, which are the lower secondary air operation and shutdown form, upper secondary air layout form, velocity, and others. It was found that increasing lower secondary air is beneficial for improving the slagging of the furnace arch. Adopting the opposed arrangement of the upper secondary air or increasing the secondary air speed can promote the flue gas mixing and improve the flue gas residence time, so as to improve the combustion efficiency. Lu Yanning [15] et al. studied the combustion of secondary air and recycled flue gas under the front and rear walls under different mixing ratios, and found that the flue gas recirculation air distribution mode can improve the combustion uniformity and burnout rate, and can effectively reduce the generation of nitrogen oxides.

At present, the research on the air distribution optimization of the biomass grate boiler mostly focuses on the relevant influencing factors of primary and secondary air. However, the new burnout air at the tail of the grate has not been involved. This study is motivated to employ the numerical model for evaluating the effect of adding grate tail burnout air for cold flow field on a grate boiler. The effect of different tail over-fired air supply parameters, including outlet velocity, pipe diameter, and incidence angle are investigate and optimized. Based on the comprehensive simulation results, the optimal operation angle of the over-fired tail air is recommended for the boiler under study. This study can provide theoretical guidance for the engineering transformation and efficient operation of a biomass boiler.

2. Methodology

2.1. Physical Properties of the Grate Boiler

This paper takes a 130 t/h grate biomass boiler as the research object. The furnace section size (Width × Depth) is 6.480 m × 9.2 mm. The drum center line is 23.9 mm. The primary air enters the furnace through a small hole on the furnace drainage wall through the primary air chamber below the grate. The secondary air is arranged on the front and back walls of the furnace throat. The secondary air on the front wall is divided into upper, middle, and lower levels. The biomass fuel is mainly Chinese fir leftovers, and the auxiliary fuels include pine bark, sawdust, bamboo, and other agricultural and forestry wastes. The design biomass fuel characteristics are in Table 1. The main design parameters of the grate boiler are in Table 2.

Table 1. The designed biomass fuel characteristics.

%					Q (kJ/kg)
C	H	O	N	S	
36.25	6.16	28.97	0.38	0.13	10,332

Table 2. The main design parameters of grate boiler.

Design Parameters	
Main steam flow (t/h)	130
Main steam pressure (MPa)	9.18
Main steam temperature (°C)	538
Feed water temperature (°C)	220
The primary air temperature (°C)	200
The secondary air temperature (°C)	200
The ratio of the primary and secondary air	3:7

Figure 1 shows the three-dimensional physical model of the furnace according to the actual size of the boiler structure. The overall grid division of the furnace adopts the structured grid division method, and the secondary air at the front and rear walls and the exhaust air nozzle at the tail of the grate are locally densified. This paper simulates the flow field under the original working condition, establishes 2.1 million, 2.99 million, and 4.11 million grid models, respectively, for simulation, and compares them with the actual operation data of the boiler. Finally, the number of grids is determined to be 2.99 million.

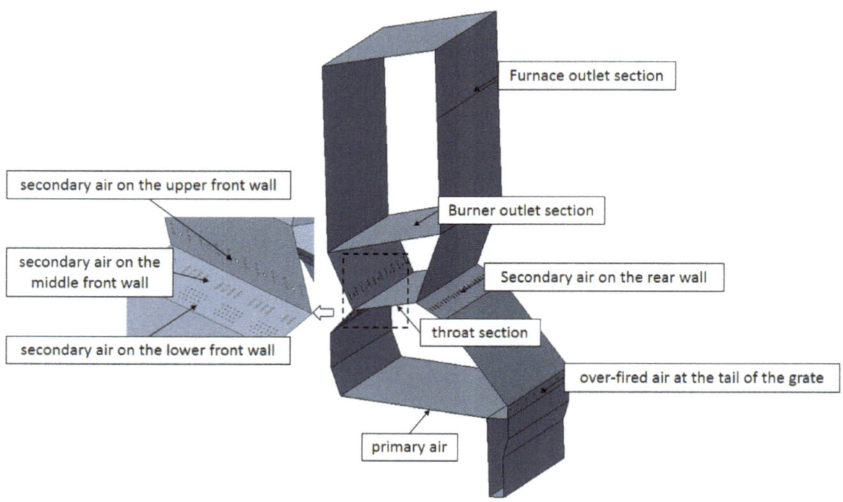

Figure 1. The sketch of the grate boiler with a steam generation rate of 130 t/h.

2.2. Numerical Model

The description of gas flow in the whole furnace is based on the solution of mass conservation equation and momentum conservation equation. The mass conservation equation and momentum conservation equation are separately presented in equations as the following:

$$\frac{\partial \rho}{\partial t} + \nabla \cdot \left(\rho \vec{v} \right) = 0 \tag{1}$$

$$\frac{\partial}{\partial t}\left(\rho \vec{v}\right) + \nabla \cdot \left(\rho \vec{v} \vec{v}\right) = -\nabla p + \nabla \cdot \left(\overline{\overline{\tau}}\right) + \rho \vec{g} + \vec{F} \tag{2}$$

where, p is static pressure, $\rho \vec{g}$ is gravitational volume, $\vec{F}$ is external volume force, and $\overline{\overline{\tau}}$ are stress tensors.

This is because the simulation area of the furnace is large, and the internal fluid flow is very strong. So, the standard k-epsilon model is selected for calculation.

$$\frac{\partial}{\partial t}(\rho k) + \frac{\partial}{\partial x_i}(\rho k u_i) = \frac{\partial}{\partial x_j}\left[\left(\mu + \frac{\mu_t}{\sigma_k}\right)\frac{\partial k}{\partial x_j}\right] + G_k + G_b - \rho\varepsilon \tag{3}$$

$$\frac{\partial}{\partial t}(\rho \varepsilon) + \frac{\partial}{\partial x_i}(\rho \varepsilon u_i) = \frac{\partial}{\partial x_j}\left[\left(\mu + \frac{\mu_t}{\sigma_\varepsilon}\right)\frac{\partial \varepsilon}{\partial x_j}\right] + C_{1\varepsilon}\frac{\varepsilon}{k}(G_k + C_{3\varepsilon}G_b) - C_{2\varepsilon}\rho\frac{\varepsilon^2}{k} \tag{4}$$

where, C_2, C_{1z}, and C_{3z} are constants, σ_K and σ_ε are turbulent Prandtl numbers of k and ε, and G_k and G_b are turbulent kinetic energy due to average velocity gradient and buoyancy.

In this paper, the gas phase combustion reaction in the furnace adopts the dense discrete phase model. The tracking of solid particles adopts the Lagrangian discrete phase model regarding the gas phase as a continuous medium. The contact force between the particles and grate is solved by the Hertz-Mindlin non-sliding model, and the interphase drag model adopts the Gidaspow model. The turbulence model in the furnace adopts standard K-ε Model. In the model, the second-order discrete format is used for the calculation of each physical quantity, and the standard wall function is used for each wall.

2.3. Assumptions and Boundaries

In this paper, the finite volume method is used for discretization. The simple algorithm is used for the pressure velocity coupling algorithm. The second-order discrete format is used for the calculation of each physical quantity. Each solid wall was analyzed using the method of the standard wall function. The other assumptions and boundaries of the model are as the following:

(1) In the original air distribution system working condition (0), the primary air ratio is 30%; the front wall upper, middle, and lower secondary air speeds are, separately, 32 m/s, 50 m/s, and 36 m/s.
(2) The back wall secondary air speed is 57 m/s.
(3) Model port setting: The flow inlet: the lower primary air outlet of the grate.
(4) The speed inlet: each secondary air outlet of the furnace; the pressure outlet: furnace outlet.
(5) The total air volume of the air distribution system is constant.
(6) The maximum ratio of burnout air is limited at 7.3%.

In order to reduce the carbon content in the slag and to improve the combustion efficiency of the boiler, an exhaust air distribution system is installed at the tail of the furnace row. The air distribution system in this paper consists of the primary air system, secondary air system, and tail burnout air system. The tail burnout air is exhausted from the secondary air and is under the condition of a certain total air volume. When the number of burnout air pipes is fixed, the outlet velocity and the pipe diameter of the burnout air will affect its proportion in the total air volume. At the same time, the changes of these two factors will also affect the flow field in the furnace and the movement of unburned carbon particles on the grate tail. Therefore, the original and the air distribution system with trail burnout air of different proportions are separately simulated and compared in this paper. The specific air distribution proportion is shown in Table 3.

Table 3. Air distribution condition of boiler cold flow field.

Numbers	Primary Air (%)	Lower Secondary Air on Front Wall (%)	Middle Secondary Air on Front Wall (%)	Upper Secondary Air on Front Wall (%)	Secondary Air on Back Wall (%)	Tail Burnout Air on Grate (%)	Outlet Velocity of Grate Burnout Air (M/S)
0	30	8	25	14	23	-	-
A30	30	8	25	14	21.35	1.65	30
A40	30	8	25	14	20.8	2.2	40
A50	30	8	25	14	20.25	2.75	50
A60	30	8	25	14	19.7	3.3	60
B60	30	8	23	14	20	5	60
C60	30	8	22	13	19.7	7.3	60

Notes: 1. A/B/C refers to the air distribution system with tail burnout air and the pipe diameter, which is ϕ 76 mm/ϕ 89 mm/ϕ 108 mm. 2. A30 is defined as trail burnout air pipe diameter, and is ϕ 76 mm, and outlet velocity is 30 m/s.

3. Analysis and Discussion

In order to investigate the influence of outlet velocity, pipe diameter, and incident angle of the exhaust air at the tail of the furnace on the cold flow field in the furnace; and the movement of unburned carbon particles on the grate tail, the cold flow field distribution in the furnace and the position distribution of unburned carbon particles on the tail grate were simulated and analyzed by changing these three factors.

3.1. The Air Distribution System without Tail Burnout Air

Figures 2 and 3 show the velocity distribution of the furnace center section and the grate tail section in the 0 working condition. It can be seen from the Figures 2 and 3 that the high-speed secondary air jet from the front and back walls does not change significantly within a certain distance after entering the furnace. After the flow field in the furnace is stable, an obvious vortex is formed near the throat. On the lower right side of the furnace, the air velocity at the grate tail is very slow; the turbulence degree and the turbulence intensity is very small, which is not conducive to the combustion of unburned carbon at the grate tail. This is the main reason for the high carbon content in the slag.

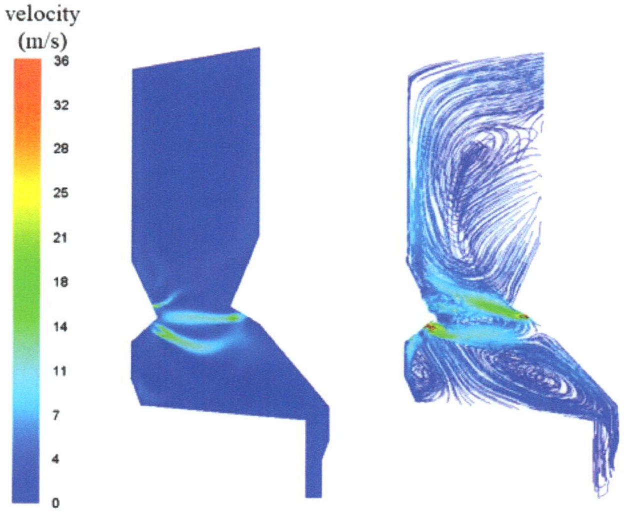

Figure 2. The velocity distribution of the furnace center section in 0 working condition.

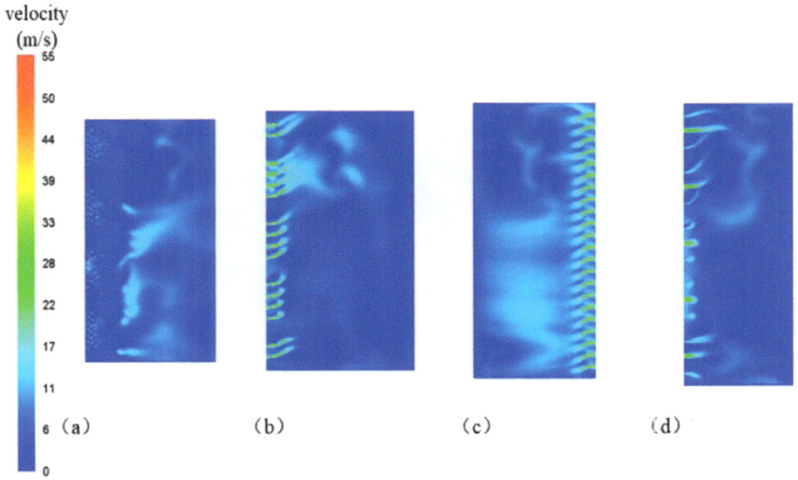

Figure 3. The velocity distribution of the grate tail section in 0 working condition. (**a**) Lower secondary air on front wall; (**b**) Middle secondary air on front wall; (**c**) Secondary air on back wall; (**d**) Upper secondary air on front wall.

3.2. Different Outlet Velocities of Grate Burnout Air

In order to explore the influence of outlet velocity of burnout air on the cold flow field in the furnace, the flow field distributions of different outlet velocities (30 m/s, 40 m/s, 50 m/s, and 60 m/s) of burnout air with pipe diameters of ϕ 76 mm are numerically simulated, and the following results are obtained.

Figures 4 and 5 show the changes of the flow field inside the furnace and the flow field at the tail of the furnace row when the velocity of the tail burnout air changes. When the pipe diameter is fixed, the turbulence degree of the flow field in the grate tail, and the radiant heat transfer and convective heat transfer in the furnace improve with the increase of the wind speed. In Figure 5d, when the outlet speed of the burnout air at the grate tail reaches 60 m/s, the secondary air on the rear wall deflects, and the front and back wall above begin to form vortices. This phenomenon can avoid air brushing of the wall. Moreover, it is beneficial to reduce the corrosion and slagging of the water-cooled wall, and improve the thermal efficiency of the boiler. The average velocity of the gas flow at the grate tail (an average velocity at 120 mm above the end of the grate along the boiler depth direction) is shown in Table 4. Under the same pipe diameter of the burnout air, the gas velocity at the tail of the furnace row changes in direct proportion to the velocity of the burnout air. When the outlet velocity of the burnout air increases from 30 m/s to 60 m/s, the average velocity of the gas flow at the tail of the furnace row increases by 91%.

Table 4. Variation of flow field at the grate tail at different flow velocities.

Numbers	Burnout Air Pipe Diameter (mm)	Average Velocity of Gas Flow at the Grate Tail (m/s)
A30	76	2.50
A40	76	2.78
A50	76	3.40
A60	76	4.79

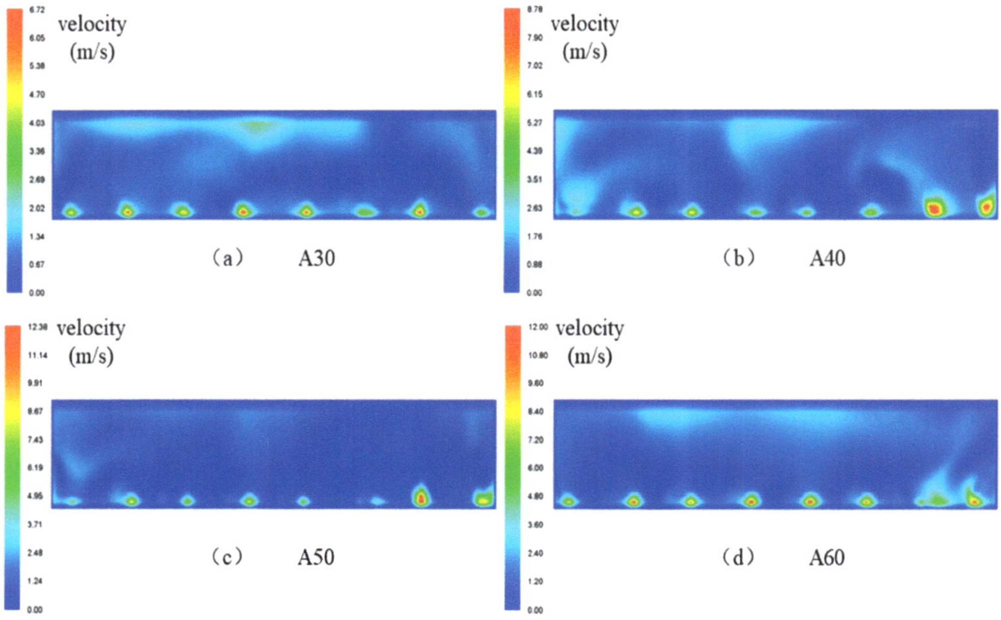

Figure 4. Velocity nephogram of vertical section at the grate tail of A30–A60.

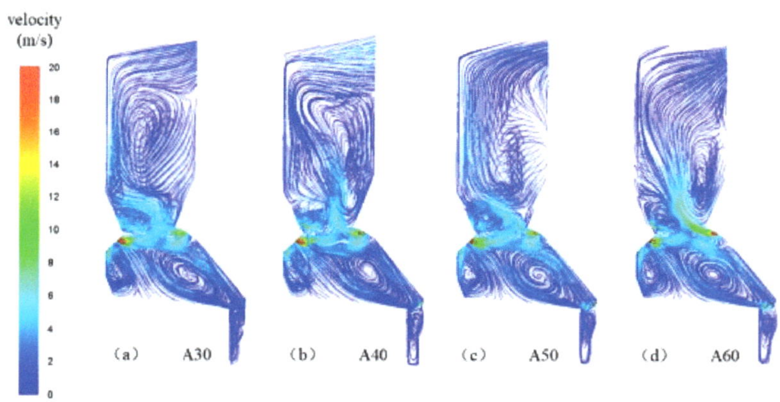

Figure 5. Flow line distribution in furnace of A30–A60.

3.3. Different Pipe Diameters of Grate Burnout Air

Furthermore, the influence of pipe diameter change on the flow field distribution in the furnace is simulated and analyzed when the wind speed of the burnout air is 50 m/s and 60 m/s. Figures 6 and 7 show the velocity cloud diagram and flow field diagram in the furnace at the longitudinal section of the grate tail under the conditions of fixed wind speed and diameter, changing from ϕ 76–ϕ 108 mm. The average speed of the gas flow at the tail of the furnace row and the speed increase ratio are given in Table 5.

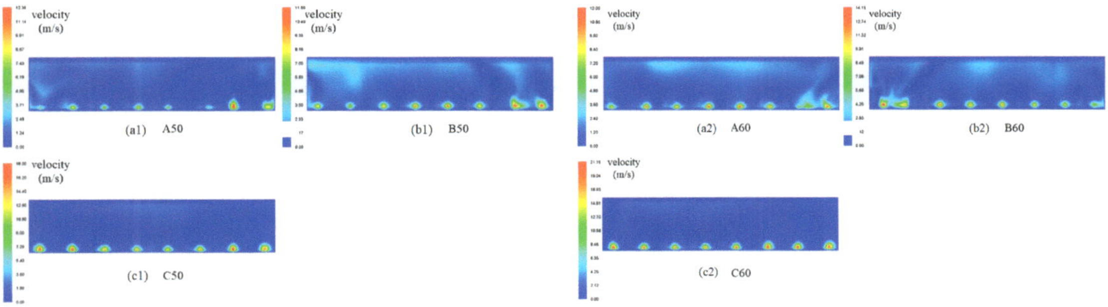

Figure 6. Velocity nephogram of vertical section at the grate tail.

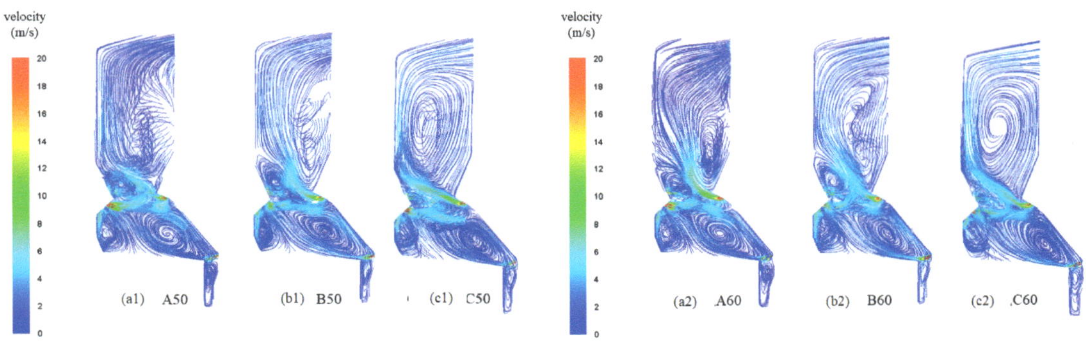

Figure 7. Flow line distribution in furnace.

Table 5. Variation average speed of the gas flow at the tail of the furnace at different pipe diameters.

Numbers	Burnout Air Pipe Diameter (mm)	Average Velocity of Gas Flow at the Grate Tail (m/s)	Ratio (%)	Ratio of Particle Quantity Reduction (%)
A50	76	3.4	0	4.76
B50	89	4.53	33	5.36
C50	108	5.08	49	7.53
A60	76	4.85	0	7.74
B60	89	5.45	12	8.33
C60	108	6.00	24	10.42

It can be seen from Figures 6 and 7 and Table 4 that the gas flow velocity at the tail of the furnace row improves with an increase in the pipe diameter of the burnout air. At the same burnout air speed, the larger the pipe diameter, the higher the increase proportion of the tail air velocity is. At the same pipe diameter, the proportion of the increase of the tail gas flow velocity is larger when the burnout air flow velocity is 50 m/s, compared to 60 m/s. This shows that it is not the higher the velocity of the burnout air, but rather, the better the velocity of the tail flow. In a certain flow rate range, increasing the diameter of the burnout air can better accelerate the flow rate of the tail gas, improve the combustion efficiency in the furnace, and reduce the unburned particles at the tail of the furnace row.

From Table 5, it can be seen that when the outlet speed of the burnout air is constant, increasing the diameter of the air duct can effectively improve the average speed of the gas flow at the tail of the furnace row. The disturbance degree of the flow field in this area increases with the increasing of the average velocity of the gas flow at the grate tail, which leads to enhancement of the combustion of unburned carbon on the grate tail. Meanwhile, the reduction in the number of particles on the tail grate decreases significantly, indicating that more particles are far away from the end of the grate and they enter the high-temperature zone of the furnace for combustion, while the residence time of particles on the grate increases. With the increase of air duct diameter, the particle density on the tail grate also decreases significantly. The oxygen content of unburned carbon in the slag increases, and the combustion intensity is enhanced. Finally, the carbon content in the slag decreases and the boiler efficiency improves.

3.4. Different Jet Angles of Grate Burnout Air

The change of the injection angle of the burnout air at grate tail will change the movement path of the air jet after entering the furnace, and affect the flow field distribution in the furnace. Therefore, under the conditions of C60, six working conditions with the incident angle of burnout air changing from 10° to 15° are simulated in this paper. The flow line distribution in the furnace is shown in Figure 8.

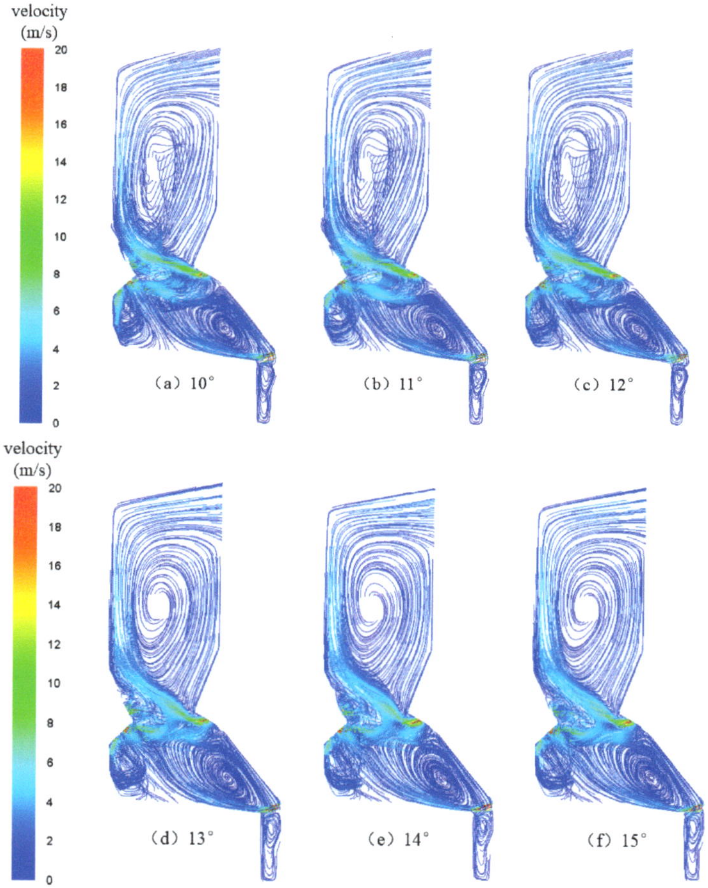

Figure 8. Flow line distribution in furnace with burnout air incident angle from 10–15°.

As can be seen from Figure 8, with the increase of the incident angle of the burnout wind, the upper half and the lower half of the furnace form vortices, respectively. The turbulence in the upper half enhances the radiation and convection heat transfer between the high-temperature flue gas and the water wall, improves the boiler efficiency, and reduces the possibility of slagging and corrosion of the water wall. The vortex under the rear arch increases the temperature of the rear arch, promotes the combustion at the tail of the furnace row, and reduces the carbon content in the slag. When the incident angle is in the range of 10–12° to 13–15°, the streamline distribution in the furnace is basically similar. Therefore, under the condition of C60, setting the incidence angle of burnout air between 12–14° is conducive to strengthening the tail flow speed and improving the combustion efficiency in the furnace.

It can be seen from Table 6 that when the incidence angle of the burnout wind changes in the range of 10–15°, taking the value at the incidence angle of 15° as the initial quantity, the average speed of the gas flow at the tail of the furnace row and the proportion of the reduction of the number of particles basically increase to a certain extent. Therefore, when the outlet speed of the burnout air is 60 m/s and the pipe diameter is 108 mm, in order to simultaneously improve the airflow speed at the tail of the furnace row and the number of particles moving to the higher temperature of the furnace, the incident angle of the burnout air can be set between 11–13°.

Table 6. Variation incident angle of the gas flow at the tail of the furnace under C60.

Incident Angle (°)	Average Velocity of Gas Flow at the Grate Tail (M/S)
10	6.96
11	7.03
12	6.60
13	6.86
14	5.87
15	6.00

4. Conclusions

In this paper, the influence of the exhaust air on the flow field at the tail of the furnace is deeply studied through numerical simulation, which establishes the foundation for further thermal simulation and provides a reference for the design and optimization of air distribution in actual operation. The following conclusions can be drawn from the above simulation of the boiler cold flow field:

(1) Adding burnout air at the tail of the grate can improve the airflow velocity at the tail of the grate, strengthen the disturbance of the flow field, and enhance the radiation and convection heat transfer in the furnace. It can create a good combustion condition for the unburned carbon on the grate tail. The high-speed burnout air can blow some particles back to the high-temperature combustion zone and prolong the residence time of particles on the grate. This phenomenon is due to the improvement of the combustion efficiency in the furnace and the reduction of carbon content in the slag.

(2) When the number of burnout air ducts at the tail of the furnace row is certain, increasing the wind speed of burnout air or expanding the diameter of burnout air can effectively improve the airflow velocity at the tail of the furnace row, increase the disturbance degree of the flow field in this area, and increase the proportion of particle reduction on the grate tail. Therefore, under the assumption of the maximum proportion of burnout air volume, the outlet velocity and the pipe diameter of burnout air should be increased as much as possible under the boundary conditions.

(3) When the down dip angle of the burnout wind changes, the air velocity at the tail of the furnace row increases within 1 m/s, but the upper flow field is improved to a certain extent. In order to obtain a greater air velocity and to increase the turbulence degree of the flow field at the grate tail, the down dip angle of the burnout wind can be set at between 11–13°.

The model presented in this work can contribute to previous works through accurate three-dimensional predictions in particle movement and the furnace cold flow field of the biomass boiler with the tail burnout air system. The simulation of a particle trajectory is complicated. In this paper, only single-sized unburned carbon particles are simulated. Further studies are necessary for investigating the operation of multi-sized particles. Furthermore, relevant experiments should be carried out to verify the results of the numerical simulations in this paper, and the tail burnout air system should be improved accordingly.

Author Contributions: Conceptualization, T.L., Z.H.; methodology, S.C., W.C.; software, W.C.; validation, Y.L., Z.H.; formal analysis, W.C.; investigation, Y.L.; resources, T.L., Z.H.; data curation, S.C., W.C.; writing—original draft preparation, S.C.; writing—review and editing, S.C., T.L.; visualization, Z.H.; supervision, T.L., Z.H.; project administration, Z.H. All authors have read and agreed to the published version of the manuscript.

Funding: The authors acknowledge financial support from the China Postdoctoral Science Foundation (2020M671983) and the Shandong Postdoctoral Innovation Project (202103077).

Conflicts of Interest: The authors declare no conflict of interest. The funders (company, etc.) had no role in the design of the study; in the collection, analyses, or interpretation of data; in the writing of the manuscript, or in the decision to publish the results.

References

1. Wang, F.; Liu, X.; Chen, L.G. Research status and development prospect of energy and high value utilization of biomass resources. *Trans. Chin. Soc. Agric. Eng.* **2021**, *37*, 219–231.
2. International Energy Agency. World Energy Outlook 2018. 2018. Available online: https://www.iea.org/weo2018/ (accessed on 13 November 2018).
3. Sánchez, J.; Curt, M.D.; Robert, N.; Fernández, J. Chapter two biomass resources. In *The Role of Bioenergy in the Emerging Bioeconomy, Resources, Technologies, Sustainability and Policy*; Lago, C., Caldes, N., Lechon, Y., Eds.; Academic Press: Cambridge, MA, USA, 2019; pp. 2–111. [CrossRef]
4. Ismail, T.M.; Abd El-Salam, M.; El-Kady, M.A.; El-Haggar, S.M. Three dimensional model of transport and chemical late phenomena on a MSW incinerator. *Int. J. Therm. Sci.* **2014**, *77*, 139–157. [CrossRef]
5. Demirbas, A. Potential applications of renewable energy sources, biomass combustion problems in boiler power systems and combustion related environmental issues. *Prog. Energy Combust. Sci.* **2005**, *31*, 171–192.
6. Lv, Y.; Jiang, D.L.; Zhao, W.J.; Liu, J.Z. Technology of Biomass Direct Combustion Power Generation and Combustion Analysis. *Power Syst. Eng.* **2011**, *27*, 4–7.
7. Zhang, J.H. Numerical Simulation of Combustion Characteristics and NOx Emission of Biomass Grate Furnace. Master's Thesis, South China University of Science and Engineering, Guangzhou, China, 2020.
8. Sun, R.; Ismail, T.M.; Ren, X.H.; El-Salam, M.A. Numerical and experimental studies on effects of moisture content on combustion characteristics of simulated municipal solid wastes in a fixed bed. *Waste Manag.* **2015**, *39*, 166–178. [CrossRef] [PubMed]
9. Sun, R.; Ismail, T.M.; Ren, X.H.; El-Salam, M.A. Influence of simulated MSW sizes on the combustion process in a fixed bed: CFD and experimental approaches. *Waste Manag.* **2016**, *49*, 272–286. [CrossRef] [PubMed]
10. Gómez, M.A.; Porteiro, J.; de la Cuesta, D.; Patiño, D.; Míguez, J.L. Numerical simulation of the combustion process of a pellet-drop-feed boiler. *Fuel* **2016**, *184*, 987–999. [CrossRef]
11. Kaer, S.K. Numerical modeling of a straw-fired grate boiler. *Fuel* **2004**, *83*, 1183–1190. [CrossRef]
12. Zhou, A.; Xu, H.; Tu, Y.; Zhao, F.; Zheng, Z.; Yang, W. Numerical investigation of the effect of air supply and oxygen enrichment on the biomass combustion in the grate boiler. *Appl. Therm. Eng.* **2019**, *156*, 550–561. [CrossRef]
13. Hu, Y.M.; Wu, C.B.; Zhu, X.C.; Shao, Y. Optimization and numerical simulation of secondary air induction into refuse incinerator. *J. Environ. Eng.* **2009**, *3*, 951–955.
14. Liu, R.M.; Liu, Y.K.; Wang, Z.H.; Liu, Y.-Z. CFD simulation and optimization of secondary air distribution for waste combustion in grate furnace. *J. Zhejiang Univ. (Eng. Sci.)* **2017**, *51*, 500–507.
15. Lu, Y.; Zhang, H.; Xu, Y.; Zhu, Y.; Wan, K.; Shao, Z.; Wang, Z. Numerical simulation of effects of flue gas recirculation on biomass combustion in grate boiler. *J. Zhejiang Univ. (Eng. Sci.)* **2019**, *53*, 1898–1906.

energies

MDPI

Article

Reliability Analysis and Life Prediction of Aging LNG Unloading Arms Based on Non-Destructive Test Data

Duc-Vu Ngo [1], Jong-Kwon Lim [2] and Dong-Hyawn Kim [3,*]

[1] Department of Ocean Science and Engineering, Kunsan National University, Gunsan 54150, Republic of Korea
[2] VARM Brain Co., Ltd., Seoul 05634, Republic of Korea
[3] School of Architecture and Coastal Construction Engineering, Kunsan National University, Gunsan 54150, Republic of Korea
* Correspondence: eastlite@kunsan.ac.kr; Tel.: +82-63-469-1862

Abstract: Unloading arms (ULAs) among seaport infrastructures are susceptible to deterioration posed by the effects of harsh marine environmental conditions. During infrastructure's service life, the deterioration of structural integrity may increase the risk of failure of infrastructure, and should be taken into account during structural reliability assessment. In this study, a simple non-destructive test (NDT) was employed to examine the structural deterioration of ULAs which were installed over 30 years ago. Then, these aging ULAs were modeled by the finite-element program, using non-destructive test data to update the thickness dimensions of structural members. Next, a reliability assessment was conducted based on the stress distribution of the main structural components under external loads, which are calculated by their relation to wind speed. Moreover, the time-dependent reliability index curve was also built by considering the deterioration function to predict the failure probability of the particular components during the remaining lifetime. The study revealed that the present condition of the ULA system was satisfactory for current loading conditions. A reliability index predicted with deteriorations factors may be a rational and appropriate approach for the assessment of aging structures needed for efficient infrastructure management.

Keywords: reliability assessment; time-dependent reliability index; aging unloading arm; non-destructive testing; wind speed

Citation: Ngo, D.-V.; Lim, J.-K.; Kim, D.-H. Reliability Analysis and Life Prediction of Aging LNG Unloading Arms Based on Non-Destructive Test Data. *Energies* **2022**, *15*, 9408. https://doi.org/10.3390/en15249408

Academic Editors: Li Tian and Jianyong Han

Received: 12 October 2022
Accepted: 9 December 2022
Published: 12 December 2022

1. Introduction

Structural capacity is generally associated with the geometry and mechanical properties of a structural component. Unfortunately, during its service life, these factors may degrade due to environmental conditions, variation in load intensity over time, and quality of periodic maintenance. The periodic safety evaluation and damage assessment of existing aging structures considering the factors that influence structural deterioration is essential. In fact, the resistance deterioration of aging structures is a stochastic process and is related to the load history (e.g., long-term load intensity may result in fatigue causing cracking), environmental conditions, operation conditions, and protective measures (e.g., the loss of section thickness for structural elements by corrosion [1]). Wang et al. [2] proposed that the deterioration of the structural model consists of two components: gradual deterioration caused by environmental actions and shock deterioration caused by severe attacks of significant loads. Moreover, there is a lot of research published in the past two decades regarding the assessment of aging structures [3–7].

Reliability analysis is a frequently used tool in terms of estimating the probability of structural safety over time or after suffering the sudden natural events such as hurricanes, earthquakes, etc. [8,9], aimed at giving quantitative information to ensure that structures maintain performance within an acceptable safety level during their remaining service life. Concerning probabilistic approaches to assessments of deteriorating structures, there are many uncertainties that may directly threaten the safety of the in-service structures. A lot of research has been conducted on developing approaches for reliability analysis. Mori and Ellingwood [10] developed methods using structural reliability principles to evaluate the time-dependent reliability of reinforced or pre-stressed concrete structures. This work was further used by Enright and Frangopol to predict the probability of safety of deteriorating bridges [11]. Rodolfo Mussini et al. [12] developed a computational structural reliability model method in conjunction with the failure-assessment diagram method and the user-defined probability of detection curves of non-destructive testing is used. Bhargava et al. [13] conducted analysis to estimate the time-dependent failure probability for a corrosion-affected RC beam and investigated the influence of variability in degradation functions on structural failure probability. Liu et al. [14] studied different failure modes under hazards that cause both progressive and shock deteriorations of structural performance. In their study, a cumulative-time approach was presented for the probabilistic structural deterioration process. Moreover, the life-cycle cost assessment and optimization approach is a practical method to predict the time-to-repair and the number of repair operations on the degrading structures [15].

This study was conducted on the ULAs of an LNG terminal (hereinafter referred to as "A") that has existed for more than 30 years in a certain area in Korea. Its storage capacity has continuously expanded over the past three decades. Seaports have largely created their sources of revenue from cargo handling, and marine unloading arms (ULAs) are an essential part. With high working intensity, infrastructure facilities, especially ULA systems, may have deteriorated. The assessment of the failure probability of this ULA system needs to be conducted. The simulation-based reliability analysis approach is considered as a suitable tool for doing this. A key factor in this approach is the model of the degraded structure, which must be accurately simulated. Ignoring the single and multiple deteriorations involved can overestimate or underestimate the safety of your infrastructure, which can lead to significant economic losses or sudden accidents. The scope of this paper is to perform a reliability analysis of existing aging ULAs subjected to stochastic wind load. NDT was performed to examine the loss of section thickness and structural integrity. In reliability analysis, the yield stress was used to define a limit state as well as random variables such as member thickness, elastic modulus, and wind speed. The wind speed distribution was derived from data collected in "A" harbor and it can be fitted into the lognormal distribution. The RSM (Response Surface Method)-based FORM (First Order Reliability Method) was used to calculate the reliability index. Then, a system reliability assessment of existing ULAs was given. Finally, a time-dependent reliability index of the most dangerous component was constructed to evaluate the failure probability of the ULA during the remaining lifetime.

2. Reliability Analysis

2.1. First-Order Reliability Method (FORM)

In reliability analysis, the limit state of interest must be defined first. This is the boundary between the safe and unsafe responses in the design parameter space. The limit-state function (LSF) can be written in the form of Equation (1), where R and S are the distributions of the resistance and load, respectively.

$$\mathrm{g}(X) = R - S \tag{1}$$

Considering the LSF of a structure, the failure probability P_f can be defined as [16]:

$$P_f = \int \cdots \int_{g(X)<0} f_X(x_1, x_2, \ldots, x_n) dx_1 dx_2 \ldots dx_n \tag{2}$$

where $f_X(x_1, x_2, \ldots, x_n)$ is the joint probability density function for the uncertain input variables of the structure $x_1, x_2, \ldots, x_n$ and $g(X) < 0$ represents the structural failure region. The probability of failure in Equation (2) is solved by integral multiples and, obviously, it is a difficult challenge. FORM provides an approximate method. In the FORM process, the LSF $g(X) = 0$ was transformed into a corresponding limit-state surface $g(u) = 0$ in standard space (U-space). This process was conducted by Rosenblatt transformation [17]. In the next step, the most probable failure point on $g(u) = 0$ has to be determined in U-space. This is the point in the failure set with the largest probability density, that is, the closest point on the failure surface to the origin of the U-space. The shortest distance between the limit-state-boundary surface and the origin in the U-space is called the reliability index β.

Once the reliability index is obtained, the probability of failure P_f is often represented in terms of the reliability index β given by:

$$P_f = 1 - \Phi(\beta) \tag{3}$$

where the notion Φ is the cumulative distribution function (CDF) of the normally distributed random variable with mean of 0 and standard deviation of 1.

2.2. Response Surface Method (RSM)

For complex and implicit LSF, the RSM [18] provides a solution that allows expressing it in explicit form. The structural LSF in the RSM can be approximated using a quadratic polynomial function as follows:

$$g(X) \cong a + \sum_{i=1}^{n} b_i x_i + \sum_{i=1}^{n}\sum_{j=i}^{n} c_{ij} x_i x_j \tag{4}$$

where x_i (i = 1, 2, ... , n) is the space variable and a, b_i, c_{ij} are the coefficients. At the beginning of process, the mean point of the random variables is chosen as the initial point. An acceptable error margin (ε) is established so that performing the analysis is as simple as possible. The design test point of the structure of each iteration step can be determined by Equation (5) [19], where X_i^C is the center point, σ_{X_i} is the standard deviation of X_i, h_i is the width, I_i is the scattering index, and i is the variable number.

$$X_i = X_i^C \pm h_i \sigma_{X_i} I_i \tag{5}$$

The analysis will be stopped at the kth iteration if the Equation (6) can be established, then it is possible to evaluate structural reliability.

$$\beta^k - \beta^{k-1} \leq \varepsilon \tag{6}$$

2.3. Reliability of Aging Structure

In realistic structures, the deterioration of structures will lead to stochastic degradation of resistance and an increased load effect. As a result, performance evaluation of structures degraded over time has become more challenging for structural engineers owing to the stochastic nature of material properties and their random change over time. The relationship between resistance and load in reliability analysis of degraded structures can be seen in Figure 1. Note that there will be uncertainty both about the precise resistance and the rate of resistance loss with time. The load change with time is not considered in this paper. Instead, the load effect is established as a stochastic process to represent its randomness.

It is evident from Figure 1 that the failure probability of a structure increased with an increase in resistance loss ratio. Therefore, in the reliability analysis, it is necessary to consider the factors affecting the degradation of structural resistance In realistic structures, there will be many resistance variables. Typically, these are random variables, and some may deteriorate with time. For example, for steel structures at seaports, the resistance of the structure is mainly dominated by the steel properties and moment of inertia including thickness (or cross-section) of the structural components. The probabilistic nature (uncertainty) of steel properties is represented by the probability density function. The uncertainty of structural thickness, besides being expressed by the probability density functions, also has its mean value defined as a function of time; see Figure 1. The structural thickness over time (residual thickness) can be inferred from the function of the amount of material loss due to corrosion. The bi-logarithmic model, Equation (7), is typically adopted to describe corrosion loss (*C*) as a function of the time of exposure, where *A* and *B* are two experimental parameters used differently by researchers. *B* can be called an environmental parameter. It depends on the atmosphere condition, the effect of protection level, etc., and is normally less than unity. A larger *B* value indicates a more severe environment, less effective protective measures, and vice versa. For the marine atmosphere, *B* = 0.69 [20]. *A* is the corrosion loss in the first year.

$$C(t) = At^B \tag{7}$$

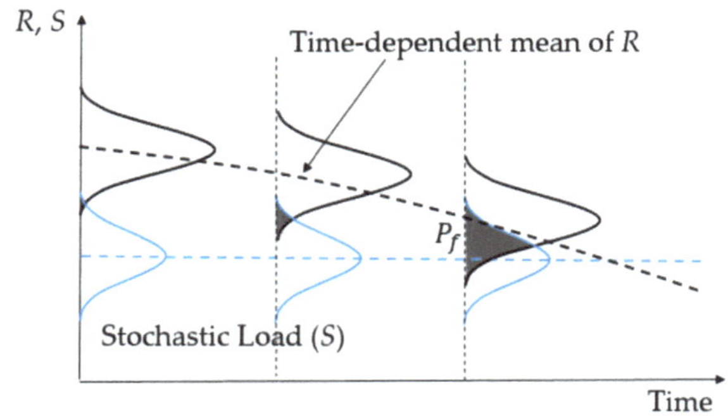

Figure 1. Stochastic process of load effect $S(t)$ and the deteriorating structural resistance $R(t)$.

2.4. System Reliability

Considering the correlation between components in the same system, the system failure probability can be conveniently calculated from the failure probabilities of the individual components. Basically, structural systems are of two types. The first is a structural system which consists of individually operated components, that is, if any of its components fail or lose function, the remaining components in the system still perform the function. Therefore, it can be seen as a functionally parallel system. In this case, the system failure probability can be calculated by the following equation:

$$P_{f,ULAs} = \prod_{1}^{n} P_{f,ULAi} \tag{8}$$

where $P_{f,ULAi}$ is the failure probability of components in the same system. The second is a structural system which consists of components that work in mutual support as a distributed system. Normal operation of the distributed system is possible only when all its components are not damaged. This means that if any of its structural components fail or lose function, the operation is discontinued. This failure probability of system can be determined based on the failure probability of each component $P_{f,i}$ by the equation below:

$$P_{f,ULA} = 1 - \prod_{1}^{n} \left(1 - P_{f,i}\right) \tag{9}$$

2.5. Target Reliability Index

The target reliability indices used by researchers are different and should be determined before performing structural reliability analysis. Such decisions will be based on the consequences of the failure, reference rules, inspection, and repair processes, etc. Ideally, the target reliability level so selected should minimize the total cost (i.e., optimize the cost of strengthening or replacement), but still comply with a certain minimum safety level. The target reliability indices corresponding to several reference rules were illustrated in Figure 2. Accordingly, the target reliability index applied in the design standards varies from 2.0 to 4.7, and is applied to general structures such as bridges, pipelines, seaport facilities, skyscrapers, etc. ISO [21] and Eurocode [22] standards recommend target reliability indices in the range of 3~4.7 for general structures. In the case of bridges, the damage to human life and property is high, so a target reliability of 3.0 or higher for the US and 3.1 or higher for Korea is applied. In this study, the ULA system was selected to perform the assessment. The failure of the ULA system here had no risk of human injury, minor economic consequences, and an insignificant impact on the environment. Therefore, it might be overly prudent and wasteful to apply reliability index levels corresponding to the reference rules listed above. Sotberg et al. [23] proposed a failure probability level, shown in Table 1, for the limit-state-based designs of pipelines. The low safety class in Table 1 can be used as a reference for target reliability evaluation. Accordingly, a failure probability level in the range of $10^{-2} - 10^{-3}$ corresponding to the target reliability index of 2.33–3.09 of the ultimate limit state was proposed as a reference for evaluation of reliability and safety assessment of unloading arms in this study. Specifically, in this study, the target reliability index was assumed to be 2.6, corresponding to failure probability 4.66×10^{-3} for performance evaluation of aging structures in the next section. As mentioned in Figure 2, the defined target reliability level is reasonable compared to the target reliability level in Japan, where earthquakes occur frequently.

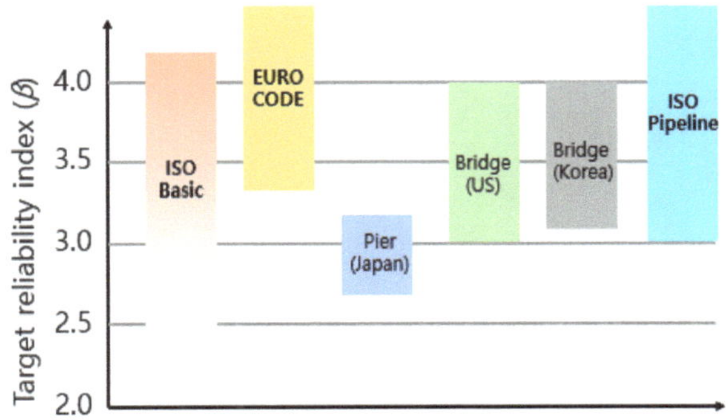

Figure 2. The target reliability indices corresponding to several reference rules.

Table 1. Failure probability level.

Limit States	Safety Classes		
	Low	Normal	High
SLS	10^{-1}–10^{-2}	10^{-2}–10^{-3}	10^{-2}–10^{-3}
ULS	10^{-2}–10^{-3}	10^{-3}–10^{-4}	10^{-4}–10^{-5}
FLS	10^{-3}	10^{-4}	10^{-5}
ALS	10^{-4}	10^{-5}	10^{-6}

3. Numerical Analysis

3.1. ULA

The structure selected for this study was the unloading arm (ULA) system installed in "A" Port 30 years ago. As shown in Table 2, it is a system composed of six marine arms divided into three main types. Here, the LNG (Liquefied Natural Gas), BOG (Boil-off gas), and Bunker unloading arms are denoted by the letters (L), (BOG), and (B/C), respectively. Since this system base was installed on a slab made of steel located about 14.5 m above sea level, it is hardly influenced by sea wave effects. Figure 3a is a picture of the ULAs taken at

"A" LNG terminal. When the components are disassembled as in Figure 3b, it is also very convenient to perform non-destructive testing (NDT) to evaluate structural condition. The ULA was operated using a hydraulic system. The main components of this ULA are made of carbon steel, including some basic components such as risers, baseplates, anchor bolts for connection to the ground, inner and outer arm supports, pipelines, and operating systems as shown in Figure 4. Its structural system and pipeline system are isolated from each other. Connected by a swivel joint, the swivel joint is the heart of the ULA and plays an important role during operation. The ULA is designed so that pipelines and swivel joints do not bear the additional load, and the load is very light. All weights and external loads are carried by structural supports. This design allows for easy replacement of deteriorated structural systems and better preservation of pipelines and operating systems, extending the life of the ULA. Only structural systems are evaluated in this study.

Table 2. Six models of the ULA system.

System	Model	Mass (Metric Ton)
LNG arm (L)	2101A, 2101B, 2101C, 2101D	52.000
BOG arm (BOG)	2102A	52.000
Bunker arm (B/C)	2103A	49.000

(**a**) (**b**)

Figure 3. (**a**) Six models of the ULA system located at the "A" LNG terminal and (**b**) image of disassembling ULA components during a major restoration.

Figure 4. Finite-element model with main components of a typical unloading arm of this study.

3.2. NDT

The ULA system suffered from a complex marine environment that significantly deteriorated the structure's performance over its service life. Ultrasonic equipment was used to measure the thickness of structural components to determine the extent of corrosion, erosion, and damage. The thickness of the test piece was calculated from this measurement.

This experiment was performed using a technical device, the SONOWALL 70 ultrasonic wall thickness meter. A total of 22 sections thicknesses were calculated along the entire structure. For each section, thickness measurements were performed at four points, as shown in Figure 5b. The minimum thickness measured in each section was taken as its representative thickness.

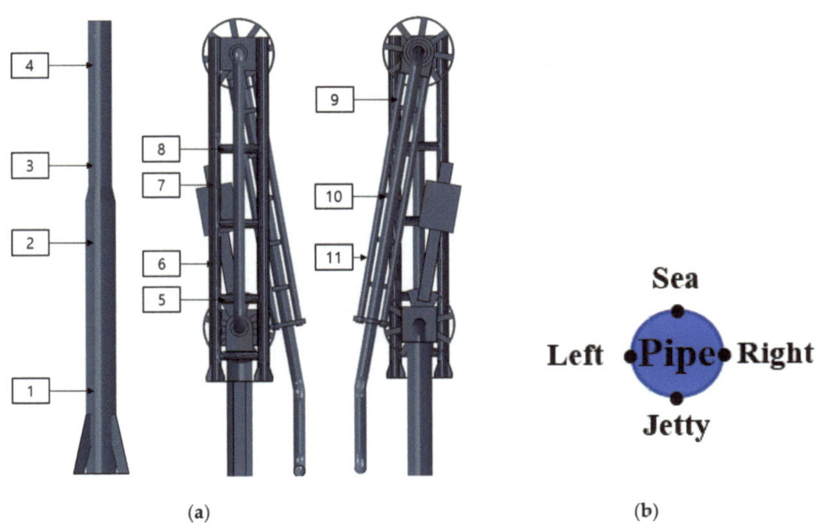

Figure 5. (**a**) Sample points of the NDT used to define the deterioration function of structure over time and (**b**) thickness-inspection positions of each section in the NDT.

Then, the aging function was determined in a time-dependent reliability analysis using the 11 positions of the NDT belonging to the major bearing components of the ULA, as shown in Figure 5a. The measured thickness of these locations is given in Table 3.

Table 3. Measured thickness from the NDT at sample points (unit: mm).

Component	Point	2101A	2101B	2101C	2101D	2102A	2103A
Riser	1	12.76	13.29	13.35	14.42	13.42	13.57
	2	12.82	12.42	13.22	13.41	13.65	13.15
	3	21.69	21.3	21.7	16.45	22.11	20.45
	4	14.21	13.3	13.44	19.63	13.2	12.82
Outer arm support	5	12.92	12.48	12.51	13.4	12.89	12.64
	6	13.34	13.58	12.96	13.08	13.36	13.26
	7	9.73	9.53	9.24	9.84	10.31	10.22
	8	12.31	12.88	12.45	13.27	13.34	12.76
Inner arm support	9	9.39	11.79	9.50	9.11	9.06	9.24
	10	14.72	14.38	14.99	14.37	15.02	14.89
	11	9.95	9.82	9.66	9.93	9.13	9.60

3.3. Wind Analysis

In this study, the foundations of the ULA system were located about 14.5 m above sea level (A.S.L.), as shown in Figure 6. Therefore, the main loads on these structures were the permanent load (self-weight) and the wind load. Due to the uncertainty of the wind speed of the harbor area, the model was established by statistical methods and considered the wind load as one of the random variables in the reliability analysis. The wind-speed data were collected in "A" harbor in Korea as described in Table 4. The data used in this study are the observation data produced by the Automatic Weather Station (AWS) at 10.97 m A.S.L. for 10 min interval from January 2001 to December 2020.

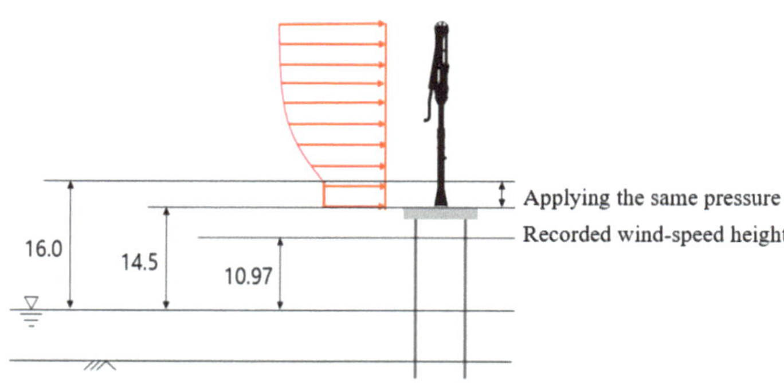

Figure 6. Schematic of the loads on the ULA system.

Table 4. Wind-speed data at "A" harbor.

Year	Wind Speed, 10 Min Mean (m/s)	Year	Wind Speed, 10 Min Mean (m/s)
2001	5.16	2011	9.18
2002	10.62	2012	11.94
2003	5.22	2013	7.75
2004	10.08	2014	11.15
2005	8.77	2015	8.57
2006	11.34	2016	15.49
2007	11.6	2017	8.27
2008	8.69	2018	12.43
2009	8.62	2019	13.24
2010	14.06	2020	13.78

Wind speed increases with height and the expression for the variation in wind speed (V) with height (h) by the following equation bellow [24]:

$$\frac{V}{V_0} = \left(\frac{h}{h_0}\right)^n \tag{10}$$

where V_0 is the mean wind speed recorded at height h_0 = 10.97 m A.S.L, V is the mean wind speed corresponding to the height h, and n is the power law exponent. Numerically, parameter n lies in the range 1/10–1.4. A typical value of n is assumed in most cases to be 1/7 [25]. However, for tower crane structures installed on the coast, an n value of 1/4 is recommended. The wind speed at the height of 16 m was applied for the structural components under 16 m.

Probability density functions are used to determine the wind potential in a specific area. There are several distributions used in modeling wind speed in the literature as Weibull, Log-normal, and Normal distributions. Given that fitting between the selected model and actual data is a question, several probability density functions should be compared to produce a minimum fitting error. In this study, the wind-speed data were expressed by the Weibull, Log-normal, and Normal distributions. Then, the Kolmogorov–Smirnov (K–S) and chi-square (χ^2) tests were used to evaluate the performances of these distributions to find out the best fitting tool for the wind-speed data.

The two-parameter Weibull distribution is the most commonly used distribution in analyzing wind speed. The probability density function (PDF) and cumulative distribution function (CDF) of Weibull distribution are expressed as:

$$f_X(x) = \frac{k}{b}\left(\frac{x}{b}\right)^{k-1} \exp\left[-\left(\frac{x}{b}\right)^k\right] \tag{11}$$

$$F_X(x) = 1 - \exp\left[-\left(\frac{x}{b}\right)^k\right] \tag{12}$$

where x is the wind speed, and k and b are the shape and scale parameters, respectively. The Log-normal distribution is also represented by two parameters: ζ_X is the scale parameter and λ_X is the shape or location parameter. The PDF and CDF for Log-normal distribution are as follows:

$$f_X(x) = \frac{1}{\sqrt{2\pi}\zeta_X x} \exp\left[\frac{1}{2}\left(\frac{\ln x - \lambda_X}{\zeta_X}\right)^2\right] \tag{13}$$

$$F_X(x) = \Phi\left(\frac{\ln x - \lambda_X}{\zeta_X}\right)^2 \tag{14}$$

The Normal PDF $f_X(x)$ and its CDF $F(x)$ are expressed, respectively, by the following formulae with the mean and standard deviation of μ_X and σ_X:

$$f_X(x) = \frac{1}{\sigma_X\sqrt{2\pi}} \exp\left[-\frac{1}{2}\left(\frac{x-\mu_X}{\sigma_X}\right)^2\right] \tag{15}$$

$$F_X(x) = \int_{-\infty}^{x} \frac{1}{\sigma_X\sqrt{2\pi}} \exp\left[-\frac{1}{2}\left(\frac{x-\mu_X}{\sigma_X}\right)^2\right] dx \tag{16}$$

Characteristics of these distributions calculated analytically from the available data are presented in Table 5. Figure 7 is the comparison of PDFs of actual and theoretical distributions. PDFs of all applied distributions are shown by different line types over the actual distribution represented in the form of a histogram. It can be observed from the naked eyes that peak densities of Weibull, Log-normal, and Normal distributions seem in line but are not in line with the peak density of the actual distribution. These qualitative comparisons also show that each PDF follows the form of a histogram, and the fitting of the two models Weibull and Normal is somewhat close. The selected distribution will be concluded based on the comparison results of goodness-of-fit statistics tests among three statistical distributions.

The K–S test gives absolute deviations between the actual distribution function (S_n) and the specified theoretical CDF (F_X). The empirical CDF for actual wind-speed data can be calculated by Equation (16) and the K–S statistics can be calculated by using Equation (17). The χ^2 test can be represented mathematically by Equation (18) [26]. The χ^2 value can be called as the overall error of probability distribution. Therefore, the lower the value of χ^2, the more accurate the model results will be.

$$S_n(x_i) = \frac{1}{N}(n_i) \tag{17}$$

where N is the size of x_i range, and n_i is the number of points smaller than x_i.

$$KS = |S_n - F_X| \tag{18}$$

$$\chi^2 = \sum \frac{(v_i - e_i)^2}{e_i} \tag{19}$$

where v_i and e_i are the observed and theoretical frequency in the ith class interval, respectively.

Figure 8 presents CDFs of the three distributions. The difference between the observation and estimation values can be seen clearly. From Figure 8, the error between the empirical CDF and theoretical CDF was obtained and shown in Figure 9. Table 6 includes the K–S test and χ^2 test values of applied distributions. Bold values indicate the best results. The results reported in Table 6 indicated that, based on the results of the K–S test, log-normal distribution was found to be the best one followed by Weibull and Normal distribution. The Weibull distribution showed better performance compared with others based on the statistics value of the χ^2 test. In this study, as a conclusion, the Log-normal distribution was chosen for the wind-speed random variable in the reliability analysis because it provides a better modeling in terms of the K–S criteria.

Table 5. Characteristics of distributions.

Distribution	Weibull	Log-Normal	Normal
Parameters	$k = 11.326$ $b = 4.296$	$\lambda_X = 2.293$ $\zeta_X = 0.297$	$\mu_X = 10.298$ $\sigma_X = 2.278$

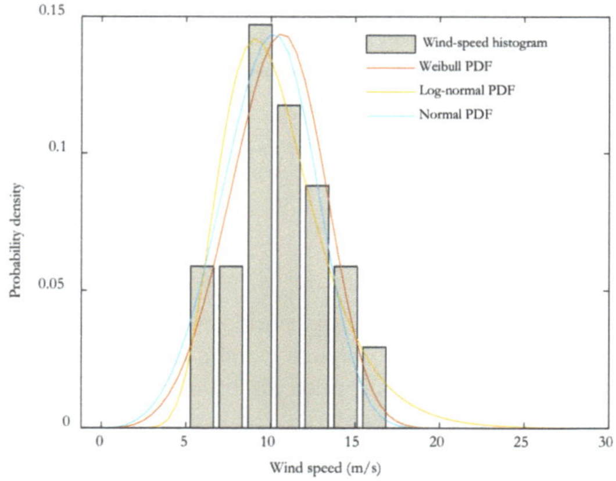

Figure 7. PDF comparison of all distributions.

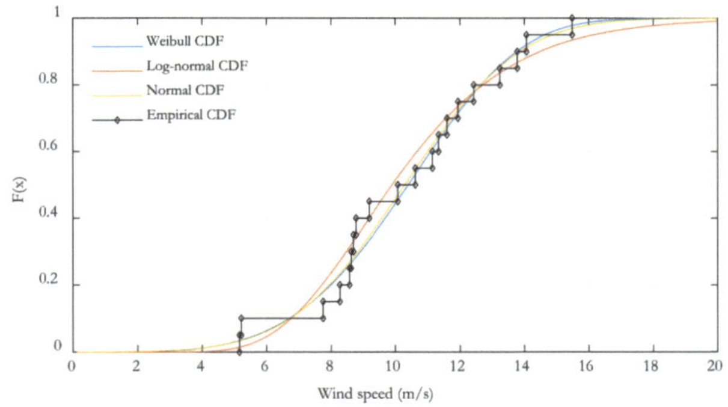

Figure 8. Comparison of CDFs.

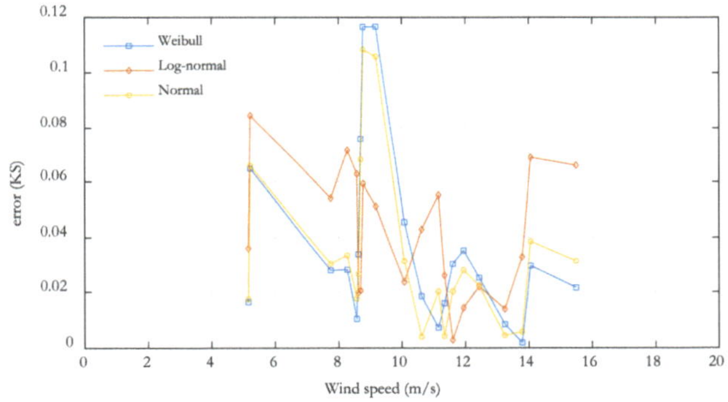

Figure 9. The relative error of CDFs between the applied distributions and actual distribution.

Table 6. Comparison of three distributions in terms of goodness-of fit test indicators.

Test	Weibull	Log-normal	Normal
K–S	0.108	**0.085**	0.117
χ^2	**2.03**	3.97	2.13

3.4. System Reliability in View of the NDT

The reliability analysis of the existing ULA system (36 age structural system) subjected to wind action is now performed. The limit-state function, Equation (1), for stress failure of the particular structural component can be written as:

$$g(X)_{36} = S_{y,36} - S_{\max,\,36}(T, E, V) \tag{20}$$

where $S_{y,36}$ and $S_{\max,\,36}$ are the yield stress and the maximum stress at time t = 36 years. T and E are the member thickness and the elastic modulus, respectively. The yield stress $S_{y,\,36}$ was defined as a random variable of which characteristic values are described in Table 7. Table 8 summarizes stochastic model assumptions of the random variables considered in the reliability analysis related to the assessment of the ultimate stress of the structural components. The notation "various" means that the value depends on the structural component being considered. Both member thickness and elastic modulus are assumed to follow a normal distribution [27,28]. Accordingly, the bias factor for thickness is taken as 1.05 with a coefficient of variation (COV) of 4.4%, whereas the bias and COV for elastic modulus are taken as 0.987 and 7.6%, respectively. The yield stress distribution is assumed as the log-normal distribution where the bias factor and COV are 1.11 and 6.8%, respectively. The distribution of wind speed was discussed in Section 3.3. Under the action of wind and self-weight, the maximum response of each structural component in terms of stress was obtained and analyzed.

Table 7. Design value of yield stress.

Component	Material Code	Yield Stress, S_y (MPa)
Riser	ASTM A516 G60	220
Base plate	ASTM A312 TP304L	170
Anchor bolts	ASTM A312 TP304L	170
Outer arm support	ASTM A516 G60	220
Inner arm support	ISO 898-1 8.8	640

Table 8. Probabilistic models of the random variables used in the reliability analysis.

Random Variable	Distribution Parameters			Type of Distribution
	Mean	Bias Factor	COV	
T (Thickness, mm)	Various	1.05	0.044	Normal
E (Elasticity Modulus, GPa)	200	0.987	0.076	Normal
V (Wind speed, m/s)	λ_X = 2.293, ζ_X = 0.297, see Table 5			Log-normal
S_y (Yield stress, MPa)	Various	1.11	0.068	Log-normal

The reliability index is a parameter that provides a measure of structural safety. As presented in Section 2, in this study, the RSM (Response Surface Method)-based FORM was used to calculate the reliability indices. The basic methodology in the form of a detailed flowchart can be summarized as shown in Figure 10. Most of the reliability index simulations converged after three iterations.

Tables 9 and 10 are the reliability indices and failure probability results, respectively, for five major structural components, namely, riser, base plate, anchor bolts, and inner and outer arm supports, corresponding to each model of the ULA. Figure 11 summarizes the reliability indices of ULAs classified by components. From Table 9 and Figure 11, it can be seen that the minimum reliability index corresponds to the riser component, which accounts for the damage caused by wind phenomena. Risers with the function of connecting the upper structure to the ground, bearing the entire weight, were particularly vulnerable. For the six ULA models, reliability ranges were 2.465–2.882 for risers, 5.136–5.203 for baseplates, 7.262–7.346 for anchor bolts, 4.049–4.088 for external arm supports, and 3.654 to 3.796 for internal arm supports, respectively. Obviously, the reliability indices depended on the particular component being considered. However, in general, the reliability indices corresponding to each ULA model increase in the order of riser, inner arm support, outer arm support, base plate, and anchor bolt. Anchor bolts have the highest reliability indices followed by the base plate, which implies that these two components are less vulnerable to wind action. This is because these two components are located on the foundation surface, convenient for maintenance and periodic inspection. In the evaluation, the deterioration state of structural components was considered in consideration of inspection data by an NDT, but the anchor bolt and base plate components were not deteriorated due to good maintenance such as early replacement. Their reliability index can here be seen as the original or design reliability index. Comparing the reliability index target with the results in Table 9, it can be concluded that, presently, all structural components are safe since their reliability values are higher than the target reliability index value (2.6). Some evaluations were also similarly given in terms of failure probability. The failure probability of the riser was the highest and the lowest for the anchor bolt component. In general, the failure probabilities of the entire ULA system are low, within the safe range ($<4.66 \times 10^{-3}$).

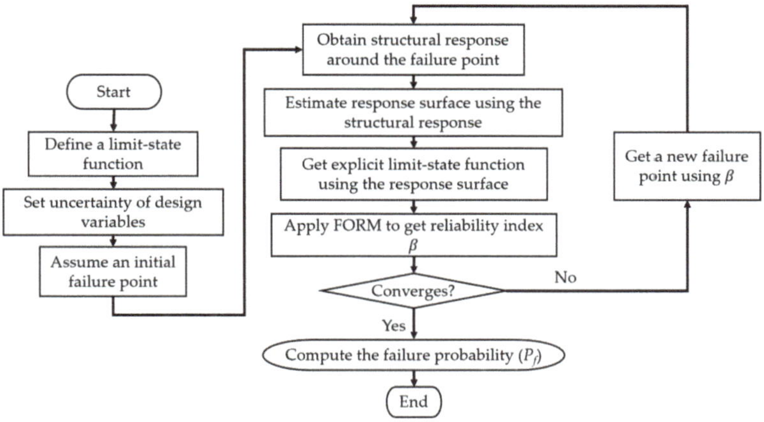

Figure 10. Flowchart for reliability analysis of an unloading arm.

Table 9. Reliability index results of structural components.

Component	2101A	2101B	2101C	2102A	2101D	2103A
Riser	2.716	2.645	2.805	2.85	2.882	2.792
Base plate	5.194	5.136	5.203	5.189	5.193	5.147
Anchor bolts	7.324	7.338	7.301	7.262	7.317	7.346
Outer arm support	4.055	4.088	4.057	4.049	4.05	4.053
Inner arm support	3.694	3.674	3.654	3.705	3.796	3.778

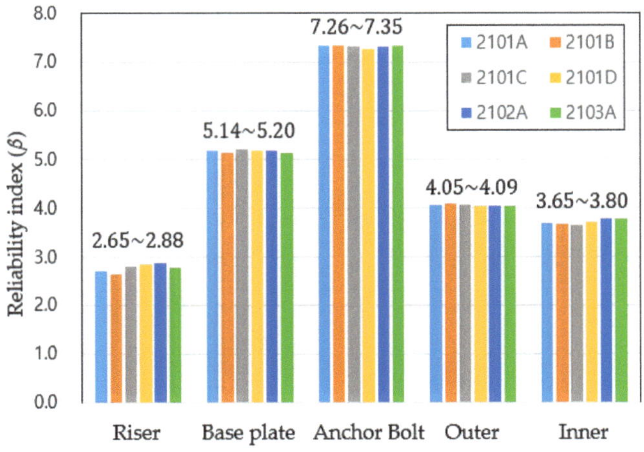

Figure 11. Summary of the reliability index of ULAs classified by components.

Table 10. Probability failure results of structural components.

Component	2101A	2101B	2101C	2102A	2101D	2103A
Riser	3.30×10^{-3}	4.09×10^{-3}	2.52×10^{-3}	2.18×10^{-3}	1.98×10^{-3}	2.62×10^{-3}
Base plate	1.03×10^{-7}	1.41×10^{-7}	9.83×10^{-7}	1.06×10^{-7}	1.04×10^{-7}	1.32×10^{-7}
Anchor bolts	1.21×10^{-13}	1.08×10^{-13}	1.43×10^{-13}	1.91×10^{-13}	1.27×10^{-13}	1.03×10^{-13}
Outer arm support	2.51×10^{-5}	2.18×10^{-5}	2.49×10^{-5}	2.57×10^{-5}	2.56×10^{-5}	2.52×10^{-5}
Inner arm support	1.10×10^{-4}	1.19×10^{-4}	1.29×10^{-4}	1.06×10^{-4}	7.36×10^{-4}	7.91×10^{-4}

It can be seen that the normal operation of ULAs is possible only when all structural members are not damaged. Thus, each individual ULC can be considered as a distributed system of its components, as illustrated in Figure 12. Hence, the failure probability of each ULA can be determined as in Equation (9), in which $P_{f,i}$ is the failure probability of each component (see Table 10). In this study, only failure probability of five main structural components (including riser, base plate, anchor bolt, and outer and inner arm support) was considered, so the overall failure probability of each ULA model can be considered to be calculated from the failure probability of the five components. The results are shown in Table 11. As seen, the highest and lowest failure probabilities corresponded to the 2101B model of the LNG arm system and 2102A model of the BOG arm system. On the other hand, it can be seen that the failure probability of the models in the LNG arm system is different even though they are all of the same model type, are installed in the same location, and perform the same functions. This difference is the result of the difference in corrosion loss from location to location even if they experienced the same corrosion environment and time. This leads to the conclusion that assessing the reliability of port infrastructures should be implemented case-by-case. The four LNG arms (LA-2101A to LA-2101D) operate on the same system and perform the same functions. However, if one of them fails, the remaining ULAs in the system still function, so they can be seen as functionally parallel systems.

Since this LNG arm system loses its function if all four ULAs in the system fail at the same time, the failure probability of this LNG arm system can be calculated by Equation (8), where $P_{f,ULAi}$ is the failure probability of ULA models in the same system (see Table 11). The BOG arm (BOG) and Bunker arm (B/C) systems consist of only one unloading arm, as in Figure 13, so the failure probability of the ULA is the same as the failure probability of systems. The results of failure probability and corresponding reliability index of systems are given in Table 12. It can be seen that the failure probability of the LNG arm system is very low. This can be explained in that four ULA models failing at the same time is extremely rare.

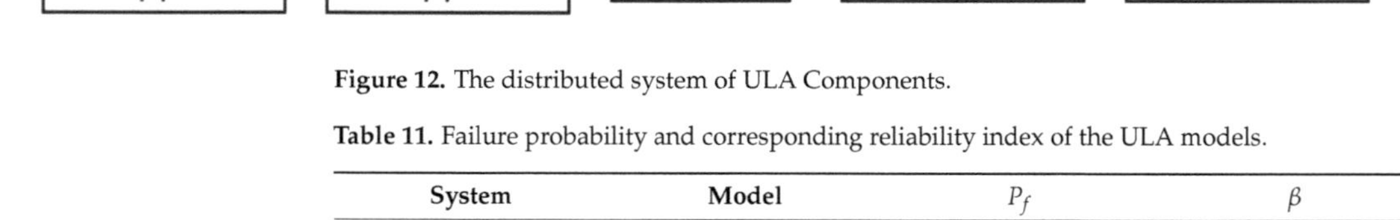

Figure 12. The distributed system of ULA Components.

Table 11. Failure probability and corresponding reliability index of the ULA models.

System	Model	P_f	β
LNG arm (L)	2101A	3.436×10^{-3}	2.703
	2101B	4.228×10^{-3}	2.633
	2101C	2.673×10^{-3}	2.785
	2101D	2.314×10^{-3}	2.832
BOG arm (BOG)	2102A	2.077×10^{-3}	2.866
Bunker arm (B/C)	2103A	2.723×10^{-3}	2.779

Table 12. Failure probability and corresponding reliability index of the ULA systems.

System	P_f	β
LNG arm (L)	8.984×10^{-11}	6.378
BOG arm (BOG)	2.077×10^{-3}	2.866
Bunker arm (B/C)	2.723×10^{-3}	2.779

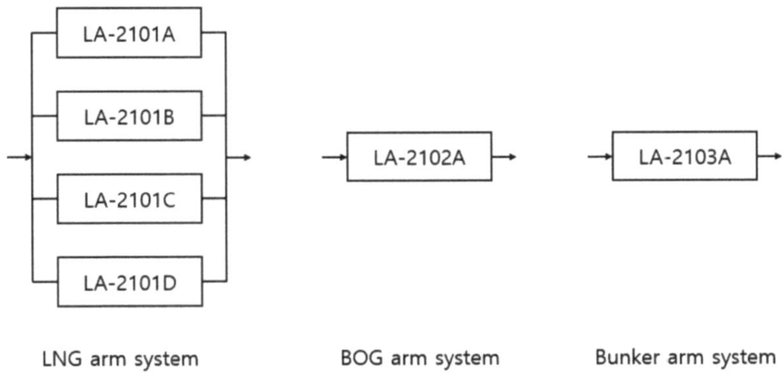

Figure 13. The parallel system of ULAs.

3.5. Time-Dependent Reliability of Riser

As an example of the application of reliability in practical engineering, in this section, the most vulnerable structural component of each ULA model is analyzed based on the proposed time-dependent reliability analysis process. The riser has been shown to be the most vulnerable component to environmental deterioration and wind action, and it has the lowest reliability index compared with other components, in the range of 2.645–2.882 (see Table 9). Accordingly, the change in the reliability index of the riser over time was investigated, and analysis was performed until the reliability index reached the target reliability index limit (2.6).

As discussed before, the deterioration of structures will lead to degradation of resistance and an increased load effect. In this study, only the mean value of the member thickness variable is considered as a mean time-dependent resistance. With other factors such as yield stress and elastic modulus, their mean values are assumed to be constant over time. In the coastal marine environment, corrosion caused by oxidation is a major concern affecting the durability of steel structures. This study assumed that general corrosion occurs in the entire outer shell. In practice, corrosion is a stochastic process managed by many variables such as location, maintenance, environmental condition, etc. Therefore, only probabilistic models can describe the thickness loss. For this reason, the remaining thickness member of structural components was assumed to follow a normal distribution with two parameters: bias and COV are 1.05 and 0.044, respectively. The mean value of the remaining thickness member at any time can be derived from the bi-logarithmic model, which is typically used to describe the general corrosion of materials, as follows:

$$T(t) = T_0 - C(t) \tag{21}$$

where $C(t) = At^B$ is the corrosion loss as a function of time and is presented in Section 2.3. Accordingly, B = 0.69 was assumed and A is the corrosion loss in the first year. With aging structures, and missing or lost recorded data, this coefficient determination is difficult and usually assumed or obtained from laboratory observations. However, this method is still of limited usefulness. This study proposes to use NDT data to determine this parameter. The results are shown in Table 13. Note that, in the Table, T_0 is the initial thickness member, and *T(36)* is the result of an NDT. With the obtained coefficients A and B, the relationship between corrosion loss (mean corrosion depth) and exposure period is illustrated in Figure 14 where the value denoted by the square is the initial corrosion loss (=0) and the values denoted by the circle are the measured corrosion loss at the time of assessment *C(36)*.

It is evident from Figure 14 that the corrosion loss increased with the exposure period, but the loss rate became slower. This could be the effect of protective measures such as anticorrosion paint, routine maintenance, etc. This means that, if a more effective protection level of a steel pipe is provided, then a lower rate of corrosion may be obtained.

Table 13. Determining the A parameter.

Components	Point	T_0	$T(36)$	$C(36)$	B	A
Riser	1	13	12.76	0.24		0.0202
	2	13	12.82	0.18		0.0152
	3	22	21.69	0.31		0.0262
	4	14.5	14.21	0.29		0.0245
Outer arm support	5	13	12.92	0.08	0.69	0.0067
	6	13.5	13.34	0.16		0.0135
	7	10	9.73	0.27		0.0228
	8	12.5	12.31	0.19		0.016
Inner arm support	9	9.5	9.39	0.11		0.0093
	10	15	14.72	0.28		0.0236
	11	10	9.95	0.05		0.0042

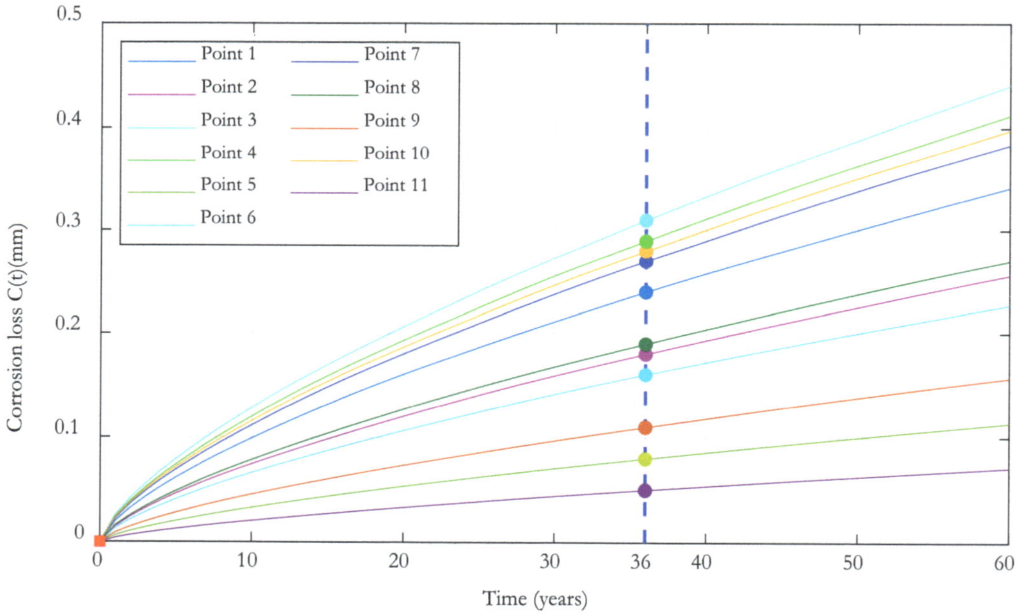

Figure 14. Time-dependent corrosion loss.

The results are presented in Table 14 and Figure 15. As illustrated in Table 14 and Figure 15, one can see how the reliability index of the initial structure changes over time under the effect of degradation. General corrosion is considered as the major factor leading to the deterioration of the structure. These reliability index curves can be divided into three phases: the current reliability index (36 age), the past reliability index (before 36 age), and the future reliability index (after 36 age). Furthermore, if the current protective measures and routine maintenance are maintained, when the target reliability index was assumed to be 2.6, the riser had around a 40.7-year lifetime for the 2101B model, and 47.3, 52, 54, 58, and 50.9-year lifetimes for 2101A, 2101C, 2101D, 2102A, and 2103A models, respectively. The corresponding remaining service life (RSL) is also summarized in Table 15. This result provides quantitative prediction useful for planning maintenance or replacement to increase structural service life.

Table 14. Reliability analysis results.

Time (Year)	2101A	2101B	2101C	2101D	2102A	2103A
0	2.832	2.769	2.916	2.962	3.028	2.875
10	2.827	2.766	2.907	2.959	3.022	2.873
20	2.807	2.739	2.888	2.944	2.991	2.863
30	2.759	2.688	2.846	2.903	2.931	2.833
36	2.716	2.645	2.805	2.850	2.882	2.792
40	2.679	2.607	2.765	2.809	2.841	2.750
45	2.627	2.554	2.705	2.741	2.782	2.687
50	2.563	2.494	2.633	2.664	2.718	2.614
55	2.482	2.431	2.547	2.582	2.649	2.527
60	2.397	2.364	2.449	2.485	2.566	2.431

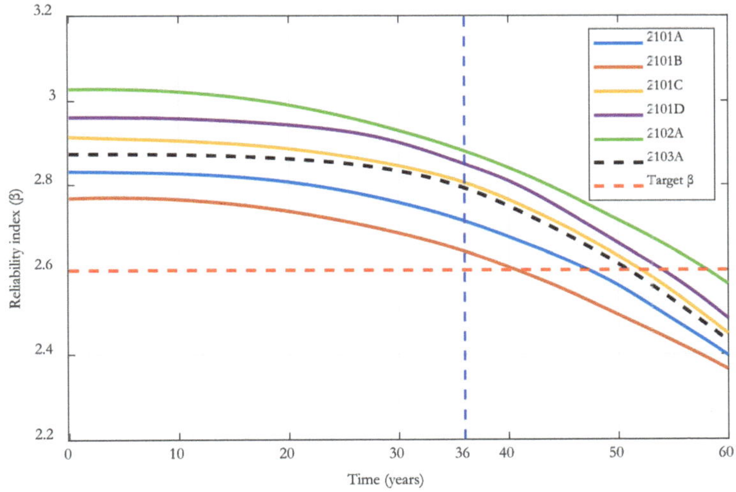

Figure 15. Reliability index as a function of time.

Table 15. Predicted remaining service life of Riser based on the assumption of target β of 2.6.

ULA Model	2101A	2101B	2101C	2101D	2102A	2103A
RSL (years)	11.3	4.7	16.0	18.0	22.0	14.9

4. Conclusions

The paper has described the process and models used for reliability analysis of the aging ULA systems in the harsh marine environment. Both the load effect (stress) and the acceptable limit (yield stress) were modeled based on stochastic processes to consider deterioration. In particular, an appropriate probabilistic model that describes the loss of material due to corrosion was proposed by combining the bi-logarithmic model $C(t) = At^B$, typically used to describe corrosion loss as a function of time, with NDT data. This method is relatively simple but still provides reliable results. The results from the reliability index analysis of the individual components of the existing ULA system indicated that, presently, all structural components are safe since their reliability values are above the target reliability index. On the other hand, a process was also established to evaluate the failure probability of a system based on the correlation between components in the same system. In addition, the results from the time-dependent reliability index analysis provided a quantitative prediction of the maximum remaining service time of the most vulnerable component. This prediction is useful for repair and replacement planning and infrastructure management. By this approach, the remaining service time of other components or other aging infrastructure is conveniently obtained.

Author Contributions: Conceptualization, D.-V.N. and D.-H.K.; methodology, D.-V.N. and D.-H.K.; software, D.-V.N.; validation, D.-H.K.; formal analysis, D.-V.N. and J.-K.L.; data curation, D.-V.N. and J.-K.L.; writing—original draft preparation, D.-V.N.; writing—review and editing, D.-V.N. and D.-H.K.; supervision, D.-H.K. All authors have read and agreed to the published version of the manuscript.

Funding: This research received no external funding.

Data Availability Statement: Not applicable.

Conflicts of Interest: The authors declare no conflict of interest.

References

1. Melchers, R.E. The effect of corrosion on the structural reliability of steel offshore structures. *Corros. Sci.* **2005**, *47*, 2391–2410. [CrossRef]
2. Wang, C.; Zhang, H.; Li, Q. Reliability assessment of aging structures subjected to gradual and shock deteriorations. *Reliab. Eng. Syst. Saf.* **2017**, *161*, 78–86. [CrossRef]
3. Ciampoli, M. Time dependent reliability of structural systems subject to deterioration. *Comput. Struct.* **1998**, *67*, 29–35. [CrossRef]
4. Vu, T.K.A.; Mark, G.S. Structural reliability of concrete bridges including improved chloride-induced corrosion models. *Struct. Saf.* **2000**, *22*, 313–333. [CrossRef]
5. Mark, G.S.; John, A.M. Spatial time-dependent reliability analysis of corrosion damage and the timing of first repair for RC structures. *Eng. Struct.* **2007**, *29*, 1457–1464. [CrossRef]
6. Li, Q.; Wang, C. Updating the assessment of resistance and reliability of existing aging bridges with prior service loads. *J. Struct. Eng. ASCE* **2015**, *141*, 04015072. [CrossRef]
7. Wang, C.; Michael, B.; Bilal, M.A. Time-dependent reliability of aging structures: Overview of assessment method. *J. Risk Uncertain. Eng. Syst.* **2021**, *7*, 3121003. [CrossRef]
8. Frangopo, D.; Dong, Y.; Sabatino, S. Bridge life-cycle performance and cost: Analysis, prediction, optimisation and decision-making. *Struct. Infrastruct. Eng.* **2017**, *13*, 1239–1257. [CrossRef]
9. Akiyama, M.; Dan, M.F.; Hiroki, I. Toward life-cycle reliability-, risk- and resilience-based design and assessment of bridges and bridge networks under independent and interacting hazards: Emphasis on earthquake, tsunami and corrosion. *Struct. Infrastruct. Eng.* **2019**, *16*, 26–50. [CrossRef]
10. Mori, Y.; Ellingwood, B.R. Reliability-based service-life as assessment of aging concrete structures. *J. Struct. Eng. ASCE* **1993**, *119*, 1600–1621. [CrossRef]
11. Michael, P.E.; Dan, M.F. Service-life prediction of deteriorating concrete bridges. *J. Struct. Eng.* **1998**, *124*, 309–317.
12. Mussini, R.; Cancela, H.; Clarke, T. Structural reliability assessment of cracked pipes: The role of probability of detection data. *Fatigue Fract. Eng. Mater. Struct.* **2019**, *42*, 664–673. [CrossRef]
13. Bhargava, K.; Mori, Y.; Ghosh, A.K. Time-dependent reliability of corrosion-affected RC beams. Part 2: Estimation of time-dependent failure probability. *Nucl. Eng. Des.* **2011**, *241*, 1385–1394. [CrossRef]
14. Liu, Y.; Dan, M.F. Time-dependent reliability assessment of ship structures under progressive and shock deteriorations. *Reliab. Eng. Syst. Saf.* **2018**, *173*, 116–128. [CrossRef]
15. Sajedi, S.; Huang, Q. Reliability-based life-cycle-cost comparison of different corrosion management strategies. *Eng. Struct.* **2019**, *186*, 52–63. [CrossRef]
16. Haldar, A.; Mahadevan, S. *Reliability Assessment Using Stochastic Finite Element Analysis*; John Wiley & Sons: New York, NY, USA, 2000; pp. 56–60.
17. Murray, R. Remarks on a multivariate transformation. *Ann. Math. Stat.* **1952**, *23*, 470–472.
18. Gerhart, I.S.; Christian, G.B.; Ulrich, B.; Winai, O. On efficient computational schemes to calculate structural failure probabilities. *Probabilistic Eng. Mech.* **1989**, *4*, 10–18. [CrossRef]
19. Kim, D.-H.; Lee, S.-G. Reliability analysis of offshore wind turbine support structures under extreme ocean environmental loads. *Renew. Energy* **2015**, *79*, 161–166. [CrossRef]
20. Feliu, S.; Morcillo, M.; Feliu, S. The prediction of atmospheric corrosion from meteorological and pollution parameters—II. Long-term forecasts. *Corros. Sci.* **1993**, *34*, 415–422. [CrossRef]
21. *EN 1990*; Eurocode Basis of Structural Design. European Committee for Standardization (CEN): Brussels, Belgium, 2002.
22. *ISO 2394*; General Principles on Reliability for Structures. International Organization for Standardization: Geneve, Switzerland, 2015.
23. Sotberg, T.; Moan, T.; Bruschi, R.; Jiao, G.; Mork, K.J. The super B project: Recommended target safety levels for limit state based design of offshore pipelines. In Proceedings of the 16th International Conference on Offshore Mechanics and Arctic Engineering, Yokohama, Japan, 13–18 April 1997.

24. James, F.M.; Jon, G.M.; Anthony, L.R. *Wind Energy Explained: Theory, Design and Application*; Wiley: New York, NY, USA, 2002.
25. Sunday, O.O.; Muyiwa, S.A.; Samuel, S.P. Analysis of wind speed data and wind energy potential in three selected locations in south-east Nigeria. *Int. J. Energy Environ. Eng.* **2012**, *3*, 7. [CrossRef]
26. Eugene, C.M.; Matthew, L.; Richard, M.V.; Laurie, G.B. Probability distributions for offshore wind speeds. *Energy Convers. Manag.* **2011**, *52*, 15–26. [CrossRef]
27. Sonja, K.; Jasmina, M.; Goran, S.; Djordje, V.; Branko, T. Uncertainty in the determination of elastic modulus by tensile testing. *Eng. Sci. Technol. Int. J.* **2022**, *25*, 100998. [CrossRef]
28. Paul, E.H.; Daniel, B.; Ibrahim, A.A.; Bilal, M.A. Uncertainties in material and geometric strength and load variables. *Nav. Eng. J.* **2002**, *114*, 139–166. [CrossRef]

energies

Article

A Study of the Segment Assembly Error and Quality Control Standard of Special-Shaped Shield Tunnels

Peinan Li [1,*], Zeyu Dai [1], Xi Wang [2,*], Jun Liu [3], Yi Rui [2], Xiaojun Li [2], Jie Fan [4] and Peixin Chen [4]

1 College of Environmental Science and Engineering, Donghua University, Shanghai 201620, China; 2201977@mail.dhu.edu.cn
2 College of Civil Engineering, Tongji University, Shanghai 200092, China; ruiyi@tongji.edu.cn (Y.R.); hylw2022@126.com (X.L.)
3 College of Urban Railway Transportation, Shanghai University of Engineering Science, Shanghai 201620, China; junliurs@sues.edu.cn
4 Shanghai Tunnel Engineering Co., Ltd., Shanghai 200032, China; zyhylw2021@163.com (J.F.); pnhylw2021@163.com (P.C.)
* Correspondence: lipeinan@dhu.edu.cn (P.L.); xiwang_chn@foxmail.com (X.W.)

Abstract: Large-section special-shaped shield tunnels feature many advantages, such as versatility and a large space utilization rate in energy transmission and public transport; however, guaranteeing the quality of segments' assembly is difficult. Based on the quasi-rectangular shield tunnel project of Hangzhou Line 9 in China, this study investigated the formation mechanism and control measures of lining segment assembly defects. By quantifying the manufacturing error and positioning error, a simulation program for segment assembly is developed to calculate the error. Furthermore, considering the relative accumulative error between the upper and lower T blocks, the finite element model of key blocks (T-LZ block) is established to perform mechanical analysis, based on which the relative error control standard of the key block under the corresponding working conditions is proposed. The results show that the assembly quality can be effectively improved by assembling the LZ block first and applying corresponding error control measures, and the displacement of the segment along the rZ direction should be carefully controlled during the construction. The error caused by normal assembly will not damage the LZ block, and the corresponding control standard under the action of multi-degree-of-freedom error (extreme case) is 9.8 mm. Using this method to predict the assembly quality of segments can provide a basis for actual construction control of segment assembly.

Keywords: quasi-rectangular shield tunnel; lining segments; assembly quality; error analysis; control standard

Citation: Li, P.; Dai, Z.; Wang, X.; Liu, J.; Rui, Y.; Li, X.; Fan, J.; Chen, P. A Study of the Segment Assembly Error and Quality Control Standard of Special-Shaped Shield Tunnels. *Energies* **2022**, *15*, 2512. https://doi.org/10.3390/en15072512

Academic Editor: Dimitrios Katsaprakakis

Received: 16 February 2022
Accepted: 27 March 2022
Published: 29 March 2022

1. Introduction

At present, the construction of large-section special-shaped tunnels has become an important part of energy and transportation infrastructure construction. Special-shaped tunnels used as heat exchangers in the form of energy tunnels can sustainably meet the thermal energy needs of buildings across city districts [1–3]. Compared with circular tunnels, special-shaped tunnels can provide a larger available space and energy utilization rate for energy equipment. With an equal effective use space as a circular tunnel, the special-shaped tunnel can save more than 35% of the underground space [4,5]. Due to its special geometry, a large-section special-shaped shield tunnel is prone to lining segment damage during assembly, which not only affects the safety and durability of the tunnel structure but also increases the maintenance cost during the service period of the tunnel [6]. Engineering practice shows that assembly defects are closely related to segment damage. The damage is aggravated, and the stability of the tunnel structure is significantly affected under a special ground load [7–10]. In the process of shield tunnel construction, assembly defects will accumulate. If there are no corresponding control measures, the number of assembly defects increases ring by ring, and finally some serious problems, such as an inability to assemble segments, are induced.

Chen and Mo [11] assessed the segment damage of a shield tunnel and proposed that the biggest cause of segment damage is assembly defect. Reducing segment manufacturing and assembly error can effectively abate segment damage. Some scholars also believe that segment damage is mostly caused by longitudinal dislocation derived from assembly defects [12–14]. The essential result of segment damage caused by assembly defects is that it affects the contact state of segments [15]: the smaller the contact area between segments, the worse the bearing capacity. Cavalaro [16] proposed the use of contact deficiency as an index to measure assembly defects, and established 2D and 3D analytical and finite element models to quantitatively analyze the rotation limit of joints. Xu and Zhang [17] performed a statistical study on the large-scale lining segment cracking in a channel project crossing a river. Combined with numerical analysis and engineering experience, they proposed that the assembly defect is one of the main causes of segment cracking [18]. Most of the above studies only performed mechanical analysis on the mechanism of segment damage, and studied the control range of segment damage through a structural test and failure mode during actual construction. The obtained control standards are mainly based on structural service requirements (mainly about result control, for example, controlling the opening of joints within a certain value), and did not include the issue of error control. The formation mechanism of assembly defects and the quantitative relationship between various factors have been scarcely analyzed.

Based on the research on the assembly defects of shield tunnels, the main causes of assembly defects can be classified as design error, manufacturing error, or assembly positioning error [19,20]. Jeang [21] and Wu [22] believe that when the control details are unknown, the Gaussian function can be used to describe the distribution form of error values. Cavalaro [23] conducted in-depth research on the cumulative mechanism of the manufacturing error and assembly positioning error of shield tunnel segments, and proposed error control measures and indicators based on numerical analysis. Such studies havee mainly focused on the error control of circular shield tunnels, and are limited to the only the calculation of the two-dimensional and single degree of freedom. However, the special-shaped shield tunnel differs from the circular shield tunnel in terms of the stress and assembly of the segment structure. At present, no scholars have performed in-depth research on the segment assembly error control of a large-section special-shaped shield tunnel. Therefore, the segment assembly error and quality control in the construction of a quasi-rectangular shield tunnel is a novel and significant topic.

In previous studies, integrated analysis of the formation mechanism and the mechanical damage of segments has hardly ever been performed, both of which are closely correlated in engineering. There is no mature experience for the assembly control measures of the quasi-rectangular shield tunnel segment in China Hangzhou Metro Line 9. From the perspective of error control, this paper applies geometric calculation to assembly simulation research, obtaining the assembly sequence of key segments, the assembly quality of an integral ring, and its corresponding control measures. Then, the corresponding key block assembly error control standards (limit values of angle and displacement) are generated from the numerical analysis of the stress of key block segments. Finally, through the combined geometric and mechanical analysis, guidance for practical construction is provided.

2. Materials and Methods

2.1. Project Overview

A 11.83 m × 7.27 m quasi-rectangular earth pressure balance shield machine was used for the construction of the China Hangzhou Line 9. Considering the rationality and expansibility of the tunnel structure in the deep underground environment, the quasi-rectangular section contains 4 arcs, the lining ring is divided into 11 pieces, and the ring width is 1.2 m. The general section (column set up in the middle) is made of concrete segments, with an outer diameter of 11.5 m × 6.94 m, an inner diameter of 10.6 m × 6.037 m, and a thickness of 0.45 m. The quasi-rectangular shield machine is shown in Figure 1. The lining structure is classified as (ZR1) type and (ZR2) type, which are assembled by staggered joints (Figure 2). A 6-degree-of-freedom serial-type assembly machine was used for the assembly, and the accuracy of the assembly machine was 0.5 mm [24]. The burial depth at the top of the tunnel on the entry and exit line ranges from 2.5 to 10.46 m.

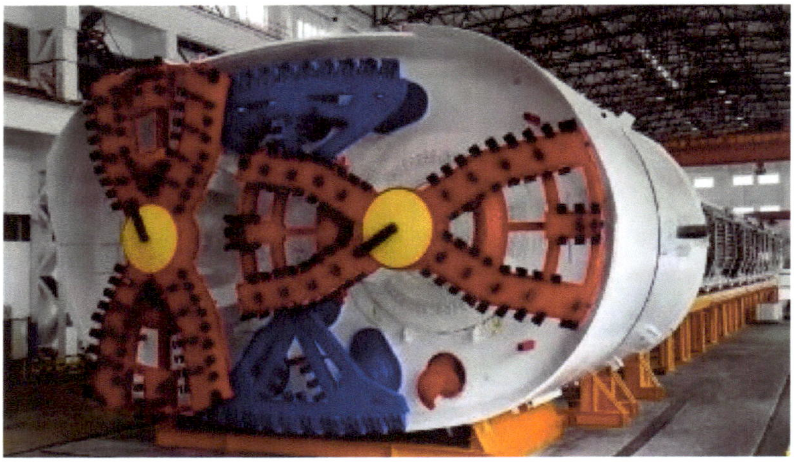

Figure 1. Quasi-rectangular EPB shield machine.

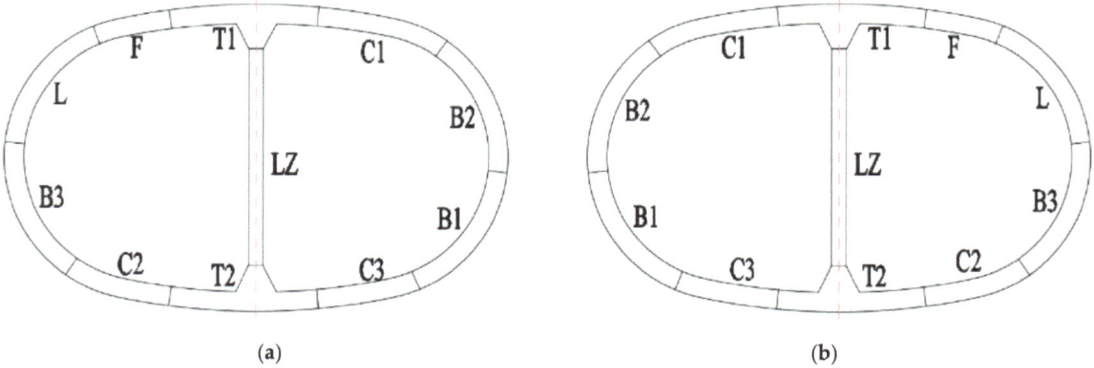

Figure 2. Segmentation diagram of segment. (**a**) (ZR1) segment; (**b**) (ZR2) segment.

2.2. Technical Framework

In this study, geometric simulation and mechanical calculation are integrated to jointly analyze the assembly error and control standard of the segments of a quasi-rectangular shield tunnel. Combined with the assembly quality analysis, the assembly measures are further optimized to improve the ring-forming quality of the segments. Finally, the accuracy of the simulation results is verified by the field-measured data. The overall technical framework is shown in Figure 3.

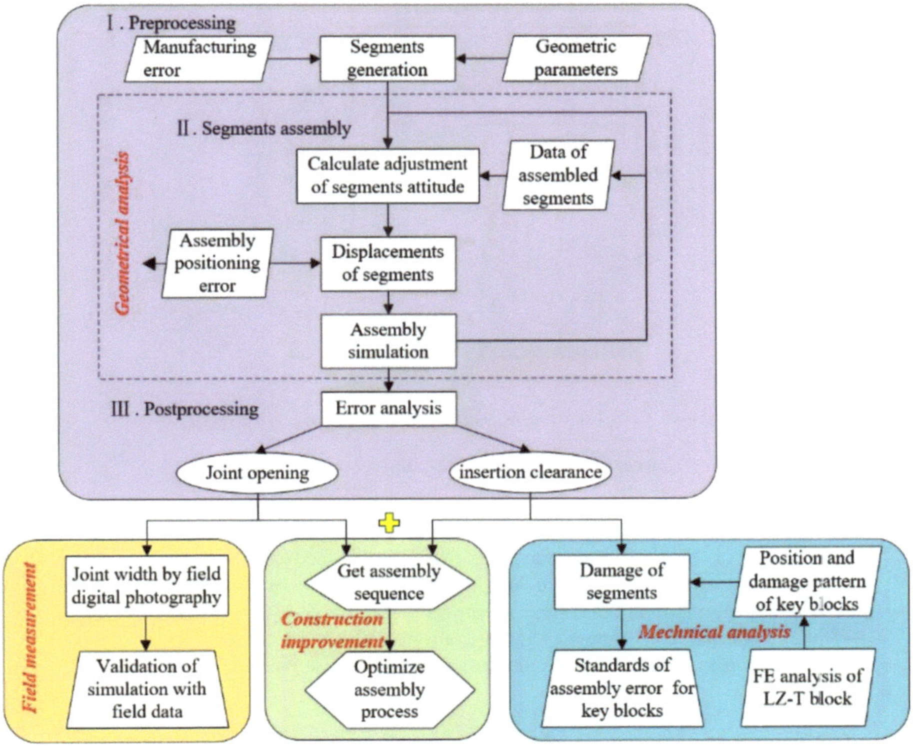

Figure 3. Overall technical route framework.

2.3. Geometric Simulation Method of Assembly Error

2.3.1. Coordinate System and Segment Displacement Mode

The block diagram of the tunnel segment of this project is shown in Figure 2. The lining ring has a total of 11 pieces, consisting of an F piece, an L piece, 3 B pieces, 3 C pieces, and 2 T pieces and a column LZ block, between which the staggered seams are used for assembly.

To facilitate the expression of segment displacement and its error, two local coordinate systems are defined: segment centroid coordinate system (hereinafter referred to as the centroid coordinate system) and the contact surface coordinate system.

(1) *Centroid Coordinate System*

The centroid coordinate system of the segment is defined because it can easily represent and calculate the displacement of the segment in 6 degrees of freedom (T, R, Z, rT, rR, rZ). The calculation result of the segment displacement is more intuitive.

The centroid coordinate system is established at the centroid of the segment, as shown in Figure 4. It is established at the lifting arm nut position of the current assembly block (it is very close to the centroid position), and the direction is radial R, shield propulsion direction Z, and tangential T (counterclockwise is positive, column downward is positive). The 6-DOF displacement of the segment is expressed in this coordinate system.

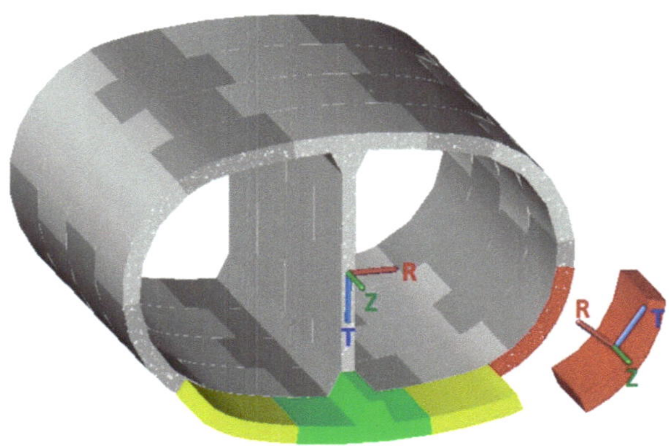

Figure 4. Coordinate of the segments.

(2) Contact Surface Coordinate System

The contact surface coordinate system is defined to make the displacement representation of the contact surface more intuitive, as shown in Figure 5. The displacement of the contact surface is expressed in this coordinate system, and T, R, and Z represent the tangential, radial, and longitudinal directions of the contact surface, respectively. The circumferential contact surface is marked as S_{Z+}, S_{Z-} surface, and the longitudinal contact surface is marked as S_{T+}, S_{T-} surface. The T1 block and T2 block have S_{R+} and S_{R-} planes, and their definitions are the same and are not listed.

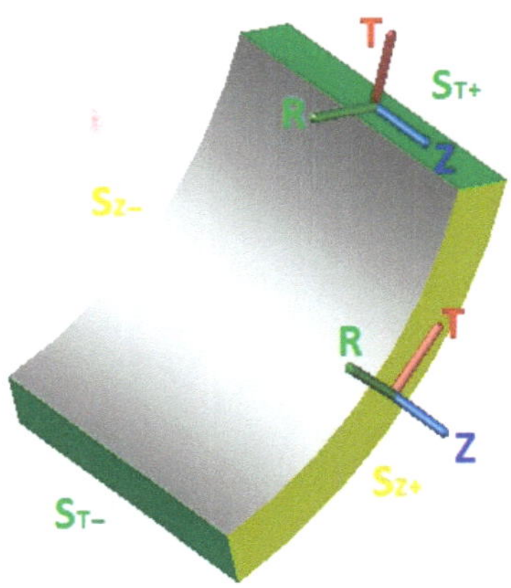

Figure 5. Definition and coordinates of the surface.

(3) 6-DOF Displacement Mode of Segment

The 6-DOF displacements of the segment include Z_0, T_0, R_0, rZ_0, rT_0, and rR_0, wherein Z_0, T_0, and R_0 are translational displacements; rZ_0, rT_0, and rR_0 (r represents rotation) are rotational displacements about the Z, T, and R axes of the centroid coordinate system, respectively; and the rotation direction complies with the right-hand rule. The subscript 0 is added to distinguish it from the displacement of the contact surface coordinate system. The displacements of adjacent segments correlate with the displacements on the contact surface, which should be converted from the centroid displacement during simulation to compute the adjustment of segments and calculate the joint dislocation and included angle.

The displacement of the contact surface is the relative displacement of adjacent contact surfaces, including displacements on longitudinal contact surfaces (S_{T+}, S_{T-}, S_{R+} and,

S_{R-}) and displacements on circumferential contact surfaces (S_{Z+}, S_{Z-}) The displacements (6 directions) of the contact surface S_2 relative to S_1 are represented in the coordinate system of the S_1 surface, as shown in Figure 6.

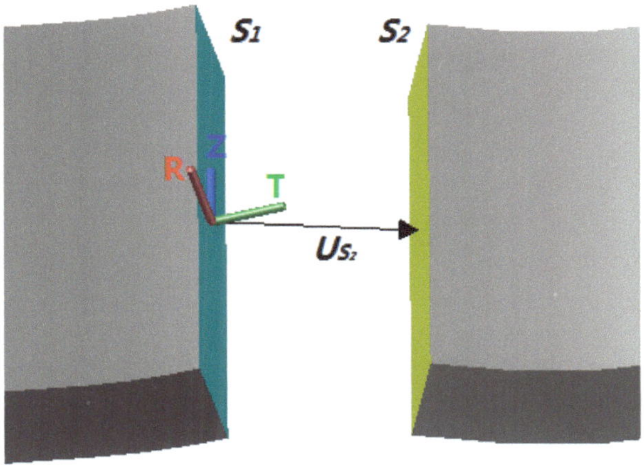

Figure 6. Motion of the interface.

(4) *Displacement Range of Contact Surface*

Based on the principle that the bolt can normally penetrate the bolt hole and the nut can be tightened, the relative displacement range of the contact surface is calculated as the evaluation index of the joint connection quality (the probability of contact surface displacement exceeding the limit) in the post-treatment. The results are shown in Table 1.

Table 1. Displacement range of the contact surface.

Direction	Longitudinal Contact Surface (S_{T+}, S_{T-})				Circumferential Contact Surface (S_{Z+}, S_{Z-})			
	Segment		LZ Segment		T Segment		F Segment	
	Lower Limit	Upper Limit	Lower Limit	Upper Limit	Lower Limit	Upper Limit	Lower Limit	Upper Limit
T (mm)	0	7.5	0	6	−5	5	−5	5
R (mm)	−4	4	−6	6	−5.77	5.77	−5.77	5.77
Z (mm)	−4	4	−6	6	0	10	0	10
rT (°)	−0.358	0.358	−0.625	0.625	−0.63	0.63	−0.63	0.63
rR (°)	−0.467	0.467	−0.341	0.341	−0.10	0.10	−0.17	0.17
rZ (°)	−1.719	2.149	−0.988	0.988	−0.20	0.20	−0.34	0.34

2.3.2. Calculation Assumptions and Parameter

The assembly simulation objective is the assembly defects caused by the segment production error and assembly positioning error. Defects include joint dislocation, opening, LZ and F block insertion gap, and relative displacement between contacting surfaces.

(1) *Calculation Assumptions*

(i) Assuming that the segment is a rigid body, the deformation during assembly is not considered.

(ii) Since the assembly errors are small, the corner displacement values of the segments caused by translation and rotation are generally within several centimeters, so the coordinate system changes caused by the segment displacement are ignored.

(iii) Segment manufacturing and assembly errors follow a Gaussian distribution.

(2) *Manufacturing Error*

The manufacturing error Δ_p is recorded as the difference between the actual manufacturing size and the design size of the segment. The arc error Δ_p^T and width error Δ_p^Z are considered separately in the simulation calculation.

During construction or design, the error control value given is generally the tolerance *Tol*. In our study, it is assumed that the error follows a normal distribution, and the exceedance probability of the absolute value $|\Delta|$ exceeding the tolerance *Tol* is 2.5%. Its standard deviation σ can be derived using $\sigma = Tol/2.24$.

The allowable manufacturing error of the width of a single segment Tol_p^T is ± 0.4 mm, and the arc length Tol_p^Z is ± 1.0 mm. the manufacturing errors along the arc and width directions are obtained from $\sigma_p^T = Tol_p^T/2.24$ and $\sigma_p^Z = Tol_p^Z/2.24$.

(3) *Positioning Error*

The design position of the current segment is defined as L_D, the optimal target position in the assembly is $\hat{L}$, and the actual position is L. The displacement from the actual position L of the current segment to the optimal target position Δ_L is recorded as the positioning error:

$$\Delta_L = L - \hat{L} \tag{1}$$

where Δ_L, L, and $\hat{L}$ are vectors containing six components: *T*, *R*, *Z*, *rT*, *rR*, and *rZ*.

The positioning errors in 6 directions ($\sigma_L^T, \sigma_L^R, \sigma_L^Z, \sigma_L^{rT}, \sigma_L^{rR}, \sigma_L^{rZ}$) are calculated, respectively. In the process of segment assembly, because it us always in contact with adjacent segments, the positioning error along the translation direction is 0.

The standard deviation of the positioning error in the rotation direction is:

$$\sigma_L^r = Tol_L^r/2.24 \tag{2}$$

Based on the principle that the maximum stagger formed by the corner is d_r = 1 mm, Tol_L^r is calculated according to the following formula:

$$\sigma_L^{rZ} = Tol_L^{rZ}/2.24 = d_r/(2.24 \times l_{rZ}) \tag{3}$$

where l_{rZ} is the maximum rotation radius of the segment along the *rZ* direction, that is, the distance from the farthest point on the segment to the *Z* axis.

The algorithms to compute $\sigma_L^{rT}, \sigma_L^{rR}$ are similar to σ_L^{rZ}. The standard deviation of the positioning error of each segment is shown in Table 2.

Table 2. Standard deviation of the location error.

Segment	σ_L^T (mm)	σ_L^R (mm)	σ_L^Z (mm)	σ_L^{rT} (°)	σ_L^{rR} (°)	σ_L^{rZ} (°)
T	0	0	0	0.0426	0.0170	0.0170
C3	0	0	0	0.0426	0.0168	0.0168
C2	0	0	0	0.0426	0.0167	0.0167
B	0	0	0	0.0426	0.0189	0.0189
L	0	0	0	0.0426	0.0186	0.0186
C1	0	0	0	0.0426	0.0144	0.0144
F	0	0	0	0.0426	0.0301	0.0301
LZ	0	0	0	0.0426	0.0098	0.0098

2.3.3. Simulation Procedures

(1) *Generation of Segments*

According to the design size, the manufacturing error parameters are added, and the numerical model of the segment is established for assembly calculation. The manufacturing error of the segment in a certain direction is simulated by applying displacements on corresponding contact surfaces: for an arc error with a size, the S_{T+}, S_{T-} surfaces are applied with an initial displacement Δ, and the manufacturing error in the width direction is treated in a similar way.

(2) *Segments Assembly*

The segment assembly simulation is divided into the following four steps:

(i) According to the actual position and assembly strategy of the assembled segment, the optimal target position $\hat{L}$ of the current segment and its displacement $\hat{U}$, where $\hat{U} = \hat{L} - L_D$ relative to the design position L_D are calculated.
(ii) According to the actual position of the assembled segment, the variation range of the positioning error Δ_L is calculated, and the random Δ_{rand} is generated according to the distribution function of the positioning error.
(iii) The displacement U of the actual position L of the current segment relative to the design position L_D is calculated.
(iv) The actual position L of the current segment is calculated. Thus, the assembly of the current segment is completed, and the assembly of the next segment is initiated.

Each step is described in detail below:

(i) Calculation of the optimal target location

The six components of the $\hat{U}$ or L_D range are calculated, respectively. Based on the engineering practice, the strategy is as follows: the displacement in the rR and Z directions is calculated according to the circumferential seam fitting principle. The displacement in rT, rZ, T, and R directions is calculated according to the longitudinal seam fitting principle. The calculation sequence ensures the quality of the circumferential joint first, and then the quality of the longitudinal joint; and controls the rotation first, and then the translation.

The principle of circumferential seam fitting refers to the contact between the current segment and the protruding point of the previous segment, and the minimum relative displacement of the circumferential seam should be achieved, as shown in Figure 7. Under the principle of circumferential seam fitting, the calculation formula of the $\hat{U}$ value and Δ_L range is:

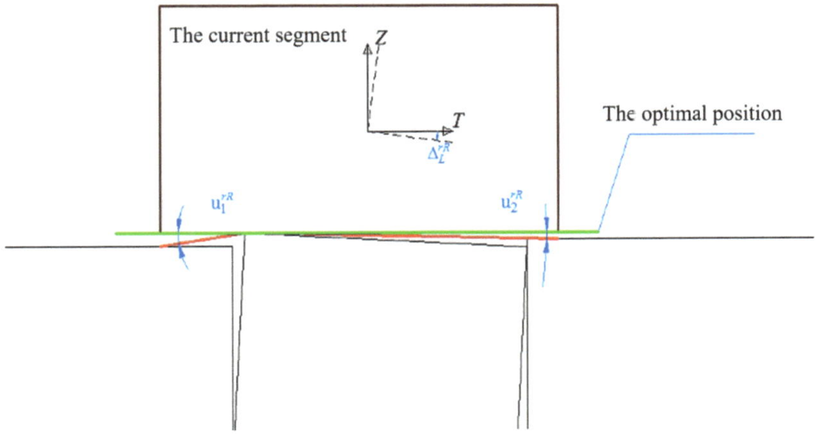

Figure 7. Schematic diagram of the ring-seam lamination principle.

$$\hat{U}^Z = max\left\{ (U^Z_{S'_{Z+}})_{i'} - (U^Z_{S_{Z-}})_i \right\}, \Delta^Z_L \in (0, +\infty) \quad (4)$$

$$\hat{U}^{rR} = 0, \Delta^{rR}_L \in \left(-u^{rR}_1, +u^{rR}_2 \right) \quad (5)$$

$$u_1^{rR} = \min\left\{\frac{\hat{U}^Z - max\left\{ (U^Z_{S'_{Z+}} - U^Z_{S_{Z-}})\Big|_{T=t,-\frac{d}{2}<R<\frac{d}{2}} \right\}}{t_{max} - t}\right\}, t_1 < t < t_{max} \quad (6)$$

$$u_2^{rR} = \min\left\{\frac{\hat{U}^Z - max\left\{ (U^Z_{S'_{Z+}} - U^Z_{S_{Z-}})\Big|_{T=t,-\frac{d}{2}<R<\frac{d}{2}} \right\}}{t - t_{max}}\right\}, t_{max} < t < t_2 \quad (7)$$

where:

$\left(U^z_{S_{z-}}\right)_i$ is the Z-direction displacement of point i on the S_{Z-} plane of the current segment, and point i is an arbitrary point on the S_{Z-} plane of the current segment;

$\left(U^z_{S'_{Z+}}\right)_{i'}$ is the Z-direction displacement of point i' on the S_{Z+} plane of the current segment, and point i' is an arbitrary point on the S_{Z+} plane of the current segment;

t_1 and t_2 are the T-direction coordinates of the left and right end faces of the current segment, respectively;

d is the segment thickness;

$t_{\max}$ is the coordinate of $\hat{U}^Z$ in the T direction;

The T and R coordinates are all in the centroid coordinate system of the current segment, and the same below.

The longitudinal seam bonding principle refers to the longitudinal seam bonding between the current segment and the adjacent assembled segment in the ring, to minimize the relative displacement of the longitudinal seam, as shown in Figure 8. Under the principle of longitudinal joint fitting, the calculation formula of the value and range is:

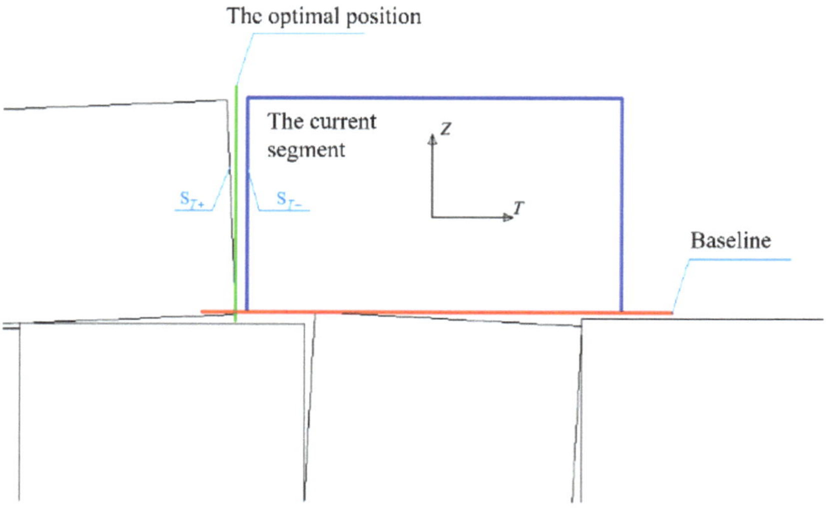

Figure 8. Schematic diagram of the longitudinal-seam lamination principle.

$$\hat{U}^{rT} = (U^{rT}_{S'_T} - U^{rT}_{S_T})\Big|_{Z=0,R=0} \times \cos(\alpha), \Delta^{rT}_L \in (-\infty, +\infty) \quad (8)$$

$$\hat{U}^{rZ} = (U^{rZ}_{S'_T} - U^{rZ}_{S_T})\Big|_{Z=0,R=0}, \Delta^{rZ}_L \in (-\infty, +\infty) \quad (9)$$

$$\hat{U}^T = Ft(max\{(U^T_{S'_T})_{i'} - (U^T_{S_T})_i\}, (U^R_{S'_T} - U^R_{S_T})\Big|_{Z=0,R=0}), \Delta^T_L \in \begin{cases} (0, +\infty), & T < 0 \\ (-\infty, 0), & T > 0 \end{cases} \quad (10)$$

$$\hat{U}^R = Fr(max\{(U^T_{S'_T})_{i'} - (U^T_{S_T})_i\}, (U^R_{S'_T} - U^R_{S_T})\Big|_{Z=0,R=0}, \Delta^R_L \in (-\infty, +\infty) \quad (11)$$

$$Ft(T,R) = T \times \cos(\alpha) + R \times \sin(\alpha) \quad (12)$$

$$Fr(T,R) = -T \times \sin(\alpha) + R \times \cos(\alpha) \quad (13)$$

where:

$\left(U^z_{S_T}\right)_i$ is the difference between the T-direction displacement of the corresponding point on the S_{T+} or S_{T-} plane of the adjacent segment and the point i on the S_{T-} or S_{T+} plane of the current segment. Point i is an arbitrary point on the S_{T-} or S_{T+} plane of the current segment, and the same definition also applies to similar expressions.

d is the segment thickness;

Ft and Fr are the conversion function from the contact surface coordinate system to the centroid coordinate system;

α is the rZ angle between the contact surface coordinate system and the centroid coordinate system.

Since the segments on the S_{T-} plane and S_{T+} plane of the LZ block and F block were assembled at the time of insertion, the calculation principle of the LZ block or F block is that the central axis is aligned with the central axis of the contact surface of adjacent segments and inserted in the middle. In other words, the relative displacement along the R direction on the S_{T+} and S_{T-} plane of the LZ block or F block with adjacent segments should be the smallest, and the other components of the relative displacement of the end faces on both sides are equal. The values of $\hat{U}^Z$, Δ^Z_L, and $\hat{U}^{rR}$ and range of Δ^{rR}_L are still calculated according to Equations (14)–(17), and the calculation formulas of the other components are as follows. The formula is given by taking the LZ block as an example. The calculation of the F block is similar to the LZ block.

$$\hat{U}^{rT} = \left(\frac{(U^{rR}_{S'_{R+}} - U^{rT}_{S_{T-}}) + (-U^{rR}_{S''_{R+}} - U^{rT}_{S_{T+}})}{2}\right)\Bigg|_{Z=0,R=0}, \Delta^{rT}_L \in (-\infty, +\infty) \quad (14)$$

$$\hat{U}^{rZ} = \frac{U^T_{S'_{R+}} + U^T_{S''_{R+}}}{l_{LZ}}\Bigg|_{Z=0,R=0}, \Delta^{rZ}_L \in (-\infty, +\infty) \quad (15)$$

$$\hat{U}^{T} = \left(\frac{(U^{R}_{S'_{R+}} - U^{T}_{S_{T-}}) + (-U^{R}_{S''_{R+}} - U^{T}_{S_{T+}})}{2}\right)\Bigg|_{Z=0,R=0}, \Delta^{T}_L \in (-\infty, +\infty) \quad (16)$$

$$\hat{U}^{R} = \left(\frac{\left(-U^{T}_{S'_{R+}} - U^{R}_{S_{T-}}\right) + \left(U^{T}_{S''_{R+}} - U^{R}_{S_{T+}}\right)}{2}\right)\Bigg|_{Z=0,R=0}, \Delta^{R}_L \in (-\infty, +\infty) \quad (17)$$

where:

s'_{R+} is the S_{R+} plane of the adjacent T1 block corresponding to the S_T plane of the LZ block;

s''_{R+} is the S_{R+} plane of the adjacent T2 block corresponding to the S_{T+} plane of the LZ block;

l_{LZ} is the length of the LZ block.

(ii) Generation of the positioning error

Based on the Box–Muller method proposed by Scott D.W. [25], which uses a uniformly distributed pseudo-random function to generate random numbers with an arbitrary distribution, a random number from a normal distribution with upper and lower limits is generated to describe the change in the positioning error Δ_L.

(iii) Calculation of the actual displacement of segments

The displacement U of the actual position L on the current segment relative to the design position L_D is calculated according to the following formula.

$$U = \hat{U} + \Delta_{rand} \tag{18}$$

(iv) Calculation of the segment position after assembly

The actual position L of the current segment is calculated according to the following formula to complete the assembly of the current segment:

$$L = L_D + U \tag{19}$$

In Equations (18) and (19), $\hat{U}$, U, $\hat{L}$, L, L_D, and Δ_L contain components in six directions: T, R, Z, rT, rR, and rZ.

(3) Post Processing

Through simulation calculation, the segment attitude and the displacement of the contact surface of each segment are obtained. According to the segment attitude data, the relative displacement overrun frequency of the segment contact surface, the joint opening, LZ, and F block insertion clearance can be calculated as the basis for the assembly quality evaluation. Considering the subsequent mechanical analysis and field test application, the calculation formulas of the insertion gap and joint opening are provided here. The contact surface of each segment only outputs the data of the S_{T-} plane and $S_{Z\text{-}}$ plane, and the contact surface data output by LZ is the data of the S_{T-} plane and S_{T+} plane.

(i) Joint opening

There is a displacement difference in the Z-direction between the each point on the $S_{Z\text{-}}$ surface and the corresponding point on the previous segment S_{Z+} surface. The maximum value of the difference is the opening of the circumferential seam, which is recorded as stretch Z:

$$\text{stretch } Z = max\{(U^Z_{S_{Z-}})_i - (U^Z_{S'_{Z+}})_{i'}\} \tag{20}$$

There is a displacement difference in the T-direction between the each point on the $S_{T\text{-}}$ plane and the corresponding point on the previous segment S_{T+} plane. The maximum value of the difference is the opening of the circumferential seam, which is recorded as stretch T:

$$\text{stretch } T = max\{(U^T_{S_{T-}})_i - (U^T_{S'_{T+}})_{i'}\} \tag{21}$$

The joint opening is easy to measure and intuitive in engineering, so it is used as a parameter to reflect the assembly quality.

(ii) The insertion gap between the LZ and F block

The insertion clearance of the LZ and F block is the T-direction component $U^{OP^T}_{S_{T-}}$ of the displacement relative to the $S_{T\text{-}}$ plane of adjacent segments. When it is positive, it means that the LZ and F block can be inserted normally. When it is negative, it means that the clearance of adjacent segments is too small, and it is difficult to insert the LZ or F block. This value can be used to measure the difficulty of inserting the LZ and F block. The LZ insertion gap is shown in Figure 9, and the definition of the block F insertion gap is similar.

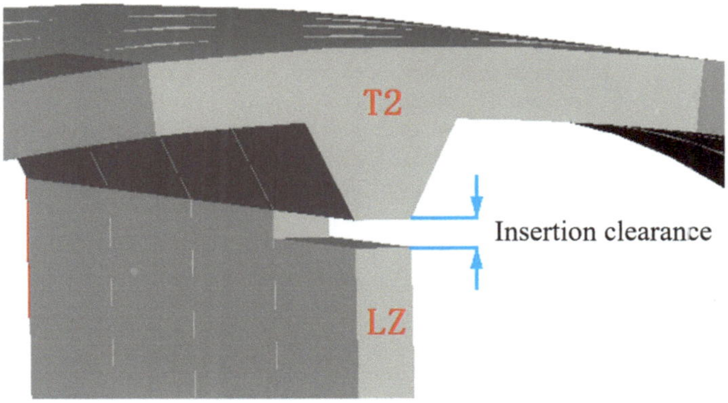

Figure 9. Schematic diagram of the LZ insertion gap.

2.4. Establishment of the Key Block Error Control Standard

2.4.1. Finite Element Method Modeling

The error control standard of the key block (T block and LZ block) is mainly obtained by mechanical analysis based on finite element analysis. The numerical model is simplified with the following assumptions for the T block and LZ block: (1) Only the main reinforcement of the LZ block is established, without considering stirrup and structural reinforcement; (2) since the main action of the T block on the LZ block is vertical compression, when considering the friction between the T block and the LZ block, the tangential action of the bolts can be ignored, and the bolts and bolt holes are omitted; and (3) the edge chamfer of the bolt handhole is not considered and is simplified to a right angle.

The concrete and embedded parts of the numerical model are simulated by the C3D8R solid element, and the reinforcement is simulated by a truss element. The mesh of the model is shown in Figure 10.

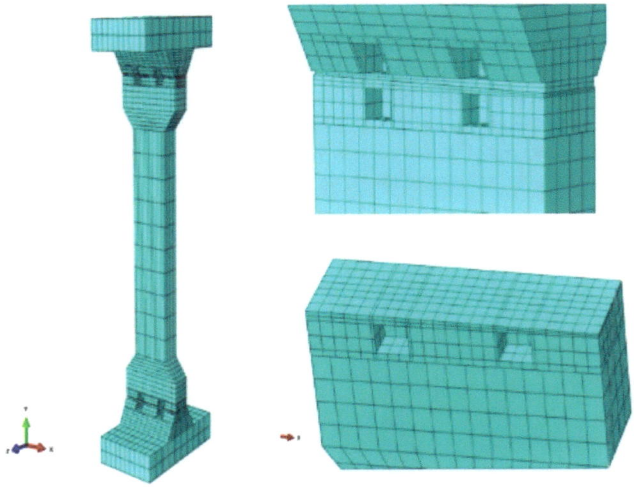

Figure 10. Grid division of the finite element model.

The "face to face" contact is set to simulate the contact between the T block and LZ block during vertical loading. Referring to the shear stiffness experimental data of the circular tunnel joint, the friction coefficient between the poured concrete with fine steel formwork is 0.5. Since the end of the T block and both ends of the LZ block are in contact with each other, which is a dangerous area where local crushing may occur, a dense grid is required. Considering that it is difficult for the whole part to perform a uniform transition when meshing, fine mesh parts and coarse mesh parts are established, respectively. They are connected by binding constraints to reduce the size of the element. The binding relationship between the end of embedded parts and the upper part of the T block and the LZ block is shown in Figure 11. The main reinforcement of the LZ block (14C20+4C16) is embedded in the concrete through the embed restraint [26].

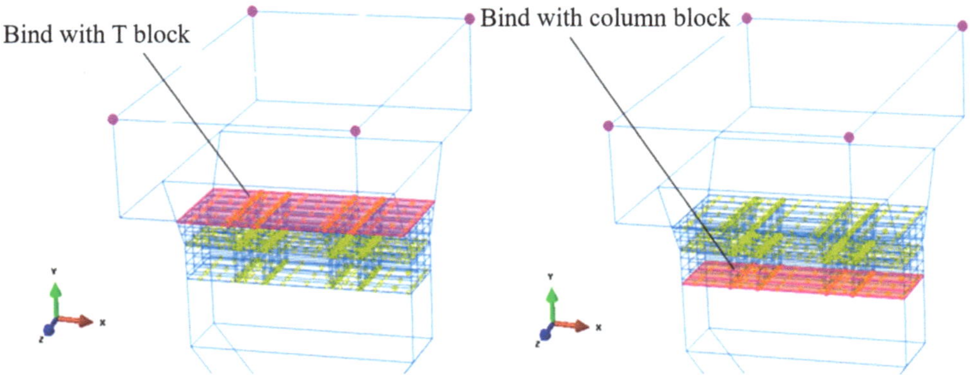

Figure 11. Connection of parts under different mesh densities.

2.4.2. Constitutive Relation and Boundary Conditions

In the numerical model, the concrete damage constitutive model in the finite element software material library is adopted. The model assumes that concrete materials are mainly damaged by tensile cracking and crushing, and the elastic stiffness matrix of concrete is reduced by the introduction of a damage factor. For concrete under compression, the damage theory establishes the damage relationship based on the relationship between inelastic strain and the compression damage factor. For tension, the theory holds that the concrete crack width is the integral of the tensile strain and the crack zone width. Assuming that the crack zone width is constant, the tensile crack width–damage factor curve can be obtained [27–29], as shown in Figure 12. In addition, considering that the main reinforcement of the LZ block in the numerical model is HPB400 hot-rolled reinforcement and the yield strength is 400 MPa, the constitutive relationship of the reinforcement adopts the linear elastic ideal plastic model.

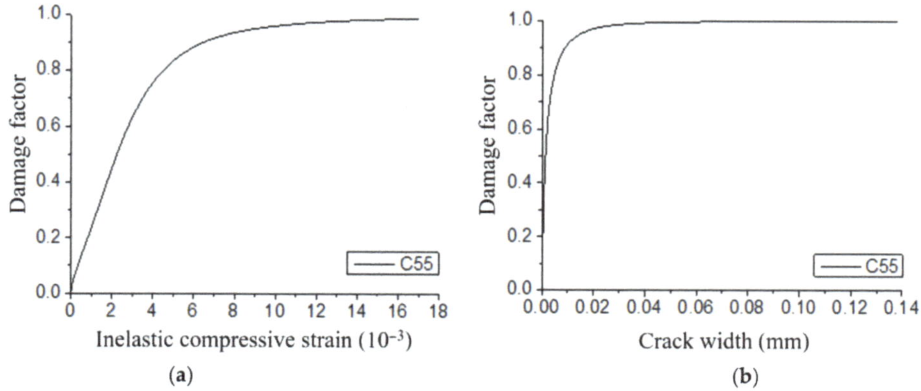

Figure 12. Definition curve of the damage factor: (**a**) Damage factor to inelastic compressive strain; (**b**) Damage factor to crack width.

Since the LZ block is mainly subjected to vertical axial force, the calculation considers the working condition of the loading vertical force when the LZ block has a certain angle or translational error with the end face of the T block after assembly. The axial force calculation results of the uniform and equal stiffness body model in the design data are directly quoted. The model considers the weakening of the longitudinal joint stiffness and the transfer of joint internal force, and the design limit buried depth of 17 m is adopted for the calculation. At this time, the maximum axial force of the LZ block is 1842 kN.

By establishing the reference point at the T block position and coupling the T block element, the unit displacement in the error direction of the reference point is given, and the constraint reaction at this time is calculated, and the boundary spring stiffness in three directions can be obtained, as shown in Table 3.

Table 3. Spring stiffness at the boundary of the T segment.

k_{rT}	k_{rZ}	k_T
2.3×10^4 kN·m/rad	1.7×10^3 kN·m/rad	7500 kN/mm

2.4.3. Limit for the Load Damage and Crack Resistance of the Structure

From the perspective of normal use, the assembly error control index of the T block is analyzed from two aspects: compression failure and tensile crack.

(1) From the perspective of compression, compression failure is not allowed except for the contact surface (20 mm concrete elements), that is, the total compression strain should not exceed 0.0033 (Figure 12a). Finite element software defines the uniaxial compression curve through stress inelastic strain, and determines the damage state through the relationship between the inelastic strain and compression damage factor. Therefore, the functional relationship between the stress and damage factor is established through inelastic strain, namely the position of equivalent stress in the uniaxial stress–strain curve can be accessed through the damage factor. Therefore, it is determined that the inelastic strain of concrete in compression failure is 0.002 and the damage factor for compression stress is 0.46.

(2) From the perspective of applicability and durability, the width of the concrete tensile crack should not exceed 0.2 mm. Figure 12b establishes the relationship curve between the tensile crack width and the tensile damage factor. From the curve, the corresponding tensile damage factor of 0.2 mm is 0.996.

2.4.4. Control Standard Calculation, Loading Steps, and Working Conditions

When the translational degrees of freedom Z, T and rotational degrees of freedom rR act alone, the end face of the T block and the end face of the LZ block are still parallel, so the structure is under uniform stress. In such cases, only the maximum allowable value of the bolt hole gap is taken, that is, the translational displacement value is 6 mm and the rotational displacement value is 0.625°. Due to the 20 mm wedge at the upper end of the LZ block, the influence of rT on the force of the LZ block under the positive and negative direction error is different. According to the defined direction, the front end of the T block tilts downward towards the tunnel propulsion direction, which is considered the positive direction of the rT rotation. The loading conditions of the different components of the error are shown in Table 4.

Table 4. Loading conditions of the DOF for different errors.

Freedom Direction	Working Condition	Error Value	Loading Steps
Z	Working condition *Z*	6 mm	I. Lift the upper T block → II. Rotate and move T block or LZ block to the position with a given error → III. Apply small vertical load to establish contact → IV. Apply vertical design load
T	Working condition *T*	6 mm	
rZ	Working condition *rZ*-1	1.2°	
	Working condition *rZ*-2	1.4°	
	Working condition *rZ*-3	1.4°	
	Working condition *rZ*-4	1.8°	
rR	Working condition *rR*	0.625°	
rT	Working condition *rT*-1	0.5°	
	Working condition *rT*-2	0.6°	
	Working condition *rT*-3	0.7°	
	Working condition *rT*-4	0.8°	
	Working condition *rT*-5	−0.5°	
	Working condition *rT*-6	−0.6°	
	Working condition *rT*-7	−0.7°	
	Working condition *rT*-8	−0.8°	

3. Results

3.1. Analysis of Multi-Ring Assembly Error

Multi-ring assembly simulation considers the error accumulation of segments. The assembly simulation is carried out according to the method proposed in this paper, during which the assembly quality under the condition of error accumulation is studied.

3.1.1. Optimization of the Assembly Sequence

In special-shaped shield tunnels, such as quasi-rectangular shield tunnels, there are many available segment assembly sequences due to the large number of segments. The analysis of assembly error in this study can be used as a reference for the optimization of the segment assembly sequence of the quasi-rectangular shield.

For the quasi-rectangular shield tunnel, there are two optional assembly sequences:

(i) T2→C3→C2→B1→B3→B2→L→C1→T1→LZ→F

(ii) T2→C3→C2→B1→B3→B2→L→C1→T1→F→LZ

Since the two groups of segments (ZR1) and (ZR2) in the straight-line section of the project are symmetrical, the analysis results of the segments in the (ZR1) group are also applicable to (ZR2).

(1) *T1 Block Attitude Adjustment Method*

In scheme i, block LZ is assembled before block F, and the lining segments have not yet formed a ring when assembling LZ. Therefore, LZ insertion will change the attitude of block T1 and the distance between blocks T1 and T2, so the difficulty of LZ insertion is reduced. In scheme ii, before assembling LZ, other segments are composed to form a ring. However, there are some risks in scheme i: after LZ insertion, the contact state of the original lining segments is changed. It may increase the difficulty of inserting block F, thus affecting the assembly quality of the overall lining.

For scheme ii, the simulation should be carried out according to the process described in Section 2.3.3. In scheme i, when inserting LZ, the attitude of the T1 block is adjusted to reduce the clearance between the S_{R+} surface and the S_{T-} surface of the longitudinal contact between T1 and LZ. The simulation of this working condition is shown in Figure 13.

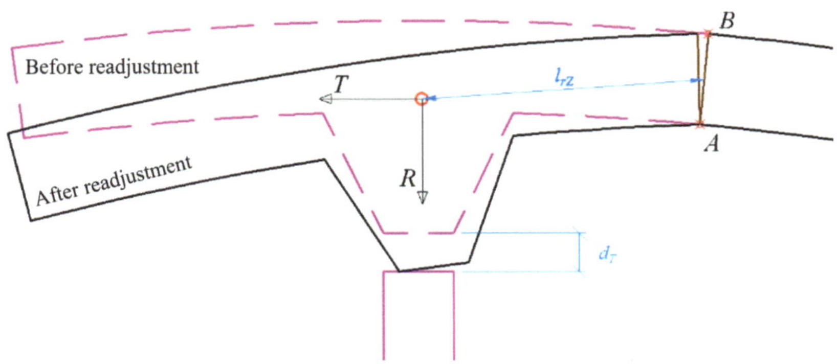

Figure 13. Segment attitude adjustment of the T1 block before installing the LZ block.

Before inserting LZ, the clearance $d_T = U_{OP}^T$ between the LZ and T1 block is estimated. The T1 block is applied with rZ rotation displacement d_T / l_{rZ}, where l_{rZ} is the distance from the centroid of the T1 block to the $S_{T\text{-}}$ plane. As shown in Figure 13, if $d_T > 0$, the rotation center is point A. If $d_T < 0$, the rotation center is point B. This method is used to simulate the T1 block attitude before inserting LZ to reduce the clearance $|d_T|$ between them.

(2) ***Calculation Results and Analysis***

From the perspective of multi-ring assembly simulation, the segments of (ZR1) and (ZR2) groups are assembled alternately. The number of rings calculated is 500. The displacement range of the longitudinal contact surface and the circumferential contact surface of the segment is shown in Table 1.

The simulation results of the key block insertion clearance and joint opening (error frequency distribution) are shown in Figure 14.

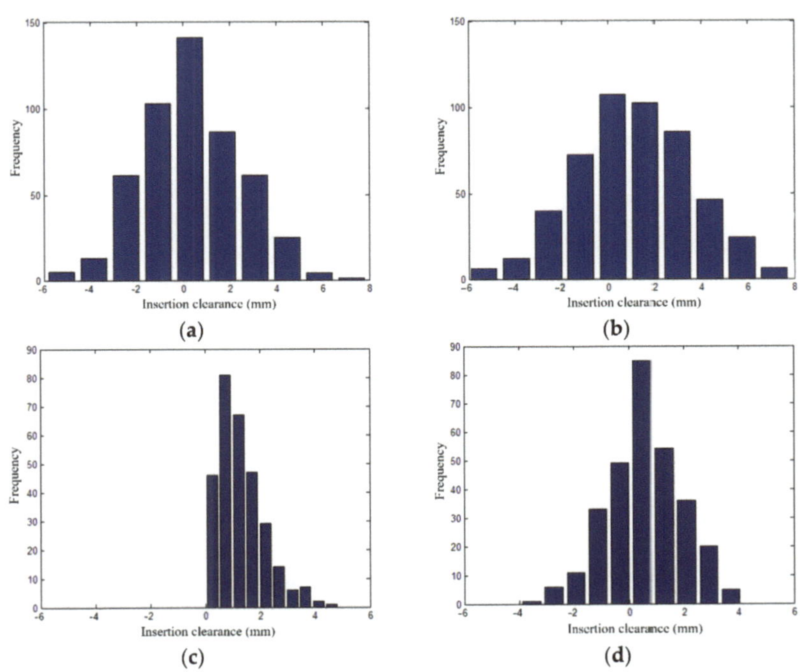

Figure 14. *Cont.*

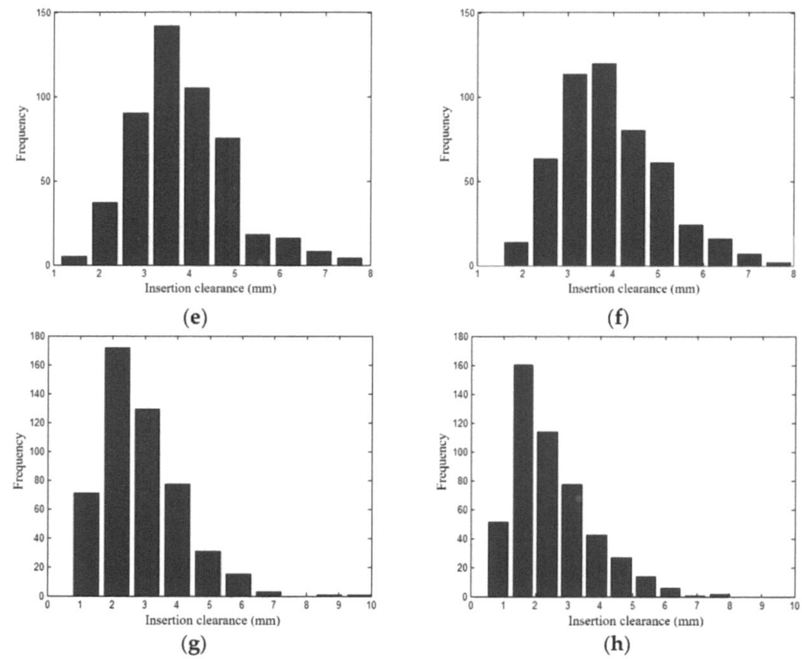

Figure 14. Error frequency distribution of the insertion clearances and joint openings under different assembly sequences. (**a**) Insertion clearance of block F under scheme i; (**b**) Insertion clearance of block F under scheme ii; (**c**) Insertion clearance of block LZ under scheme i; (**d**) Insertion clearance of block LZ under scheme ii; (**e**) Maximum opening of the circumferential seam under scheme i; (**f**) Maximum opening of the circumferential seam under scheme ii; (**g**) Maximum opening of the longitudinal joint under scheme i; (**h**) Maximum opening of the longitudinal joint under scheme ii.

As can be seen from Figure 14, although assembling LZ block under scheme i first readjusts the upper T block, it does not significantly affect the insertion of block F and the joint quality. A comparison of the key parameters of the two schemes is shown in Table 5. Considering that the exceedance probability of 2.5% is generally defined as the limit probability in the project, all data distribution ranges are within the 0.025 quantile and 0.975 quantile ($\mu_{0.025}$, $\mu_{0.975}$).

Table 5. Comparison of the key parameters of two assembly schemes.

Assembly Scheme	Insertion Clearance of Block F (mm)	Insertion Clearance of Block LZ (mm)	Longitudinal Joint Opening (mm)	Circumferential Joint Opening (mm)	Probability That the Insertion Clearance of Block F Is Less than −2 mm (%)
i	(−3.7, 4.5)	(0, 3.5)	(0, 5.9)	(0, 6.5)	13.2
ii	(−4.0, 4.2)	(−2.0, 3.0)	(0, 5.7)	(0, 6.3)	16.4

It can be seen from the calculation results that the LZ block is assembled first to eliminate its negative clearance and ensure the quality of the assembly, and the scheme has little effect on the insertion clearance of the capping F block. Considering that the probability that the insertion clearance of block F is less than −2 mm is relatively small, it is considered that assembling the LZ block first will not affect the assembly quality of block F. In addition, it should be noted that since the LZ block will cause secondary displacement of the T1 block, the opening amount of the longitudinal seam (up to 10 mm) and circumferential seam (up to 8 mm) may be large.

Compared with the single loop simulation results [30], this result has slightly increased. This indicates that the error accumulation has little effect on the insertion clearance and longitudinal seam opening of block F and LZ when the influence between the rings is considered.

From the simulation results of the multi-ring assembly, it can be concluded that the scheme of assembling the LZ block first is better for error control. During the construction of the quasi-rectangular shield section, the actual assembly scheme is scheme i, namely LZ first and then F.

3.1.2. Optimization of the Assembly Measures

The assembly quality control measures are introduced. The assembly simulation calculation is carried out according to the assembly sequence, namely the LZ block first and then the F block, and the segments of the (ZR1) and (ZR2) groups are assembled alternately. The manufacturing error, standard deviation of the positioning error, and relative displacement range of the contact surface are the same as those above, and the number of rings during the analysis of the assembly error is 500. After the improvements, the assembly construction measures are as follows:

① According to the assembly sequence of the LZ block first and then the F block, the rotation angle of the T1 block in the *rZ* direction should be adjusted before inserting the LZ block;

② Based on the above calculation and analysis, the standard deviation of the positioning error in the *rZ* direction is reduced to 0.5 times;

③ The displacement in the Z and *rR* directions should be in accordance with the principle of the fitting circumferential seam, and the displacement in the *T*, *R*, *rT*, and *rZ* directions should be assembled in accordance with the principle of the fitting longitudinal seam;

④ According to the engineering experience and relevant requirements in the bidding document, the longitudinal misalignment of the segment longitudinal joints should be controlled. When the longitudinal misalignment is greater than 4 mm, measures to level the annular surface should be performed, that is, adding gaskets on the annular surface. It is assumed that the gasket can always completely flatten the annular surface (the displacement in the Z direction is equal). The specific measures used to control the assembly error are as follows: remove the sundries on the contact surface before assembly, repair the segments and sealing strips in time, correct the shield tail clearance in time during propulsion, close the jack in time after the segments are positioned, tighten the bolts in time after the assembly of each ring of segments, and re-tighten the bolts after the shield tail.

(1) *Analysis of the Insertion Clearance between Block F and LZ*

As can be seen from Figure 15, the insertion clearance of block F is distributed between (−1.9 mm, 2.6 mm), and the insertion clearance of block LZ is between (0 mm, 2.4 mm). This result is significantly reduced compared with the previous section, which shows that the assembly quality of block F and LZ can be effectively improved with the corresponding control measures.

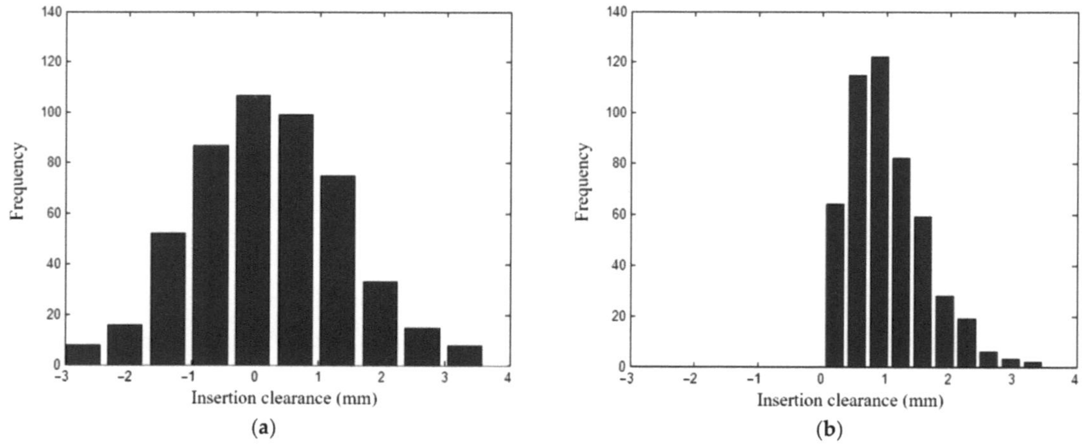

Figure 15. Error frequency distribution of the insertion clearances in key blocks after optimization of the assembly measures. (**a**) Insertion clearance of block F. (**b**) Insertion clearance of block LZ.

(2) ***Analysis of the Maximum Opening of the Circumferential and Longitudinal Joints in the Ring***

Figure 16 shows the distribution of the maximum joint opening of each ring under the consideration of construction control measures, with quantiles of 0.025 and 0.975 ($\mu_{0.025}$, $\mu_{0.975}$). It can be concluded from the calculation results that the maximum opening amount of the circumferential seam in the ring is distributed in (0 mm, 5.9 mm), and the maximum opening amount of the longitudinal seam in the ring is distributed in (1.0 mm, 4.3 mm).

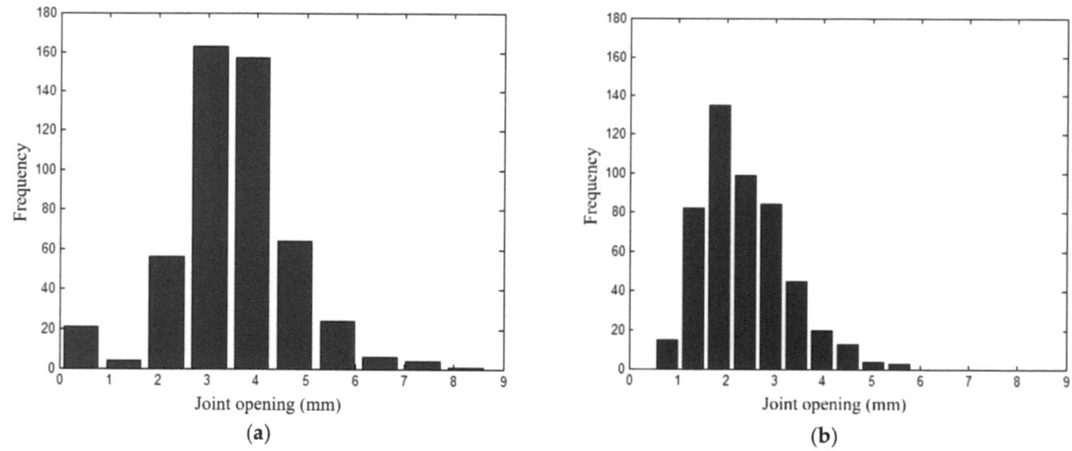

Figure 16. Error frequency distribution of the maximum joint openings in each ring after optimization of the assembly measures. (**a**) Maximum circumferential joint opening; (**b**) Maximum longitudinal joint opening.

Compared with the previous section, the error is significantly reduced, and the joint quality is greatly improved. It shows that the opening of the segment joint can be effectively controlled by taking the measures of segment ring surface leveling and improvement of the assembly accuracy in the rZ direction. The leveling measures of the segment ring surface are as follows: when the longitudinal dislocation of the longitudinal joint is greater than 4 mm after the assembly of 1 ring segment, the ring surface of the segment should be leveled using a gasket, and then the assembly of the next ring segment should be carried out. In the process of assembling 500 rings for error simulation analysis, 20 ring leveling measures are used, namely one leveling is required every 25 rings.

(3) Influence of the Positioning Error in the rT and rR Directions on the Joint Quality

To study the influence of the *rT* and *rR* corners on the joint assembly quality, based on the above construction measures, the standard deviation of the positioning error in the *rT* and *rR* directions is reduced to 0.5 times to carry out the assembly simulation. The results are shown in Figure 17.

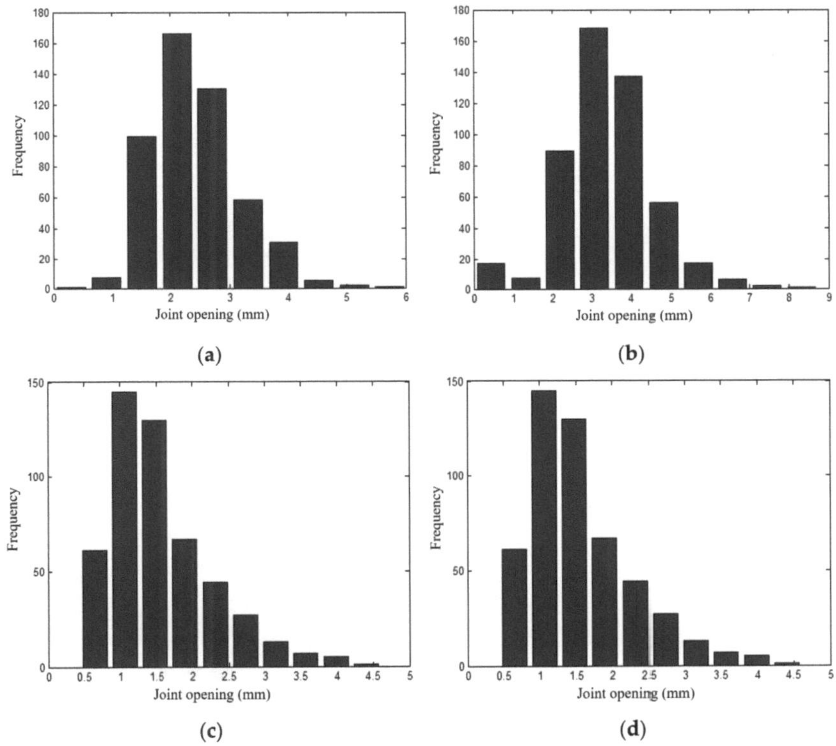

Figure 17. Error frequency distribution of the maximum joint openings in each ring after reducing σ_L^{rT} and σ_L^{rR}. (**a**) Maximum circumferential joint opening after the reduction in σ_L^{rT}; (**b**) Maximum circumferential joint opening after the reduction in σ_L^{rR}; (**c**) Maximum longitudinal joint opening after the reduction in σ_L^{rT}; (**d**) Maximum longitudinal joint opening after the reduction in σ_L^{rR}.

When the positioning error in the *rT* direction is reduced by 0.5 times, the joint quality is significantly improved. The maximum opening of the circumferential seam in the ring is distributed in the range of (1.3 mm, 4.0 mm), and the maximum opening amount of the longitudinal seam in the ring is distributed in the range of (0.6 mm, 3.3 mm), which improves the assembly quality to a certain extent. When the positioning error in the *rR* direction is reduced by 0.5 times, the joint opening does not change significantly, and the positioning error in the *rR* direction has little effect on the joint opening. It can be seen from the simulation results that if the joint quality cannot be guaranteed after controlling the positioning error in the *rR* direction, the positioning error in the *rT* direction can be further controlled.

3.2. Field Measurement and Verification

It is difficult to obtain the error parameters in the assembly of a shield tunnel [31]. The data from the construction party is often collected after the assembly [32], which generally includes the data after the segment is out of the shield tail and the deformation is stable. Because this study does not involve the change in the joint quality after the segment is stressed, all field data need to be collected before the segment comes out of the shield tail. Considering the narrow space of the construction site, the traditional measurement methods cannot obtain the joint data quickly and accurately [33]. In this paper, digital photography and image recognition technology [34] are used to collect, process, and analyze the segment joint data at the shield tail. A SONY α550 camera with 14.2 million effective pixels (fixed focus 250 mm shooting) was used to collects segment joint photos, using a laser rangefinder to measure the shooting distance. Halcon language was used to write the photo recognition program to obtain the joint opening after the segment assembly and ring forming.

3.2.1. Identification Method of the Joint Width

Using 3 calibration lines with widths of 8, 10, and 12 mm, respectively, the width of the standard calibration line was measured 18 times at a distance of 0.5~4.0 m using a measuring camera, lens, and rangefinder, and the calibration *R–D* curve was obtained. Then, through linear regression of the calibration data, the relationship between the calibration ratio *R* and distance *D* was obtained as follows:

$$R = -0.0008D^2 + 0.0278D + 0.0081 \tag{22}$$

The gray segmentation technology and sub-element edge detection technology were used to process the seam photos and obtain the pixel width of the seam, as shown in Figure 18. Combined with the *R–D* curve formula, the real width of the joint was obtained.

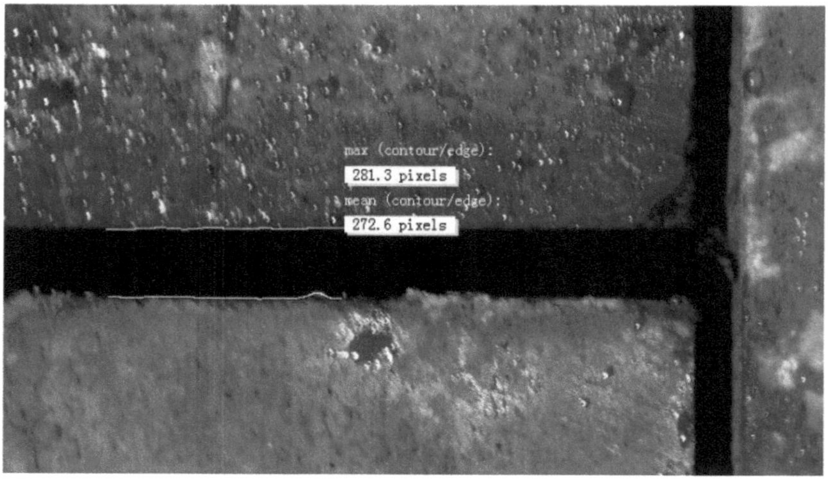

Figure 18. Segment joint width identification.

The joint width identified by digital photography includes three parts: the step width of the segment edge, the width of the rubber gasket, and the actual joint opening. There are two rubber gaskets at the longitudinal seam and one rubber gasket at the circumferential seam. Since there is no contact between the local rubber gasket at the joint opening [35], the deformation of the rubber gasket is not considered when calculating the joint opening. The calculation formula is as follows:

$$\text{circumferential joint opening} = \text{joint width (by photography)} - 12\ \text{mm} \tag{23}$$

$$\text{longitudinal joint opening} = \text{joint width (by photography)} - 10 \text{ mm} \tag{24}$$

If the result calculated by the above two formulas is negative, it means that the adjacent segments were in contact, and the opening of the joint is taken as 0 mm.

3.2.2. Field Data Acquisition Area

During field data collection, 10 longitudinal joints and adjacent circumferential joints were measured for each ring segment. Two photos at the front and back of each longitudinal seam were taken, respectively. The width and opening angle of the longitudinal seam and circumferential seam were obtained simultaneously through the calculation, as shown in Figure 19.

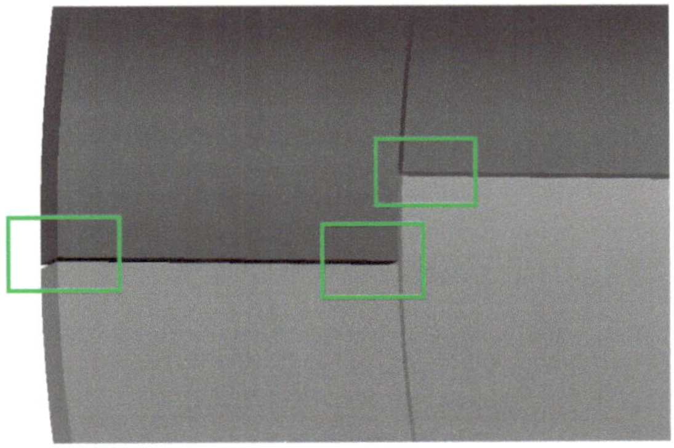

Figure 19. Shooting area for data acquisition.

3.2.3. Comparison between Field and Simulation Data

The seam data of 52 rings during assembly was measured on site, including the longitudinal joints and circumferential joints at the inner edge of the lining. The opening value of each joint is the maximum measured value. The statistical results of the measured joint opening on site are shown in Figure 20, and the distribution parameters of the calculated and measured values are shown in Table 6.

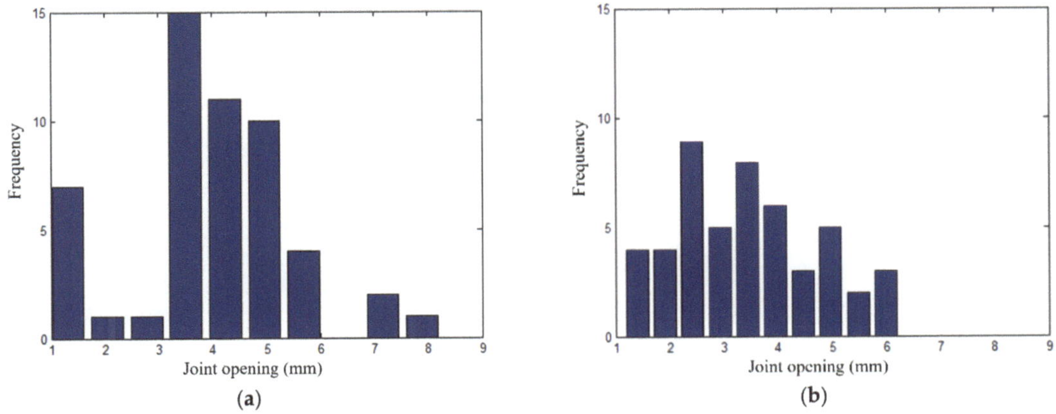

Figure 20. Error frequency distribution of the maximum joint openings in a ring measured at the project. (**a**) Maximum circumferential joint opening; (**b**) maximum longitudinal joint opening.

Table 6. Comparison between the calculated values and measured values.

Index	Statistic	Calculated Value	Measured Value
Maximum circumferential joint opening in the ring (mm)	Minimum value	0.00	0.95
	Maximum value	8.67	8.25
	Average value	3.51	3.93
	$\mu_{0.025}$	0.00	1.02
	$\mu_{0.975}$	5.94	7.25
Maximum longitudinal joint opening in the ring (mm)	Minimum value	0.53	1.20
	Maximum value	5.78	6.23
	Average value	2.34	3.61
	$\mu_{0.025}$	1.03	1.43
	$\mu_{0.975}$	4.32	6.03

As can be seen from Figure 20, the maximum joint opening in the ring presents a normal distribution in general, which is in good agreement with the calculation results. It can be seen from Table 6 that the minimum and maximum values in the comparison between the measured values and calculated values are relatively consistent. Their distribution range is also similar. For the mean value of the joint opening, the measured value is slightly larger than the calculated value. Because the error parameters used in the calculation are conservative. It can also be seen from the comparison of $\mu_{0.025}$ and $\mu_{0.975}$ that the measured value is slightly larger.

Although there is some deviation between the calculated and measured value, from the perspective of the distribution form and range, the segment joint opening obtained by the calculation method used in this paper is more consistent with the actual value, but the calculation parameters still need to be further optimized.

The statistical results of the measured joint opening of the tunnel are shown in Figure 21 for when the segment leaves the shield tail at a certain distance.

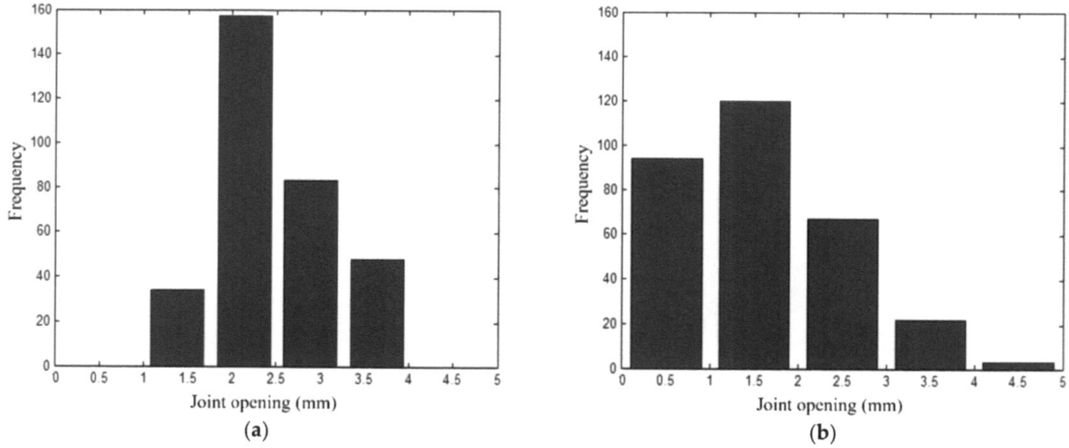

Figure 21. Error frequency distribution of the maximum joint openings in a ring measured at the forming tunnel. (**a**) Maximum circumferential joint opening; (**b**) Maximum longitudinal joint opening.

According to the measured results in Figure 21, the maximum joint opening of the tunnel can be controlled in the range of (0 mm, 4 mm). It can be seen that the segment assembly quality is guaranteed more under the above control measures.

3.3. Error Control Standard of Key Block Assembly

3.3.1. Calculation Results with Errors in the Z, T, and rR Directions

When errors occur in the *Z*, *T*, and *rR* directions, the end face of the T block is parallel to the end face of the LZ block, and the contact surface are still under approximately uniform stress. According to the calculation results, the compression damage factors under these 3 working conditions are 0.02, 0.03, and 0.04, respectively, and the tensile damage factors are 0.56, 0.49, and 0.85, respectively, which are far less than the limit value, and the reinforcement does not yield. It can be seen that errors in these three degrees of freedom have little effect on the force of the LZ block when acting alone.

3.3.2. Calculation Results with Errors in the *rZ* Direction

Figure 22 shows the curves of the maximum compression damage factor and the contact area of the *T*-LZ block with error in the *rZ* direction. It can be seen from Figure 22 that with the increase in the *rZ* direction error, the compression and tension damage factors gradually increase, the compression area gradually decreases, the final clearance continues to increase, and the LZ block is more biasedly compressed. When the error value is 1.6°, the maximum compression damage factor is 0.4, which is close to the critical damage factor of the compression failure of 0.46. In addition, the calculation shows that the crack width is far less than the limit value. When only the error in the *rZ* direction is considered, 1.6 ° can be taken as the error limit (error control standard).

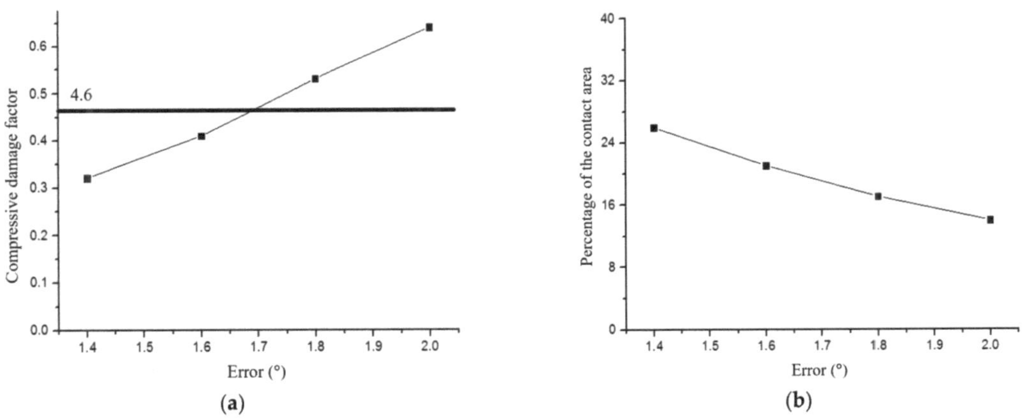

Figure 22. Maximum compression damage factor and contact area of the T-LZ block under *rZ* direction error. (**a**) Maximum compression damage factor; (**b**) Contact area of the T-LZ block.

3.3.3. Calculation Results under the *rT* Direction Error

Figure 23 shows the compression damage factor and contact of T-LZ block and under error in *rT* direction. It is found from the figure that with the increase in the error in *rT*± directions, the compression and tensile damage factor gradually increase, the compression area gradually decreases, and finally, the clearance increases. When the error value is ± 0.7°, the maximum compression damage factors are 0.67 and 0.52, respectively, which exceed the limit of the critical damage factor of 0.46 of compression failure. Through the calculation and analysis, it can be seen that the crack width is also far less than the limit value. Therefore, when the error in the *rT* direction acts alone, 0.6° can be taken as the error limit.

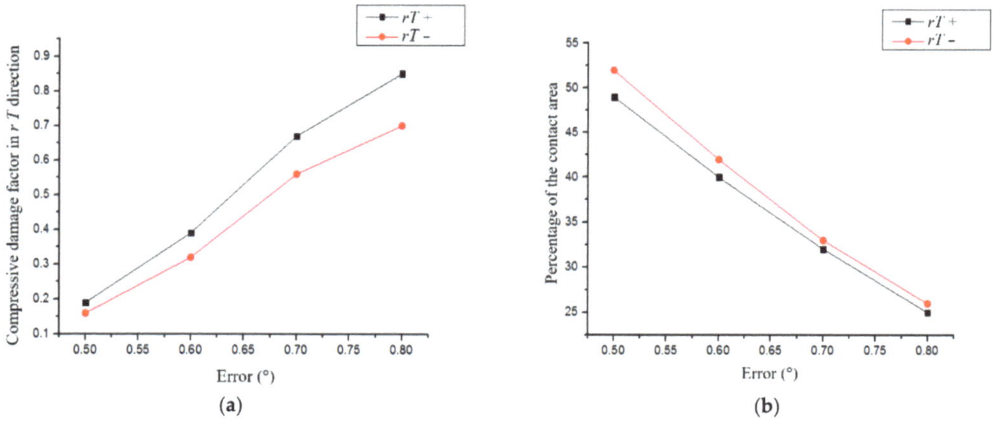

Figure 23. Maximum compression damage factor and T-LZ block contact area under *rT* direction error. (**a**) Maximum compression damage factor; (**b**) Contact area of the T-LZ block.

Table 7 presents the error limits of each direction under the error in the single degree of freedom. Considering that the rotation angle between T blocks cannot be measured directly during construction, it can only be obtained indirectly by measuring the clearance difference between the T block end face and LZ block end face (the difference between maximum clearance and minimum clearance).

Table 7. Error limits in the single degree of freedom direction.

Freedom Direction	*Z*	*T*	*rZ*	*rR*	*rT*
Error limit	6 mm	6 mm	1.6°	0.625°	0.6°
Clearance difference	—	—	9.8 mm	—	12.6 mm

3.3.4. Combination of the *rZ* Direction and *rT* Direction

According to the calculation under the separate errors in the *rZ* and *rT* directions, the clearance difference limits are 9.8 and 12.6 mm, respectively. It can be assumed that 9.8 mm is the limit value of the clearance difference under the combination of the *rZ* and *rT* directions. If it is verified that all the combinations with a clearance difference less than or equal to 9.8 mm meet the limit value of the compression and tensile damage factors, and the limit value of the clearance difference under the combination of the 9.8 mm errors in the *rZ* and *rT* directions can be determined.

The calculation results of typical combinations of clearance differences equal to 9.8 mm are shown in Table 8. From the table, it can be seen that the compression damage factor and tensile crack width meet the judgment criteria, and no reinforcement yields. Thus, the countless combinations between these four typical combinations are considered to meet the requirements. Therefore, 9.8 mm is determined as the clearance difference limit under the combination of the *rZ* and *rT* directions.

Table 8. Combined results of errors in the *rZ* and *rT* direction.

Combination	*rZ*	*rT*	Maximum Compression Damage Factor	Maximum Crack Width (mm)	Maximum Stress of Reinforcement (MPa)
1	0.2°	0.4°	0.32	0.073	269
2	0.5°	0.3°	0.36	0.062	264
3	0.9°	0.2°	0.23	0.048	275
4	1.3°	0.1°	0.41	0.082	311

4. Discussion

4.1. Analysis of Factors Affecting the Insertion Clearance of the F Block

In practical engineering, the insertion clearance of block F is often not allowed to be negative, and the tolerance value is considered to be −2 mm. Therefore, the influence of the manufacturing error and positioning error in different degrees of freedom on the insertion clearance of the F block was studied. The limit case when the standard deviation of the manufacturing error and the positioning errors in the *rT*, *rR*, and *rZ* directions is zero was simulated separately and analyze the influencing factors. The frequency distribution of the assembly error of block F with insertion clearance under the corresponding calculation conditions is shown in Figure 24, and a summary of the statistical parameters is shown in Table 9.

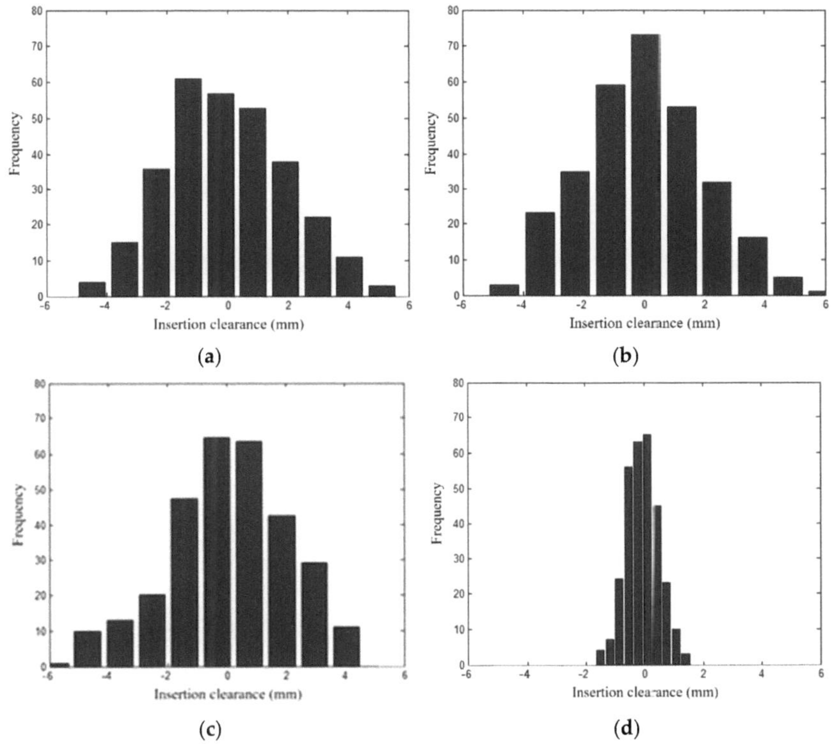

Figure 24. Influence of the manufacturing error and positioning error of each component on the insertion clearance of the F block. (**a**) No manufacturing error of the segment; (**b**) No positioning error in the *rT* direction; (**c**) No positioning error in the *rR* direction; (**d**) No positioning error in the *rZ* direction.

Table 9. Influencing factors of the insertion clearance of the F block.

Index	Statistics	No Manufacturing Error	*rT*	*rR*	*rZ*
Insertion clearance of F block	$\mu_{0.025}$ (mm)	−4.48	−3.42	−3.51	−4.50
	$\mu_{0.975}$ (mm)	3.62	3.99	3.83	3.56
	Average value (mm)	−0.24	0.03	−0.03	0.01
	Standard deviation (mm)	2.08	1.97	2.04	2.03

It can be seen from Figure 24 and Table 9 that the segment manufacturing error and positioning errors in the *rT* and *rR* directions have little effect on the insertion clearance of the F block. The main factor affecting the insertion clearance of the F block is the positioning error in the rZ direction. Therefore, the standard deviation of the positioning error in the rZ direction is reduced. The statistical relationship between the change in the standard deviation and the distribution of the insertion clearance of block F was calculated, as shown in Table 10.

Table 10. Influence of the standard deviation of the positioning error in the rZ direction on insertion clearance of the F block.

Index	Statistics	0.75 Times Smaller	0.5 Times Smaller	0.25 Times Smaller
Insertion clearance of F block	$\mu_{0.025}$ (mm)	−2.58	−2.22	−1.49
	$\mu_{0.975}$ (mm)	2.88	2.37	1.45
	Average value (mm)	0.09	0.03	0.01
	Standard deviation (mm)	1.43	1.17	0.76
	Probability of being less than −2 mm (%)	8.00	4.00	0.67

It can be seen from Table 10 that when the positioning error in the rZ direction is reduced by 0.5 times, the insertion clearance of block F is distributed between −2.22 and 2.37 mm, and the probability of being less than −2 mm is 4%. According to the calculation results, it is suggested that the positioning error in the rZ direction should mainly be controlled during the segment assembly construction, and the resulting joint dislocation and opening should be limited by 0.5 mm.

4.2. Analysis of the Most Unfavorable Attitude of the LZ Block

When there is a certain angle between the upper and lower T blocks, the LZ block attitude is uncertain. In this paper, the most unfavorable attitude of the LZ block is obtained by comparing and calculating the structural stress under the two attitudes.

Considering the error in the rT direction, it is assumed that there is a relative error θ between the upper and lower T blocks. The working condition 1 in Figure 25 shows that the end face of the LZ block is parallel to the end face of the lower T block and has an included angle θ with the end face of the upper T block. The working condition 2 shows that there is an included angle $\theta/2$ between the end face of the LZ block and the end faces of the upper and lower T blocks, respectively.

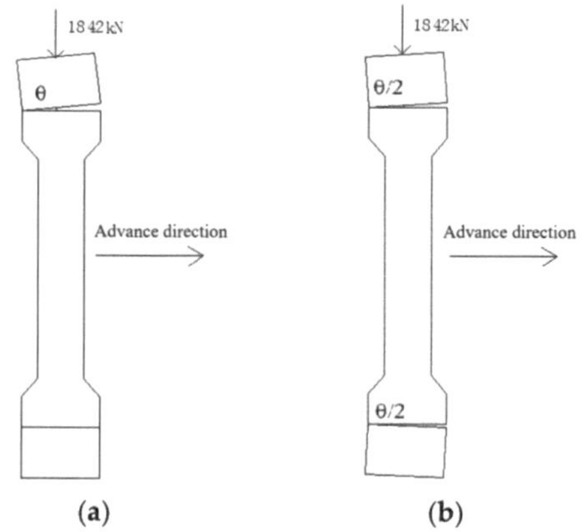

Figure 25. Calculation of the working conditions of the LZ block attitudes. (**a**) Condition 1; (**b**) Condition 2.

Through the calculation with multiple θ, it is found that the tension compression damage factor in condition 1 is larger and the damage distribution range is wider. To summarize, working condition 1 is more dangerous than working condition 2, and the LZ block can take working condition 1 as the most unfavorable attitude.

4.3. Failure Mode Analysis of the LZ Block

After analyzing the force of the LZ block when a large error occurs, the control standard of the error is put forward. The errors in two main directions (*rZ* and *rT*) are analyzed as follows. Taking the errors in the *rZ* and *rT* directions to be 1° and 2°, respectively, the distance of the contact between the T block and LZ block is calculated under the 2 directions are shown in Figure 26, where the black part is the contact region.

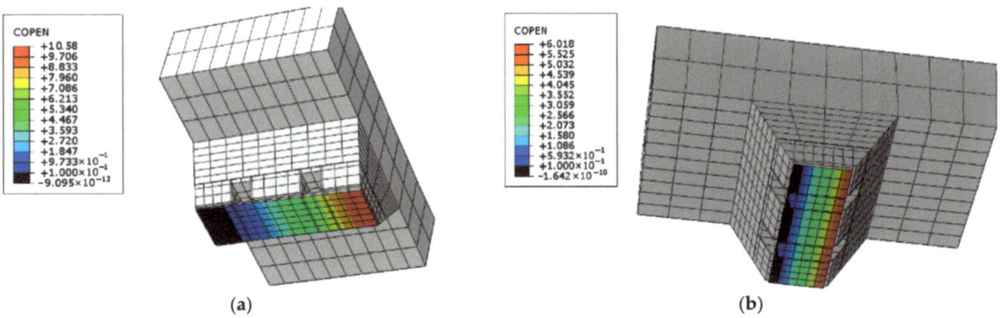

Figure 26. Space between the contact surface of the T-segment and LZ block. (**a**) Contact surface under *rZ* error; (**b**) Contact surface under *rT* error.

According to the finite element calculation, it is found that the damage position of the model is also mainly located near the contact surface of the T block and the LZ block while the damage factor of the LZ block main body is very small. From the point of view of reinforcement stress, the maximum stress of the reinforcement in the 2 working conditions is 370 and 350 MPa, which further verifies the conclusion that the LZ block will not suffer overall damage.

In summary, even under extreme errors, the LZ block will not suffer integral damage but only local damage near the contact surface. Therefore, the error control index should be put forward from the perspective of normal use requirements.

5. Conclusions

By establishing a geometric model of the segment of a quasi-rectangular shield tunnel and considering the 6-DOF displacement, this paper proposes a simulation calculation method for the segment assembly error, quantifies the manufacturing error and assembly positioning error, and compiles the assembly simulation program. A finite element model of the T-LZ block is established. Through the calculation and analysis of the relative error combination of the single degree of freedom and multi degree of freedom for the T block, the relative error control index of T block under corresponding working conditions is proposed. The conclusions are summarized as follows.

(1) The scheme in which the LZ block is assembled first can better realize error control. This scheme eliminates possible negative clearance during the assembly of the LZ block and has little impact on the insertion clearance of the capping F block. From the multi-ring assembly simulation results, the probability that the insertion clearance of the F block is less than −2 mm is even smaller. Thus, assembling the LZ block first will not affect the assembly quality of the F block.

(2) The positioning error in the *rZ* direction is the most significant factor affecting the assembly quality of the longitudinal joint and the insertion clearance of the LZ and F

blocks. From the calculation results, it can be concluded that the displacement in the *rZ* direction of the segments should be controlled first, followed by the displacement in the *rT* direction.

(3) Using photography and image recognition technology, the joint opening after segment assembly and forming was obtained. From the distribution shape and scope, the segment joint opening obtained by the calculation method recommended in this paper was found to agree with the field data. Considering that there is still a certain deviation between the calculated value and the measured value, the calculation parameters need to be further optimized.

(4) The LZ block will not suffer integral damage but only local damage on the contact surface. Errors in the *rZ* and *rT* directions will cause local contact between the T block and LZ block, resulting in the bias of the LZ block, which is the main factor causing the unfavorable stress in the LZ block. When errors in the *Z*, *T*, and *rR* directions act separately, the contact surface is still under a uniform force, and they have little effect on the force of the LZ block.

(5) Combined with the research results of the single-degree-of-freedom error and multi-degree-of-freedom error, 9.8 mm is taken as the control index of the sum of the upper and lower gap differences in the T−LZ block under the combination of multi-degree-of-freedom errors. In general, the relative error of the upper and lower T blocks meets the force requirements, and the T−LZ blocks will not be damaged by force due to assembly defects.

Author Contributions: Conceptualization, P.L.; Data curation, J.L. and J.F.; Formal analysis, X.L.; Investigation, P.C.; Methodology, P.L. and Z.D.; Project administration, Y.R. and J.F.; Resources, P.C.; Software, Z.D.; Supervision, Y.R. and X.L.; Validation, J.L.; Writing—original draft, X.W.; Writing—review & editing, Z.D. All authors have read and agreed to the published version of the manuscript.

Funding: This work supported by the "Social Development Project of Science and Technology Commission of Shanghai Municipality (21DZ1201105)", "the Fundamental Research Funds for the Central Universities (21D111320)", and "Shanghai Rising-Star Program of China (18QB1403800)".

Institutional Review Board Statement: Not applicable.

Informed Consent Statement: Not applicable.

Data Availability Statement: The data presented in this study are available on request from the corresponding author.

Acknowledgments: The research has been sponsored by the "Social Development Project of Science and Technology Commission of Shanghai Municipality (21DZ1201105)", "the Fundamental Research Funds for the Central Universities (21D111320)", and "Shanghai Rising-Star Program of China (18QB1403800)". The Experimental data used to support the findings of this study are available from the corresponding author upon request.

Conflicts of Interest: The authors declare no conflict of interest.

References

1. Loria, A.F.R. The thermal energy storage potential of underground tunnels used as heat exchangers. *Renew. Energy* **2021**, *176*, 214–227. [CrossRef]
2. Ma, C.; Di Donna, A.; Dias, D.; Zhang, J. Numerical investigations of the tunnel environment effect on the performance of energy tunnels. *Renew. Energy* **2021**, *172*, 1279–1292. [CrossRef]
3. Loria, A.F.R.; Di Donna, A.; Zhang, M. Stresses and deformations induced by geothermal operations of energy tunnels. *Tunn. Undergr. Space Technol.* **2022**, *124*, 104438. [CrossRef]
4. Zhu, Y.; Zhu, Y.; Huang, D.; Li, P. Development and application of rectangular shield tunnel technology. *Mod. Tunn. Technolgy* **2016**, *53*, 1–12.
5. Li, P.; Dai, Z.; Huang, D.; Cai, W.; Fang, T. Impact Analysis for Safety Prevention and Control of Special-Shaped Shield Construction Closely Crossing Multiple Operational Metro Tunnels in Shallow Overburden. *Geotech. Geol. Eng.* **2021**, 1–18. [CrossRef]

6. Gong, C.; Ding, W.; Mosalam, K.M.; Günay, S.; Soga, K. Comparison of the structural behavior of reinforced concrete and steel fiber reinforced concrete tunnel segmental joints. *Tunn. Undergr. Space Technol.* **2017**, *68*, 38–57. [CrossRef]
7. Zhou, P.; Lin, M.; Zhou, F.; Gong, L.; Ao, W. Bearing Characteristics and Failure Mechanism of a Novel Plate-Short Anchor Structure for Tunnel Crack Reinforcement. *Eng. Fail. Anal.* **2022**, *135*, 106160. [CrossRef]
8. Yuan, Y.; Jiang, X.; Liu, X. Predictive maintenance of shield tunnels. *Tunn. Undergr. Space Technol.* **2013**, *38*, 69–86. [CrossRef]
9. Shi, P.; Li, P. Mechanism of soft ground tunnel defect generation and functional degradation. *Tunn. Undergr. Space Technol. Inc. Trenchless Technol. Res.* **2015**, *50*, 334–344. [CrossRef]
10. Zheng, G.; Cui, T.; Cheng, X.; Diao, Y.; Zhang, T.; Sun, J.; Ge, L. Study of the collapse mechanism of shield tunnels due to the failure of segments in sandy ground. *Eng. Fail. Anal.* **2017**, *79*, 464–490. [CrossRef]
11. Chen, J.; Mo, H. Numerical study on crack problems in segments of shield tunnel using finite element method. *Tunn. Undergr. Space Technol.* **2009**, *24*, 91–102. [CrossRef]
12. Wang, M.; Dong, Y.; Yu, L.; Fang, L.; Wang, X.; Liu, D. Experimental and numerical researches of precast segment under radial dislocation conditions. *Tunn. Undergr. Space Technol.* **2019**, *92*. [CrossRef]
13. Krahl, P.A.; Palomo, I.I.; Almeida, S.J.D.C.; Siqueira, G.H.; Júnior, N.D.O.P.; Junior, L.C.M.V. Tolerances for TBM thrust load based on crack opening performance of fiber-reinforced precast tunnel segments. *Tunn. Undergr. Space Technol.* **2021**, *111*, 103847. [CrossRef]
14. Toutlemonde, F.; Quiertant, M.; Dubroca, S. Tunnel precast segments: An SFRC alternative. In Proceedings of the Fifth RILEM Symposium on Fibre-Reinforced Concretes (FRC), Lyon, France, 13–15 September 2000; Rossi, P., Chanvillard, G., Eds.; RILEM Publications: Paris, France, 2000.
15. Xu, G.; He, C.; Lu, D.; Wang, S. The influence of longitudinal crack on mechanical behavior of shield tunnel lining in soft-hard composite strata. *Thin-Walled Struct.* **2019**, *144*. [CrossRef]
16. Cavalaro, S.; Blom, C.; Walraven, J.C.; Aguadoa, A. Structural analysis of contact deficiencies in segmented lining. *Tunn. Undergr. Space Technol. Inc. Trenchless Technol. Res.* **2011**, *26*, 734–749. [CrossRef]
17. Xu, M.; Zhang, Z. Cause Analysis and Treatment of Segment Damage of Large Diameter Shield Tunnel. *Chin. J. Undergr. Space Eng.* **2013**, *9*, 1705–1712.
18. Lorenzo, S.G. In situ behaviour of an instrumented ring subjected to incipient TBM steering around a curve. *Eng. Struct.* **2021**, *249*, 113124. [CrossRef]
19. Sharghi, M.; Chakeri, H.; Afshin, H.; Daniel, D. Tunnel Lining Segments-Analysis of the Possible of Cracking and Damages during Installing Stage. *Soil Mech. Found. Eng.* **2021**, *58*, 287–294. [CrossRef]
20. Yang, Y.; Zhou, B.; Xie, X.; Liu, C. Characteristics and causes of cracking and damage of shield tunnel segmented lining in construction stage–a case study in Shanghai soft soil. *Eur. J. Environ. Civ. Eng.* **2017**, *22*, s213–s227. [CrossRef]
21. Jeang, A. Combined parameter and tolerance design optimization with quality and cost. *Int. J. Prod. Res.* **2001**, *39*, 923–952. [CrossRef]
22. Wu, C.; Tang, G. Tolerance design for products with asymmetric quality losses. *Int. J. Prod. Res.* **1998**, *36*, 2529–2541. [CrossRef]
23. Cavalaro, S.; Blom, C.; Walraven, J.; Aguado, A. Formation and accumulation of contact deficiencies in a tunnel segmented lining. *Appl. Math. Model.* **2011**, *36*, 4422–4438. [CrossRef]
24. Zhuang, Q.; Yang, Z. Six degree of freedom series segment assembly machine for quasi rectangular shield. *Mod. Tunn. Technol.* **2016**, *53*, 46–50.
25. Scott, D.W. Box–Muller transformation. *Wiley Interdiscip. Rev. Comput. Stat.* **2011**, *3*, 177–179. [CrossRef]
26. Han, J.; Liu, D.; Guan, Y.; Chen, Y.; Li, T.; Jia, D.; Yan, F.; Jia, P.; Zhao, Y. Study on shear behavior and damage constitutive model of tendon-grout interface. *Constr. Build. Mater.* **2022**, *320*, 126223. [CrossRef]
27. Huang, Z.; Zhang, C.-L.; Ma, S.-K.; Zhang, J.-B.; Zhu, Q.-X. Study of the mechanical behaviour and damage characteristics of three new types of joints for fabricated rectangular tunnels using a numerical approach. *Tunn. Undergr. Space Technol.* **2021**, *118*, 104184. [CrossRef]
28. Liu, J.; Shi, C.; Wang, Z.; Lei, M.; Zhao, D.; Cao, C. Damage mechanism modelling of shield tunnel with longitudinal differential deformation based on elastoplastic damage model. *Tunn. Undergr. Space Technol.* **2021**, *113*, 103952. [CrossRef]
29. Ye, Z.; Liu, H. Investigating the relationship between erosion-induced structural damage and lining displacement parameters in shield tunnelling. *Comput. Geotech.* **2021**, *133*, 104041. [CrossRef]
30. Hou, K.; Li, X.; Huang, D.; Wang, Y.; Hong, B. Simulation Study on Assembly Errors of Segments and Columns in Quasi-Rectangular Shield Tunnels. *Mod. Tunn. Technol.* **2016**, *53*, 269–277.
31. Jiang, X.; Zhang, Y.; Zhang, Z.; Bai, Y. Study on Risks and Countermeasures of Shallow Biogas during Construction of Metro Tunnels by Shield Boring Machine. *Transp. Res. Rec. J. Transp. Res. Board* **2021**, *2675*, 105–116. [CrossRef]
32. Lei, M.; Zhu, B.; Gong, C.; Ding, W.; Liu, L. Sealing performance of a precast tunnel gasketed joint under high hydrostatic pressures: Site investigation and detailed numerical modeling. *Tunn. Undergr. Space Technol.* **2021**, *115*, 104082. [CrossRef]
33. Xue, Y.; Li, Y. A Fast Detection Method via Region-Based Fully Convolutional Neural Networks for Shield Tunnel Lining Defects. *Comput. Civ. Infrastruct. Eng.* **2018**, *33*, 638–654. [CrossRef]
34. Zhu, H.; Li, X.; Hu, C. A Method and Device for Measuring the Opening Width of a Shield Tunnel Lining Segment Joint: CN 2011. Available online: https://patents.google.com/patent/CN102297660A/en (accessed on 10 February 2022).
35. Ding, W.; Gong, C.; Mosalam, K.M.; Soga, K. Development and application of the integrated sealant test apparatus for sealing gaskets in tunnel segmental joints. *Tunn. Undergr. Space Technol.* **2017**, *63*, 54–68. [CrossRef]

energies

Article

Study on Influence of Joint Locations and Hydraulic Coupling Actions on Rock Masses' Failure Process

Yunjuan Chen [1,2,*], Tao Gao [2], Fuqiang Yin [3], Xiaozhi Liu [2] and Jun Wang [1,2]

1 Key Laboratory of Building Structural Retrofitting and Underground Space Engineering (Shandong Jianzhu University), Ministry of Education, Jinan 250101, China; wangjun@sdjzu.edu.cn
2 Civil Engineering College, Shandong Jianzhu University, Jinan 250101, China; gaotaosju@126.com (T.G.); xiaozhisju@126.com (X.L.)
3 Shandong Provincial Institute of Land Surveying and Mapping, Jinan 250102, China; yinfuqiang@shandong.cn
* Correspondence: chenyunjuan@sdjzu.edu.cn

Abstract: Distribution of joints and fissures under hydraulic coupling condition is particularly critical to the stability of surrounding rock masses in underground engineering construction. In this paper, DDARF (Discontinuous Deformation Analysis for Rock Failure) and RFPA (Rock Failure Process Analysis) are compared and analyzed firstly based on laboratory tests. Then using preferred software RFPA, the failure process, stress state, acoustic emission characteristics and energy dissipation laws of rock masses with different joint locations are analyzed under the hydraulic coupling condition. Results show that a large tensile stress region is generated on both ends of the original joint with the micro-cracks' propagation, water pressure in cracks promotes the generation of tensile stress to a certain extent, damage effect angle increases gradually from the rock specimen with the middle joint to that with the marginal joint; the same water pressure has a certain auxiliary effect on the main crack failure when the joint is close to the middle part of the specimen, and has a dominant effect on the local crack failure when the joint is far away from the middle; the maximum water pressure shows the "U" shaped distribution. At low initial water pressure, stresses of specimens with symmetrical joint locations have similar evolution trends, while at high initial water pressure, the water pressure in cracks has significant dissipation and thus the maximum water pressure in the system does not exceed the initial value. The length of the main crack path is positively proportional to the number of acoustic emissions and the energy accumulation capacity, and evolution of the damage variable basically shows a development trend of steady growth-rapid growth-steady growth.

Keywords: jointed rock masses; hydraulic coupling; crack propagation; acoustic emission; energy accumulation and dissipation

Citation: Chen, Y.; Gao, T.; Yin, F.; Liu, X.; Wang, J. Study on Influence of Joint Locations and Hydraulic Coupling Actions on Rock Masses' Failure Process. *Energies* **2022**, *15*, 4024. https://doi.org/10.3390/en15114024

Academic Editors: Jianyong Han, Li Tian and Helena M. Ramos

Received: 17 April 2022
Accepted: 25 May 2022
Published: 30 May 2022

1. Introduction

With the rapid development of underground engineering construction, the complexity of engineering geological environment is increasing, which poses a more difficult challenge to the stability of engineering rock masses [1,2]. Due to the uneven distribution of natural joints and fissures in rock masses, as well as the fissure water, crack propagation of rock masses threatens the stability of project and even brings huge safety risks to the construction [3,4]. Therefore, it is of great significance to study the failure mechanism of rock masses under complex hydraulic coupling conditions and to carry out the study of crack propagation laws of rock masses with different joint locations.

Under the hydraulic coupling action, rock masses bear complex external geological environment, and many uncertain factors will affect rock failure, thus crack propagation would become more complex [5]. Abbass [6] analyzed the crack propagation law of layered sandstone under hydraulic coupling effect from a microscopic point of view, established five visually similar sandstone bodies with different microscopic parameters, and studied the influence of scale parameter's differences on the macroscopic failure of samples through Brazil splitting test. Yashwanth [7] monitored the acoustic emission laws of sandstone samples under the hydraulic coupling condition, analyzed the frequency of the acoustic emission, and proved that the acoustic emission activities existed discontinuously in the process of rock failure. Chen Z.Q. [8] carried out the hydraulic coupling test of sandstone under different confining pressures, analyzed the acoustic emission characteristics of rock failure by using acoustic emission technology and three-dimensional space positioning system, and obtained the specific evolution process of crack propagation in three-dimensional state. In terms of numerical simulation research, Lian Z.L. [9] and Zhang S.C. [10] used ABAQUS and developed relevant hydraulic coupling subroutine to study the crack propagation mechanism under water pressure, taking into consideration a variety of comprehensive and complex factors such as rock mechanical parameters, water pressure, ground stress and boundary conditions. Li G. [11] applied the finite difference software FLAC3D 3.00 (Fast Lagrangian Analysis of Continua) to conduct numerical analysis on a semicircle arched roadway, and analyzed the evolution laws of stresses and displacements generated by deep soft rock excavation under the influence of fissure water pressure.

For the rock failure process, the study of energy accumulation and dissipation laws can provide good guidance for the early hazard warning and disaster prevention [12–14]. Zhao Z.G. [15] used rock acoustic emission technology to conduct uniaxial whole-process loading experiments, and proposed the energy determination index K_{SN}, through which the failure time of rock specimens can be predicted. Jin Y. [16] carried out true triaxial physical hydraulic fracturing experiment on natural shale, extracted the numbers and energies of acoustic emission events to analyze the crack propagation laws. Based on the uniaxial and triaxial tests of yellow sandstone, Liu G. [17] analyzed the mechanical behavior and energy evolution laws in the rock failure process. Compared with laboratory tests, numerical simulation can also monitor the energy accumulation and dissipation laws precisely [18,19]. Hiroyuki S. [20] used the fluid-solid coupling DEM (discrete element method) to conduct a series of simulations on hydraulic fracturing process, and the simulation results were highly consistent with laboratory tests, which verified the feasibility of using numerical software to monitor acoustic emission. Yu J. [21,22] and Lin C. [23] analyzed the crack propagation process of sandstone under triaxial compression based on PFC (particle flow code), obtained the mesoscopic energy change response and studied the energy dissipation and conversion laws.

In this paper, the discontinuous deformation characteristics of rock masses will be studied, the reasonable simulation programs will be compared and determined, the influence of original joint location on rock failure will be analyzed, and the crack propagation laws, water pressure distributions, acoustic emission characteristics and energy dissipation laws will be deeply studied; at the same time, the custom acoustic emission damage constitutive model will be put forward and the damage variable evolution law will be analyzed.

2. Simulation Programs and Comparison Optimization on Rock Failure

Two kinds of software, discontinuous deformation analysis for rock failure (DDARF) and rock failure process analysis (RFPA), both can be good at simulating crack propagation and analyzing the stability in rock masses. One belongs to the discontinuous method, while the other is the continuous method; these two methods are compared in the terms of unit processing, parameters assignment, and cracking simulation as follows, and combined with the experimental study, the optimal algorithm of rock failure simulation under hydraulic coupling condition is determined.

2.1. Implementation of Crack Propagation in DDARF

DDARF has the advantage that neither the finite element method (FEM) nor the discrete element method (DEM) can replace it in studying the stability of jointed rock masses. It not only has strict theoretical basis, but also can calculate the large displacement and deformation of rock masses. DDARF is applicable to analyze any situations of continuous, discontinuous and completely discrete rock masses. It has some proprietary characteristics for simulating crack propagation [24,25]:

(1) Generation of triangular block system

Based on joints' probability distribution in underground rock engineering (such as exponential distribution or uniform distribution, etc.), the probability model of the joint network is firstly generated in the DDARF method, then the complex calculation domain is decomposed into the combination of multiple simply connected subdomain, and the triangle grids are generated in each subdomain using the traveling wave method; thereupon, the whole triangular block system can be obtained by integrating these subdomains.

(2) Weibull distribution of material heterogeneity

In the triangular block system, the non-uniformity of mechanical parameters is realized by Weibull distribution. If the number of triangular blocks is n, the linear congruence method will be firstly used to generate n random numbers, which obey uniform distribution in the interval [0, 1], then the direct sampling method is used to obtain the elastic modulus and Poisson's ratio, which also obey Weibull distribution, and then are assigned to these triangular blocks.

(3) Cracking algorithm of virtual joints

In DDARF, the block boundary is virtual joint, and the coherence algorithm is used for judging the strength limit of it. When the cohesion of virtual joint reaches its strength limit, the contact springs between blocks will fail, and tensile failure or shear failure will occur, respectively, judged by the maximum tensile stress criterion and the Mohr–Coulomb criterion, shown in Figure 1 [26]. After failure, the virtual joint will become a true joint, and the mechanical parameters are reduced from virtual joint to true joint accordingly.

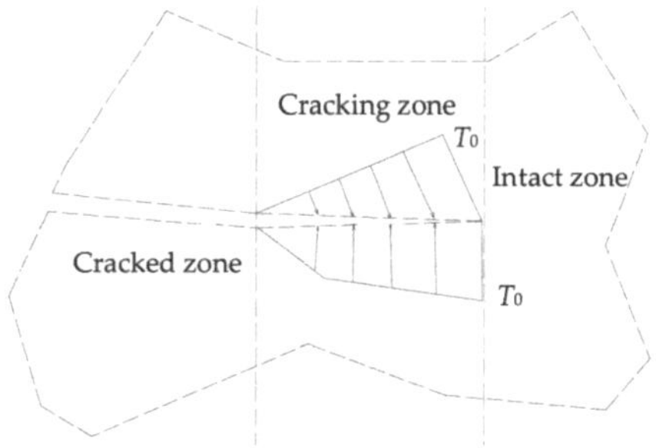

Figure 1. Crack propagation in discontinuous deformation analysis for rock failure (DDARF) [26].

2.2. Implementation of Crack Propagation in RFPA

In RFPA, it is believed that mesoscopic heterogeneity is the fundamental cause of macroscopic nonlinear failure of quasi-brittle materials, and statistical damage constitutive relation is used to study the randomness of rock masses' heterogeneity and defect distribution. RFPA is a rock failure process analysis system which uses the elastic mechanics to analyze stress and strain, the elastic damage theory and the modified Mohr–Coulomb criterion to analyze elements' deformation and failure. In the process of realizing rock masses' failure simulation, RFPA has the following characteristics [27,28]:

(1) Rock is dispersed into a numerical model composed of meso-quadrilateral elements, it is isotropic and elastic-brittle in meso-scale, and the macroscopic failure behavior of rock masses is the collective effect of mechanical behavior of these meso-quadrilateral elements.
(2) Mechanical parameters of these meso-quadrilateral elements obey Weibull distribution, which is the same as the DDARF method. Thus, the relationship between mesoscopic and macroscopic mechanical properties of rock masses is established.

(3) Phase change and damage criterion are introduced to analyze the contact between elements fails or not, and the phase transition point is judged by the modified Mohr–Coulomb criterion. On the other hand, elements' stress and strain is calculated by the finite element method (FEM, which is used as the stress analysis solver). It should be noted that the stress analysis solver is independent with the analysis of phase change. At the same time, in RFPA, it is assumed crack propagation in rock masses is a quasi-static process, and thus the effect of inertia force is ignored.

2.3. Laboratory Test and Numerical Simulation on Crack Propagation of Rock-Like Specimens

2.3.1. Laboratory Test on Crack Propagation

To study the impact of original joints' location on rock masses' crack propagation and damage rule, from the perspective of efficiency and high repeatability, sandstone is taken as the prototype, and the rock-like specimen is made based on the similarity principle [29,30]. Rock-like materials include sand, cement, water reducing agent and water with a quality ratio of 0.97:1.00:0.03:0.30; the solid materials are firstly poured into the stirrer (rated power: 2600 w), fully stirred, and then the water reducing agent is poured, still fully stirred. The role of the water reducing agent is to reduce the amount of water and cement, reduce the setting time of cement mortar and improve the plasticity of cement mortar. Joints are simulated by polyvinyl chloride sheets with a size of 0.5 mm × 15 mm. Rock-like specimens are water-cured for about two weeks, according to the preparation method of concrete members (seen Figure 2a), and then loaded for analysis after 28 days of curing.

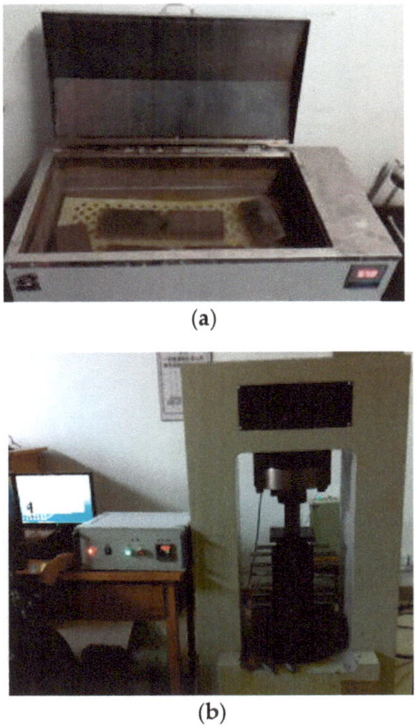

Figure 2. Preparation of rock-like specimen with (**a**) the Curing box and (**b**) the GAW-2000 rigid testing machine.

The rock-like specimen's processing device is a multi-functional mold independently developed by our own team, which can make specimens with multiple parallel or cross joints with different angles. Size of the specimen is 140 mm × 70 mm × 45mm [26], it is loaded by the GAW-2000 rigid testing machine, and firstly force control, when the machine and specimen face fully contacted, then converted to displacement control, with the loading rate 0.5 mm/min; the loading machine is shown in Figure 2b. Comprehensive material tests are carried out and the obtained physical and mechanical parameters of rock-like specimens and sandstones are shown in Table 1.

Table 1. Comparison of physical and mechanical properties of rock-like specimen and sandstone.

Materials	Compressive Strength /MPa	Tensile Strength /MPa	Elastic Modulus /GPa	Poisson Ratio	Density/(g/cm^3)
Rock-like specimen	55	5	6	0.20	2.25
Sandstone [31]	20~170	4~25	3~35	0.02~0.25	2.10~2.40

As can be seen from Table 1, the physical and mechanical test parameters of rock-like specimens are all within the variation range of sandstone. Therefore, rock-like specimens can be used to analyze the mechanical properties of sandstone to a certain extent.

In order to study the influence of original joints' location on rock masses' crack propagation, all the joints' inclination angles are set at 30°, because this angle has the most significant influence on rock masses' crack propagation [26], and joints' location is divided into five conditions from the top part of the specimen to the bottom part: the spacing of each joint is 20 mm, and the marginal joint is 30 mm away from the end of the specimen. The specific locations of joints are shown in Figure 3, and the crack morphologies of specimens after loading are shown in Figure 4.

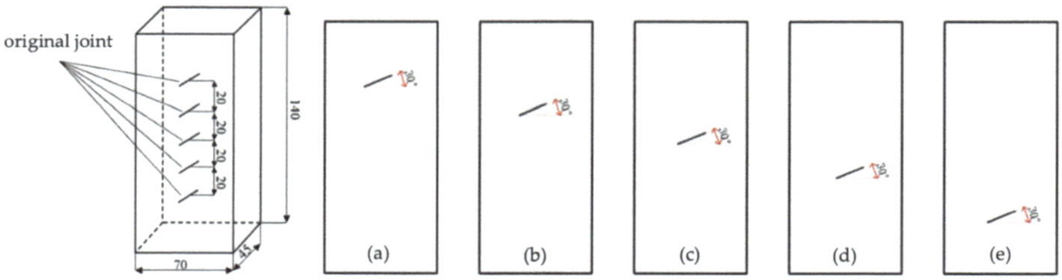

Figure 3. Original joint location and the space between joints with (**a**) top joint, (**b**) upper joint, (**c**) middle joint, (**d**) lower joint and (**e**) bottom joint.

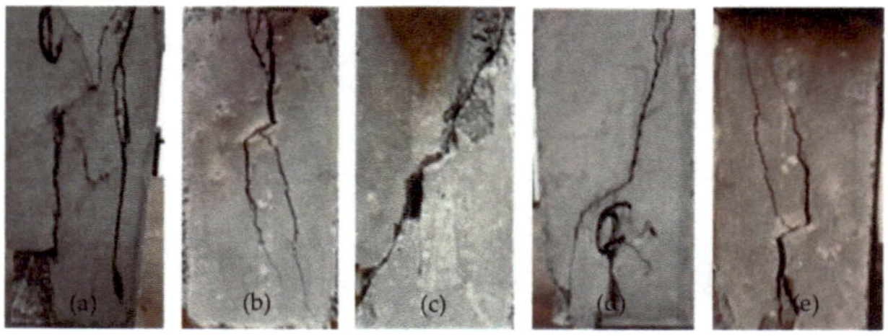

Figure 4. Crack morphology of rock-like specimens with different joint location with (**a**) top joint, (**b**) upper joint, (**c**) middle joint, (**d**) lower joint and (**e**) bottom joint.

From Figure 4, when the original joint is in the top or bottom part of the rock-like specimen (Figure 4a,e), crack will propagate along the tip of the original joint under compressive stress, and the wing cracks are rapidly expanded, the specimen will produce obvious tensile cracks, approximately parallel to direction of the maximum principal stress. When the original joint is in the upper or lower part of the specimen (Figure 4b,d), the wing cracks' propagation is still significant, secondary cracks are also produced and propagated concomitantly. In this condition, cracks are also mainly tensile cracks; while, when the original joint is in the middle part of the specimen (Figure 4c), there are no secondary cracks produced, and the final failure of the specimen is mainly caused by the wing cracks' propagation along with the original joint, which belongs to shear failure.

2.3.2. Numerical Simulation and Optimization Analysis on Crack Propagation

In order to compare and optimize the numerical simulation method in analyzing sandstone's failure process, take the original joint located in the middle of the specimen for an example; DDARF and RFPA are both used to carry out the numerical simulation. The joint angle is still 30° and the specimen size is still 140 mm × 70 mm × 45 mm. The physical and mechanical parameters are adopted in Table 1. Comparison between numerical simulation results and test results is shown in Figure 5.

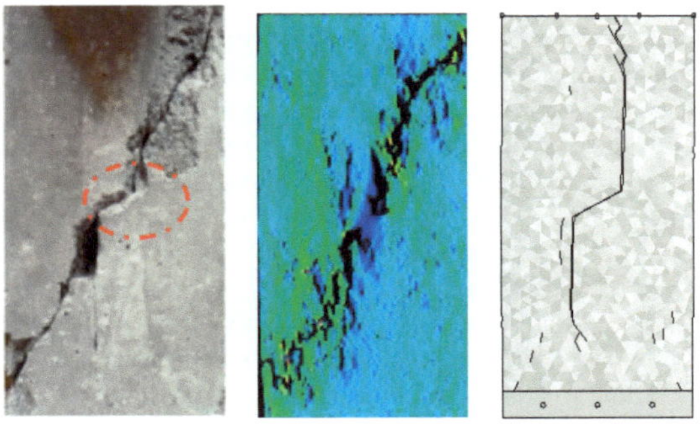

(a) (b) (c)

Figure 5. Crack propagation comparison between laboratory test and numerical simulations with (**a**) laboratory test, (**b**) rock failure process analysis (RFPA) numerical simulation, and (**c**) DDARF numerical simulation.

As can be seen from Figure 5, the laboratory result is in line with lab tests of the reference [32], when the original joint is located in the middle part of the specimen, the crack propagation of the specimen is mainly the wing cracks, and the secondary cracks can be neglected. In the laboratory test and RFPA numerical simulation, the crack propagation gives priority to shear crack. In the DDARF numerical simulation, the direction of crack propagation is basically parallel to the direction of maximum principal stress, and the tensile crack is the main one. Simulation result of RFPA is more in agreement with the laboratory test in terms of crack propagation morphology and its mechanical properties. Accordingly, in view of the superiority of RFPA in analyzing rock masses' crack propagation, influences of joints' location on rock failure and energy dissipation under hydraulic coupling conditions will be carried out based on RFPA in this paper.

3. Study on Rock Failure Process under Hydraulic Coupling Condition

3.1. Rock Failure Process with Different Joint Locations of the Same Water Pressure

The RFPA seepage stress coupling analysis module for rock failure is used to conduct numerical simulation, and a plane strain rectangular model with the size of 70 mm × 140 mm is established. A total of 9800 units are divided, and the mesh size of each unit is 1.0 mm × 1.0 mm. Shear failure of the units is determined by the Mohr-Coulomb criterion. Mechanical parameters of the model are shown in Table 2. The rock failure process is studied by locating the maximum principal stress and acoustic emission. The water pressure and crack propagation laws of rock specimens under the axial pressure and 100 m water head (1MPa water pressure) are analyzed with the original joint angle 30°.

Table 2. Mechanical parameters of rock specimens.

Mechanical Parameters	Value
Homogeneous degree/m	2
Elastic modulus/GPa	3
Poisson ratio	0.2
Density/(g/cm^3)	2.25
Frictional angle/°	45
Compressive strength vs tensile strength	10
Permeability coefficient /m·d^{-1}	0.1
Water pressure coefficient	0.1

In stress simulation, the red area represents tensile stress and the blue area represents shear stress. Through setting acoustic emission post-processing, the amount of acoustic emission in rock masses' whole failure process is recorded, and the corresponding maximum principal stress evolution diagram is intercepted to analyze the mesoscopic propagation path of cracks. In the maximum principal stress diagram, the stress state of each region in numerical simulation can be understood in real time; at the same time, the shear and tensile areas are key to accurately simulate the stress of rock masses. Therefore, those marks made to distinguish different stress states are necessary.

According to the principle of controlling variable method, the axial compressive stress is maintained, and the pore water pressure diffusion laws and stress evolution laws in the rock failure process are simulated and analyzed with different joint locations. Calculation results are shown in Figure 6.

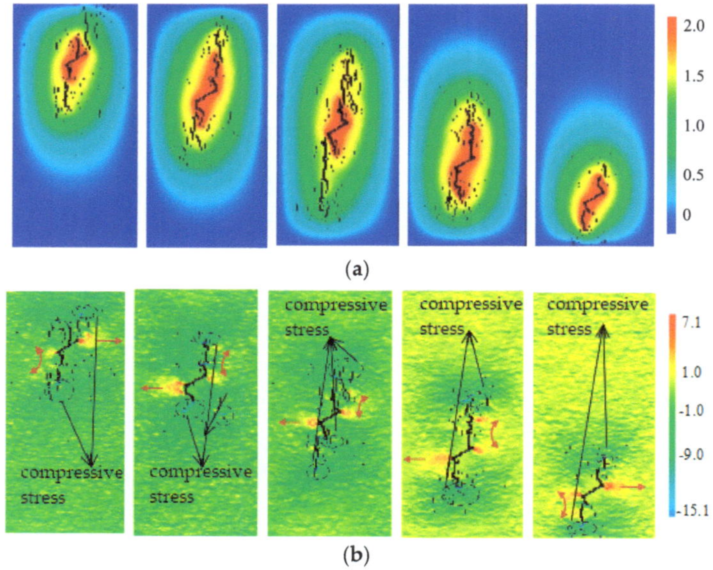

Figure 6. Distribution of water pressure and principal stress in crack propagation process with different joint locations with (**a**) water pressure diffusion during crack propagation, and (**b**) principal stress distribution during crack propagation.

As can be seen from Figure 6, water pressure mainly exists in the original joint and dissipates slowly toward the wing crack. When the original joint is in the middle part of the rock specimen, the large water pressure diffusion range is smaller than that of other joint locations. This water pressure mainly spreads uniformly around the original joint and propagated cracks and decreases unit by unit. At the same time, it can be found that the water pressure has a great promotion effect on the crack propagation, and it has a direct relationship with the tensile stress generated in the rock specimen; the decreasing direction of the water pressure gradient is consistent with the direction of the tensile stress. Water pressure is higher in the original joint areas, and the surrounding areas are affected slightly. The closer the original joint is to the end of the specimen, the smaller the water pressure range is, affected by the wing crack propagation, and the water pressure is more likely to diffuse and lead to the failure of the rock specimen earlier.

The maximum water pressure in these jointed rock specimens are shown in Figure 7. When the original joint is located in the middle part of the rock specimen, the maximum water pressure is the lowest; conversely, when the original joint is located at both ends of the rock specimen, the maximum water pressure is the highest, shown as the "U" shaped distribution. The result shows that, when the original joint is located in the middle, the water pressure has a certain auxiliary effect on the rock failure, while when the original joints are far from the center of the rock mass, the water pressure in the rock mass increases, which plays a leading role for the local rock failure.

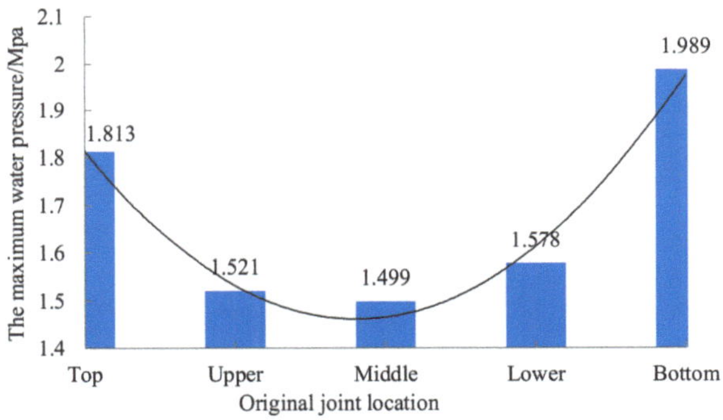

Figure 7. The maximum water pressure of rock specimen with different joint locations.

After crack initiation, expansion and coalescence, rock masses will enter the stage of rapid failure. Figure 8 shows the schematic diagram of crack microscopic evolution and stress evolution of rock masses at this stage.

As can be seen from Figure 8a, the failure of a specimen is mainly caused by the continuous propagation of wing cracks and secondary cracks. Under the action of axial compression, the new cracks generated meet with each other and penetrate to the end of the specimen, resulting in the instability and complete failure of rock masses. In order to further study the failure laws of rock specimen, the angle between the failure range of the rock specimen is defined as the failure effect angle α, which mainly reflects the influence of the failure range when they are completely destroyed. Figure 8b shows that, when the original joint is in the middle, angle α is the smallest, and approximately parallel to the axial stress direction, the closer the original joint is to the middle, the greater the stress is. In addition, it can be seen that the compressive stress areas produced by the middle joint are significantly larger than those of other locations' joints. The stress concentration phenomenon is mainly related to the location of cracks and accumulations of stress energy. At the same time, because the water pressure in cracks has a certain role in promoting the wring cracks' propagation, there are numbers of all kinds of tensile stress clouds around the main crack surface, thus making it that all the specimens' failures give priority to tensile damage finally.

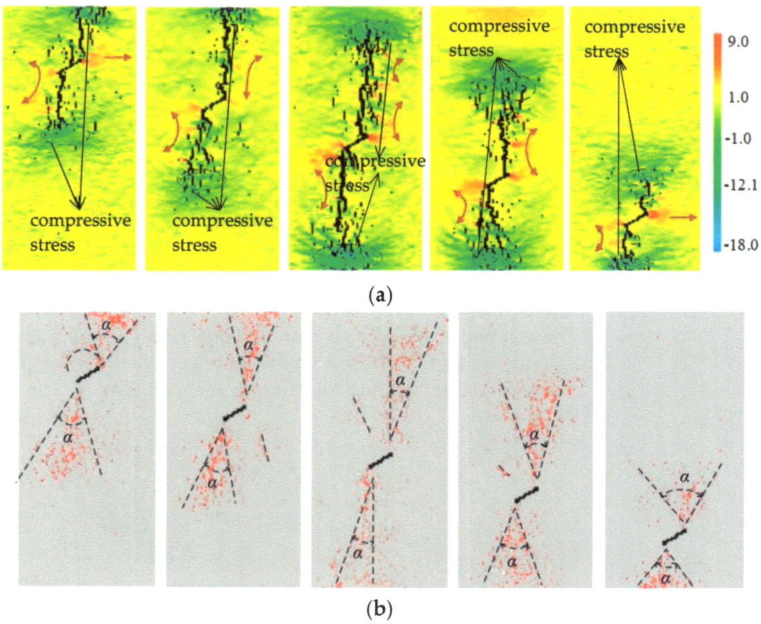

Figure 8. Stress and crack morphology evolution laws at failure stage with (**a**) stress evolution with different joint locations and (**b**) crack microscopic evolution with different joint locations.

3.2. Rock Failure Process with Different Joint Locations and Different Water Pressure

Simulations with different initial water pressures and different joint locations are done to study the influence on rock strength, results are shown in Figure 9.

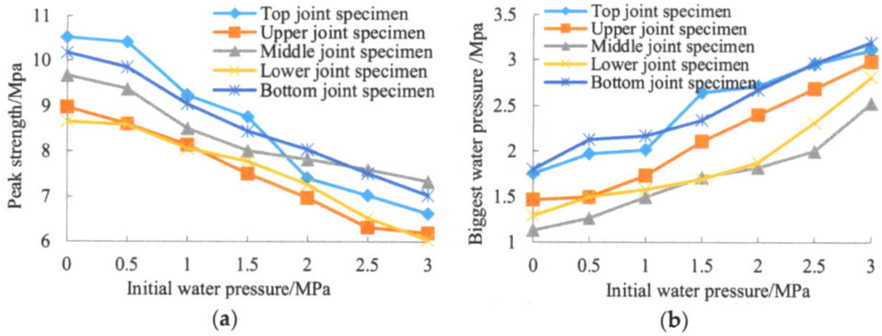

Figure 9. Influence of different initial water pressure and different joint locations on rock specimens with (**a**) peak strength vs. initial water pressure, and (**b**) biggest water pressure vs. initial water pressure.

As can be seen from Figure 9a, the peak strength of specimens with different joint locations are all inversely proportional to the initial water pressure. In the lower initial water pressure range from 0.0 MPa to 1.5 MPa, rock specimens with the symmetrical joint locations have a highly coincidence relationship, and the peak strength of rock specimens with top or bottom joint location is greater than that of other joint locations. When the initial water pressure exceeds 1.5 MPa, the peak strength of the rock specimen with top joint location is gradually smaller than that of other locations, and the peak strength decreases greatly. The peak strength intensity of specimens with upper or lower joint location is small, and changes of these two with the initial water pressure are similar. However, the peak strength drop of the specimen with top joint location increases, and the threshold phenomenon of the relative region appears. This also indicates that under the condition of relatively high initial water pressure, the damage effect of the water pressure on the rock specimen increases when the joint is far away from the middle of the rock specimen.

As can be seen from Figure 9b, under different initial water pressures, the maximum water pressure of rock specimens with different joint locations is positively correlated with the initial water pressure. The maximum water pressure of rock specimens with top and bottom joint is the largest, and their change trends with different water pressures are highly approximate. However, when the initial water pressure reaches the relatively high value of 2.0 MPa, the biggest water pressure of rock specimens with middle joint location is smaller than its initial water pressure. The reason for this may be that newer microcracks are produced because of the wing cracks' propagation, which can promote diffusing the water pressure, which indicates that under this higher initial water pressure, the axial pressure damage has the leading role for rock failure, and the influence by water pressure is not the primary factor.

4. Analysis on Energy Dissipation Laws in Rock Failure Process

4.1. Acoustic Emission Energy Analysis

When mechanical behavior occurs in the numerical simulation, the RFPA system itself can monitor and record the stress, strain signal and acoustic emission energy in the process of rock failure, thus we can extract the related data and disposal it. Under the action of hydraulic coupling of jointed rock masses, the whole system follows the first law of thermodynamics (ignoring that the system does not exchange heat with the outside world), that is, the total energy input into the system is all converted into the elastic strain and damage dissipation energy inside the rock masses, and the energy changes at each stage are analyzed through numerical simulation. Take the water head 100 m as the example, the damage acoustic emission energy in the rock failure process is studied; considering the symmetry, two cases of specimens with top and middle joints are taken here to analyze the acoustic emission laws, shown in Figure 10.

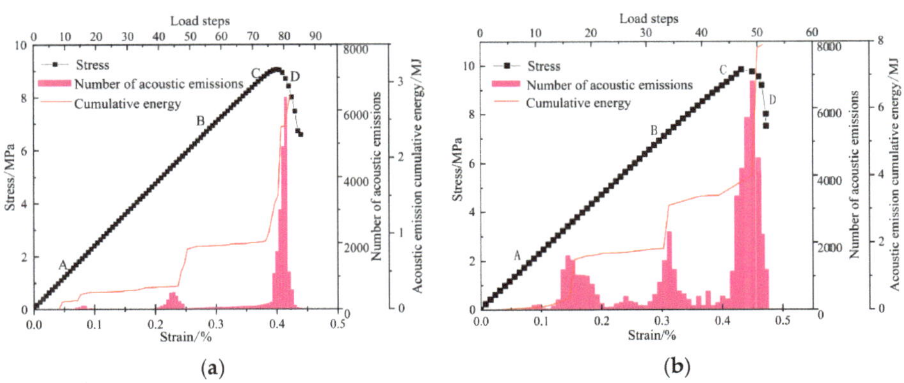

Figure 10. Stress–strain relationship and acoustic emission laws with loading steps with (**a**) the specimen with top joint, and (**b**) the specimen with middle joint.

Seen from Figure 10, the whole loading process can be divided into four stages, including the compaction stage (OA stage), the microcrack initiation and propagation stage (AB stage), the cracks' rapid growth stage (BC stage), and the failure stage (CD stage), similar to conference [32]. When the joint locates at the top part of the rock specimen, as shown in Figure 10a, the acoustic emission phenomenon is not produced at stage OA. At the stage AB, a relatively small number of acoustic emission peaks are produced, mainly due to the axial compression and the friction between micro-cracks, and the acoustic emission energy accumulation has a larger boost at this stage. When closer to the failure stage, the acoustic emission number has nonlinear rapid growth, and the cumulative energy can reach 2.83 MJ; cracks propagate rapidly, and when the axial stress reaches 9.03 MPa, the crack propagates to the specimen end and rock fails.

When the joint locates at the middle part of the rock specimen, as shown in Figure 10b, its acoustic emission law is similar with that of Figure 10a. At the OA stage, the amount of acoustic emission is relatively small. The acoustic emission phenomenon is "delayed" because the amount of acoustic emission does not increase significantly during the stage AB, which shows that the rock specimen can store lots of energy and the cumulative energy becomes larger; while with the axial stress increases, the acoustic emission amount is relatively smaller, due to the water pressure diffusing to the new wing cracks. When the specimen enters the failure stage and reaches the peak strength, the amount of acoustic emission increases rapidly and the energy accumulation also presents a nonlinear growth trend; the accumulated energy reaches 7.89 MJ, which is larger than the specimen with top joint.

These two cases show that, at the stage OA, small wing cracks occur at the original joint tip and water pressure also weakens the acoustic emission phenomenon, so the amount of acoustic emission is very small, and the energy accumulation value is almost zero. When entering the stage AB, the amount of acoustic emission gradually increases, and appears as a relatively small peak area. The closer the original joint is to the middle of rock specimen, the more the acoustic emission amount in this stage, which better proves secondary cracks are produced in the middle-jointed specimen at this stage. At the stage BC, affected by the water diffusion, the acoustic emission amount is less, which shows that the wing cracks' growth is greatly influenced by water pressure; from the perspective of the acoustic emission amount, these two cases at this stage are almost in the quiet period, while the rock specimen with the middle joint is relatively active, which is the result of secondary cracks' further propagation. Additionally, at the failure stage CD, cracks are connected and fused with each other in these two cases, and the macroscopic crack surface appears, the acoustic emission amounts all show nonlinear growth, the specimens are unstable and destroyed instantly, and the residual strength is small. The acoustic emission energy accumulation is an obvious periodic; energy growth often occurs before the sharp increase of the acoustic emission number, and the closer the original joint location to the middle of the rock specimen, the longer the crack propagation length, and the greater the accumulated energy accordingly, which indicates that the length of the main crack propagation length is proportional to the energy accumulation.

4.2. Damage Evolution Process Analysis

The essence of rock masses' deformation and failure is the accumulation and release of internal energy. In different deformation stages, the energy release of rock masses has different characteristics. Numerical simulation can be used to monitor AE, and thus the laws of acoustic emission and energy release can be studied. Assuming that the amount of acoustic emission is W_0 when the rock specimen completely fails, and the amount of acoustic emission is W_t at the time t before fails, thus the damage variable D' of rock masses at time t can be defined as:

$$D' = W_t / W_0 \tag{1}$$

According to the statistics of acoustic emission numbers in the rock failure process, evolution laws of damage variable D' in rock masses' damage process are analyzed with the load steps, as shown in Figure 11.

As shown in Figure 11, rock masses' deformation damage can be divided into four stages: (1) Initial damage stage. At this stage, the specimen is in the compaction stage, and the value of damage variable D' basically increases from zero to a tiny value. The main reason for this phenomenon is that rock masses are in the energy accumulation and storage stage at this time, undergoing the compaction of original joints and the closure of micro-pores and micro-cracks. (2) Damage stable development stage. Micro-cracks initiate and propagate at this stage, and it begins to propagate at the end of the original joint, accompanied by secondary cracks. The damage variable D' shows stable growth, and the damage ratio accounts for about 1/5 of the overall damage. The increase of damage value indicates that the rock specimen dissipates energy, but the elastic strain energy accumulation is still the main. (3) Accelerated damage stage. At this stage, the micro-cracks grow rapidly, a large number of longitudinal cracks are generated at the new cracks' ends, and the amount of acoustic emission also increases exponentially. The damage variable D' of rock specimens of all conditions increases linearly, and the increased amplitude accounts for about 60% of the whole, while the increased amplitude of middle jointed specimen accounts for about 40%. The energy accumulation of the system basically reaches the peak, and the damage development degree shows a jump unstable growth. This stage is of great significance for engineering disaster warnings. (4) Damage failure stage. With the further expansion and connection of cracks, the large-scale macroscopic crack surface is formed; with the release of the system's energy, the damage variable D' shows a gentle growth, but the middle-jointed specimen produces a second jump in growth and finally tends to be stable. When the damage variable D' reaches 1.0, the rock specimen finally loses its bearing capacity and fails.

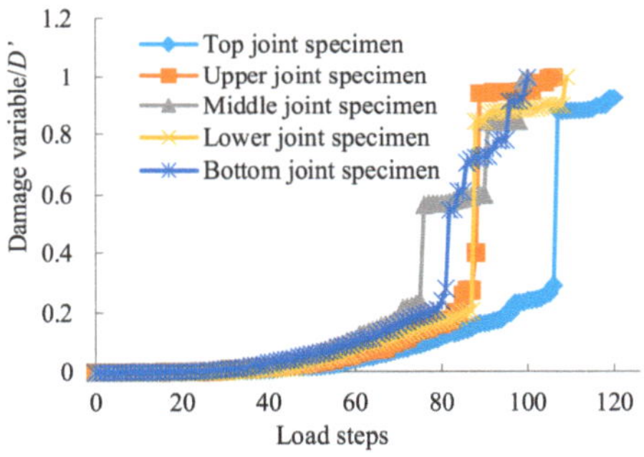

Figure 11. Evolution laws of damage variable D' with load steps.

5. Conclusions

Based on the laboratory tests, numerical simulations on crack propagation of sandstone are compared by DDARF and RFPA, indicating that RFPA is more suitable for the analysis of sandstone's failure characteristics under the hydraulic coupling condition. Based on RFPA, the following conclusions are obtained.

(1) Under the hydraulic coupling action, a large tensile stress region is generated on both ends of the original joint with the micro-cracks' propagation. Water pressure in the cracks promotes the generation of tensile stress to a certain extent, and the rock specimen is mainly subjected to tensile failure. The further away the original joint from the middle part of the rock specimen, the greater the damage effect angle is; crack paths are all approximately parallel to the maximum principal stress direction at all the different joint locations.

(2) Under the same initial water pressure, the water pressure has a certain auxiliary effect on the main crack failure when the original joint is close to the middle part of the specimen, and the water pressure has a dominant effect on the local crack failure when the original joint is far away from the middle part. At low initial water pressure, stresses of the specimens with symmetrical joint locations have similar evolution trends, while at high initial water pressure, the jointed specimens' destruction effect by water pressure is reduced, and the water pressure in cracks has significant dissipation, and thus the maximum water pressure in the system does not exceed the initial value.

(3) The length of the main crack path is positively proportional to the amount of acoustic emission and the energy accumulation capacity. In addition, for the rock specimens with all joint locations, the amount of acoustic emission and the cumulative energy

shows a nonlinear sharp increase in the failure stage. The damage variable D' can reflect the failure degree of jointed rock masses, and the evolution of damage variable D' basically shows a development trend of steady growth-rapid growth-steady growth.

Author Contributions: Y.C.: Methodology and writing—original draft. T.G.: Numerical simulation. F.Y.: Writing—review and editing. X.L.: Data curation. J.W.: Conceptualization. All authors have read and agreed to the published version of the manuscript.

Funding: This study was financially supported by the National Natural Science Foundation of China (No. 42172310 and No. 51609130), the Open Foundation of Key Laboratory of Mining Disaster Prevention and Control by Shandong University of Science and Technology (No. MDPC201909), and the Doctoral Foundation of Shandong Jianzhu University (No. XNBS1704).

Data Availability Statement: Not applicable.

Acknowledgments: The authors gratefully acknowledge Shandong Jianzhu University and Shandong Provincial Institute of Land Surveying and Mapping that have contributed to the research results reported within this paper.

Conflicts of Interest: The authors declare no conflict of interest.

References

1. Havrevoll, O.H.; Vereide, K.; Lia, L. Efficiency of pressurized rock traps for unlined hydropower tunnels. *Energies* **2021**, *14*, 4344. [CrossRef]
2. Seto, M.; Nag, D.K.; Vutukuril, V.S. Acoustic emission in coal and sandstone under triaxial compressive stress condition. In Proceedings of the Rockbursts and Seismicity in Mines, Rotterdam, The Netherlands, 31 December 1997; pp. 427–431.
3. Han, J.Y.; Liu, D.; Guan, Y.P.; Chen, Y.; Li, T.; Jia, D.; Yan, F.; Jia, P.; Zhao, Y. Study on shear behavior and damage constitutive model of tendon-grout interface. *Constr. Build. Mater.* **2022**, *320*, 126–223. [CrossRef]
4. Hashemnejad, A.; Aghda, S.M.F.; Talkhablou, M. Introducing a new classification of soft rocks based on the main geological and engineering aspects. *B Eng. Geol. Environ.* **2021**, *80*, 4235–4254. [CrossRef]
5. Riabokon, E.; Turbakov, M.; Popov, N.; Kozhevnikov, E.; Poplygin, V.; Guzev, M. Study of the influence of nonlinear dynamic loads on elastic modulus of carbonate reservoir rocks. *Energies* **2021**, *14*, 8559. [CrossRef]
6. Abbass, T.; Andre, V. Failure of layered sandstone under Brazilian test conditions: Effect of micro-scale parameters on macro-scale behaviour. *Rock Mech. Rock Eng.* **2010**, *43*, 641–653.
7. Chitrala, Y.; Moreno, C.; Sondergeld, C.; Rai, C. An experimental investigation into hydraulic fracture propagation under different applied stresses in tight sands using acoustic emissions. *J. Petrol. Sci. Eng.* **2013**, *108*, 151–161. [CrossRef]
8. Chen, Z.Q.; Li, T.B.; Chen, G.Q.; Zhang, H. Experimental study of acoustic emission characteristics of sandstone under hydro-mechanical coupling action. *Rock Soil Mech.* **2014**, *35*, 2815–2822.
9. Lian, Z.L.; Zhang, J.; Wu, H.A.; Wang, X.X.; Xue, B. A simulation study of hydraulic fracturing propagation with a solid-fluid coupling model. *Rock Soil Mech.* **2008**, *29*, 3021–3026.
10. Zhang, S.C.; Biao, F.J.; Liu, H.; Zhang, J. A numerical study of parameter influences on horizontal hydraulic fracture. *Eng. Mech.* **2011**, *28*, 228–236.
11. Li, G.; Shen, J.L.; Li, G.H.; Dai, L.P. Numerical simulation for deformation behavior of soft rock roadway under the condition of water-rock interaction. *J. Saf. Environ.* **2016**, *16*, 146–150.
12. Kharghani, M.; Goshtasbi, K.; Nikkah, M.; Ahangari, K. Investigation of the Kaiser effect in anisotropic rocks with different angles by acoustic emission method. *Appl. Acoust.* **2021**, *175*, 107831. [CrossRef]
13. Charlie, C.L.; Zhao, T.B.; Zhang, Y.B.; Wan, W. A study on the energy sources and the role of the surrounding rock mass in strain burst. *Int. J. Rock Mech. Min. Sci.* **2022**, *154*, 105114.
14. Rodrígueza, P.; Celestino, T.B. Application of acoustic emission monitoring and signal analysis to the qualitative and quantitative characterization of the fracturing process in rocks. *Eng. Fract. Mech.* **2019**, *210*, 54–69. [CrossRef]
15. Zhao, Z.G.; Song, L.L.; Jia, Z.C.; Gao, C.Z. Study on failure time prediction of rock based on acoustic emission energy judging index. *Coal Saf. Environ.* **2016**, *48*, 117–119.
16. Jin, Y.; Fan, M.; Fu, W.N.; Chen, M.; Han, H.F.; Zhou, X. Experimental study on fracture propagation behavior based on acoustic emission characteristics. *Chin. J. Rock Mech. Eng.* **2018**, *37*, 3834–3841.
17. Liu, G.; Li, Y.M.; Xiao, F.K. Study on failure mechanics behavior and damage evolution law of yellow sandstone under uniaxial triaxial and pore water action. *Chin. J. Rock Mech. Eng.* **2019**, *38*, 3532–3544.
18. Golshani, A.; Thanh, T.C. Energy analysis of hydraulic fracturing. *Ksce J. Civ. Eng.* **2009**, *13*, 219–224. [CrossRef]
19. Hajibagherpour, A.R.; Mansouri, H.; Bahaaddini, M. Numerical modeling of the fractured zones around a blasthole. *Comput. Geotech.* **2020**, *123*, 103535. [CrossRef]
20. Hiroyuki, S.; Sumihiko, M.; Tsuyoshi, I. The distinct element analysis for hydraulic fracturing in hard rock considering fluid viscosity and particle size distribution. *Int. J. Rock Mech. Min.* **2011**, *48*, 712–727.

21. Mu, K.; Yu, J.; Li, H.; Cai, Y.; Chen, X. Acoustic emission of sandstone with hydro-mechanical coupling and PFC-based modelling of energy dissipation. *Rock Soil Mech.* **2015**, *36*, 1496–1504.
22. Yu, J.; Li, T.B.; Zheng, C.T.; Cai, Y.Y.; Tu, B.X.; Mu, K. Numerical simulation research on Duncan-Chang model of coastal soft soil. *Chin. J. Rock Mech. Eng.* **2014**, *33*, 4271–4281.
23. Lin, C.; Yu, J.; Chen, X.; Cai, Y.Y.; Wen, Z.J. Mesoscopic numerical simulation on crack expansion and energy evolution of sandstone with prefabricated joint under hydro-mechanical coupling condition. *J. Huaqiao Univ. (Nat. Sci.)* **2018**, *39*, 538–544.
24. Chen, Y.J.; Li, S.C.; Zhu, W.S.; Yu, S.; Wang, W.; Zhao, C.L. Rock failure and its jointed surrounding rocks: A multi-scale grid meshing method for DDARF. *Tunn. Undergr. Space Technol.* **2014**, *43*, 370–376.
25. Li, S.C.; Chen, Y.J.; Zhu, W.S.; Zhao, C.L.; Li, L.P.; Yin, F.Q. New reinforcement algorithms in discontinuous deformation analysis for rock failure. *Geomech. Eng.* **2016**, *11*, 787–803.
26. Chen, Y.J.; Zhang, X.; Zhu, W.S.; Wang, W. Modified discontinuous deformation analysis for rock failure: Crack propagation. *Geomech. Eng.* **2018**, *14*, 325–336.
27. Ma, K.; Sun, X.Y.; Tang, C.A.; Wang, S.J.; Yuan, F.Z.; Peng, Y.L.; Liu, K. An early warning method for water inrush in Dongjiahe coal mine based on microseismic moment tensor. *J. Cent. South Univ.* **2020**, *27*, 3133–3148. [CrossRef]
28. Ma, K.; Tang, C.A.; Xu, N.W.; Liu, F.; Xu, J.W. Failure precursor of surrounding rock mass around cross tunnel in high-steep rock slope. *J. Cent. South Univ* **2013**, *20*, 207–217. [CrossRef]
29. Huang, Y.H.; Yang, S.Q.; Liu, X.R. Experimental and numerical study on the mechanical characteristics of rock-like materials. *J. Exp. Mech.* **2014**, *29*, 239–249.
30. Chen, Y.J.; Zhu, W.S.; Li, S.C.; Zhao, T.B.; Li, L.P.; Zhang, Q.Q. Similar rock specimen's creep constitutive test and realization by FLAC3D. *Res. J. Appl. Sci. Eng. Technol.* **2014**, *7*, 3015–3021. [CrossRef]
31. Li, Y.; Cai, W.B.; Zhu, W.S.; Dong, Z.; Zhang, Q. Particle flow analysis of parallel double crack evolution under uniaxial compression. *J. Cent. South Univ.* **2019**, *50*, 3035–3045.
32. Li, G.; Tang, C.A. A statistical meso-damage mechanical method for modeling trans-scale progressive failure process of rock. *Int. J. Rock Mech. Min.* **2015**, *74*, 133–150. [CrossRef]

MDPI AG
Grosspeteranlage 5
4052 Basel
Switzerland
Tel.: +41 61 683 77 34

Energies Editorial Office
E-mail: energies@mdpi.com
www.mdpi.com/journal/energies

www.ingramcontent.com/pod-product-compliance
Lightning Source LLC
Chambersburg PA
CBHW041355210926
51850CB00013B/3083
* 9 7 8 3 7 2 5 8 2 4 0 8 3 *